Quantum Mind and Social Science

There is an underlying assumption in the social sciences that consciousness and social life are ultimately classical physical/material phenomena. In this ground-breaking book, Alexander Wendt challenges this assumption by proposing that consciousness is, in fact, a macroscopic quantum mechanical phenomenon. In the first half of the book, Wendt justifies the insertion of quantum theory into social scientific debates, introduces social scientists to quantum theory and the philosophical controversy about its interpretation, and then defends the quantum consciousness hypothesis against the orthodox, classical approach to the mind–body problem. In the second half, he develops the implications of this metaphysical perspective for the nature of language and the agent–structure problem in social ontology. Wendt's argument is a revolutionary development which raises fundamental questions about the nature of social life and the work of those who study it.

ALEXANDER WENDT is Ralph D. Mershon Professor of International Security and Professor of Political Science at The Ohio State University. He is the author of *Social Theory of International Politics* (Cambridge University Press, 1999), which won the International Studies Association's Best Book of the Decade Award in 2006.

Praise for *Quantum Mind and Social Science*

"Wendt's second monograph has been eagerly anticipated. Was it worth the wait? Of course. Beautifully written and painstakingly argued, *Quantum Mind and Social Science* explores the potential impact that advances in quantum mechanics may have on the social sciences. Notwithstanding the fact that this is probably one of the best introductions to quantum mechanics I have read, the book also raises a series of pressing questions about how a careful engagement with quantum mechanics might alter how we think about social science and social practice. Do I agree with it? No. But that's not the point. This is a book of speculative grand theorizing that is sadly lacking in the social sciences today."

Colin Wight
Professor in the Department of Government and
International Relations, The University of Sydney

"Alexander Wendt, one of the leading and most original voices in International Relations, has now produced what may be his most daring effort yet. In *Quantum Mind and Social Science* Wendt argues for a new kind of physicalism that encompasses elements of mind all the way down to the quantum processes governing elementary particles. For most social scientists, all that Wendt takes us through will be a revelation. Wendt's discussion of this material is just fabulous, the best lay discussions of the issues I have seen. Whatever one thinks of the final thesis, the journey here is definitely worth the ride."

Douglas V. Porpora
Professor of Sociology in the Department of
Culture and Communication, Drexel University

"This book is very well written and engaging and introduces some very controversial new ideas. The author takes a courageous stance on a number of deep and difficult issues in philosophy of mind. Some of these ideas may ultimately not be supported, and some others may engage never-ending debates. But if even one of them turns out to be right, then the book will have made a great contribution."

Jerome R. Busemeyer
Provost Professor in the Department of Psychological
and Brain Sciences, Indiana University

Quantum Mind and Social Science

Unifying Physical and Social Ontology

Alexander Wendt

Department of Political Science
The Ohio State University

CAMBRIDGE
UNIVERSITY PRESS

University Printing House, Cambridge CB2 8BS, United Kingdom

Cambridge University Press is part of the University of Cambridge.

It furthers the University's mission by disseminating knowledge in the pursuit of education, learning and research at the highest international levels of excellence.

www.cambridge.org
Information on this title: www.cambridge.org/9781107082540

© Alexander Wendt 2015

This publication is in copyright. Subject to statutory exception and to the provisions of relevant collective licensing agreements, no reproduction of any part may take place without the written permission of Cambridge University Press.

First published 2015

A catalogue record for this publication is available from the British Library

ISBN 978-1-107-08254-0 Hardback
ISBN 978-1-107-44292-4 Paperback

Cambridge University Press has no responsibility for the persistence or accuracy of URLs for external or third-party internet websites referred to in this publication, and does not guarantee that any content on such websites is, or will remain, accurate or appropriate.

For Emma and Otto

"... the worst of all possible misunderstandings would occur if psychology should be influenced to model itself after a physics which is not there any more..."[1]

Physicist Robert Oppenheimer in 1956

---

[1] The quote is from Young (1976: 26).

Contents

Acknowledgments		*page* x
1	Preface to a quantum social science	1
	Why are we here?	1
	Introduction	2
	The causal closure of physics	7
	Classical social science	11
	The anomaly of consciousness	14
	The mind–body problem	14
	Intentionality and consciousness	18
	The threat of vitalism	21
	The anomaly of social structure	22
	Where is the state?	23
	The threat of reification	25
	As if explanation and unscientific fictions	26
	My central question, and answer in brief	28
	Re-inventing the wheel?	34
	Situating your observer	36
Part I	**Quantum theory and its interpretation**	39
	Introduction	39
2	Three experiments	43
	The Two-Slit Experiment	43
	Measurement is creative	46
	Collapse of the wave function	46
	Complementarity	48
	The Bell Experiments	50
	The Delayed-Choice Experiment	54
3	Six challenges	58
	The challenge to materialism	59
	The challenge to atomism	60
	The challenge to determinism	62
	The challenge to mechanism	63
	The challenge to absolute space and time	65
	The challenge to the subject–object distinction	66

4	**Five interpretations**	**70**
	The problem and a meta-interpretive framework	71
	Instrumentalism: the Copenhagen Interpretation	73
	Realism I: materialist interpretations	76
	The GRW Interpretation	76
	The Many Worlds Interpretation	77
	Realism II: idealist interpretations	81
	The Subjectivist Interpretation	81
	The Bohm Interpretation	85

Part II Quantum consciousness and life   91

Introduction   91

5 Quantum brain theory   **95**
- Your quantum brain   96
  - The Fröhlich tradition   98
  - The Umezawa tradition   101
- Assessing the current debate   102

6 Panpsychism and neutral monism   **109**
- Panpsychism   111
  - Background   112
  - Defining 'psyche,' aka subjectivity   114
  - Projecting subjectivity through the tree of life   116
  - ... And then all the way down   119
  - The combination problem and quantum coherence   123
- Neutral monism and the origin of time   124

7 A quantum vitalism   **131**
- The materialist–vitalist controversy   132
- Life in quantum perspective   137
  - Cognition   139
  - Will   139
  - Experience   141
- Why call it vitalism?   143

Part III A quantum model of man   149

Introduction   149

8 Quantum cognition and rational choice   **154**
- Quantum decision theory   157
  - Order effects in quantum perspective   157
  - Paradoxes of probability judgment   159
  - Quantizing preference reversals   161
- Rationality unbound?   164
- Quantum game theory: the next frontier   169

9 Agency and quantum will   **174**
- Reasons, teleology, and advanced action   175

	Free will and quantum theory	182
	The philosophical literature	183
	The Libet experiments	185
10	Non-local experience in time	189
	The qualitative debate on changing the past	191
	The Epistemological view	192
	The Ontological view	193
	The physics of changing the past	198

Part IV Language, light, and other minds — 207

Introduction — 207

11 Quantum semantics and meaning holism — 210
Composition versus context in meaning — 212
Quantum contextualism — 215

12 Direct perception and other minds — 222
The problem of perception — 223
The dual nature of light — 226
Holographic projection and visual perception — 228
Semantic non-locality and intersubjectivity — 230
   The theory of mind debate — 230
   Semantic non-locality and other minds — 233
   Three objections considered — 237

Part V The agent–structure problem redux — 243

Introduction — 243

13 An emergent, holistic but flat ontology — 247
Supervenience meets externalism — 250
Agents, structures, and quantum emergence — 255
Downward causation in social structures — 260

14 Toward a quantum vitalist sociology — 267
The holographic state — 268
The state as an organism — 273
The state and collective consciousness — 275
The politics of vitalist sociology — 281

Conclusion — 283
Night thoughts on epistemology — 284
Too elegant not to be true? — 288

Bibliography — 294
Index — 345

Acknowledgments

Writing this book has been very much a "quantum" experience. On the one hand, locked in my own personal bubble of subjectivity, I have found it quite isolating to venture into and try to get a grip on literatures with which I often had no initial familiarity, and where an iconoclastic interloper could expect little systematic help from local guides. On the other hand, my experience has also been quite holistic in the sense that innumerable individuals have joined me along the way, whether in the form of written comments on parts of the text, meetings over lunch to discuss problems I was having, countless questions at seminars, references sent that I had not seen, and a striking number of random emails from people around the world who had heard of my work and either had something substantive to say or just wanted to offer encouragement. By no means has everyone agreed with me, but their comments were overwhelmingly constructive. I was not able to incorporate all of the input I received, but a great deal of it is now enfolded in one form or another within the final product, which is participatory epistemology at its best.

The journey to this book has been a long one, over ten years depending on how you count, and over the course of this period the logistics train carrying my records of all this feedback simply broke down. Thus, much to my regret I am unable to recall every individual who helped me to clarify and formulate my ideas – or even every institution where I have presented them. All the more reason, therefore, for me to begin by expressing my heartfelt thanks to everyone I have encountered along the way, and my apologies to those whose names are now lost in the sands of time.

Those whom I do recall fall into three groups: outside help, colleagues, and family. In the first category, Stuart Hameroff has been supportive of this project from its very inception. Stu not only commented twice on early drafts of my quantum brain chapter but also invited me – at the time a complete stranger – to be a plenary speaker at the 2004 Tucson conference on "Toward a Science of Consciousness." Being at this large and hugely stimulating conference showed me that while my thesis might be crazy, at least I was in good company.

I am very grateful to Stefano Guzzini and Anna Leander, who in a 2006 volume gave me the opportunity to crystallize my argument in the form of an "auto-critique" in response to critics of my first book, *Social Theory of International Politics*. This led to the first published version of these ideas, and generated a number of subsequent comments that encouraged me to keep going with the project.

John Haslam at Cambridge University Press gave my initial proposal for the book a warm reception, and found an outstanding and diverse set of reviewers for the manuscript. Their enthusiasm – despite varying degrees of skepticism about the argument – made it possible to muster the last burst of energy I needed to take their many comments on board.

Among the colleagues who contributed extensive written comments and/or significant moral support for my work, my (again, incomplete) records include Bentley Allan, Badredine Arfi, Bear Braumoeller, Steve Brooks, Zoltan Buzas, Aldous Cheung, Bud Duvall, Karin Fierke, Caleb Gallemore, Chris Gelpi, Eric Grynaviski, Xavier Guillaume, Stefano Guzzini (again), Ted Hopf, Tahseen Kazi, Jim Keeley, Jason Keiber, Oliver Kessler, Tim Luecke, Eric MacGilvray, Sebastien Mainville, Patchen Markell, Dragan Milovanovic, Mike Neblo, Karen O'Brien, Heikki Patomaki, Sergei Prozorov, Randy Schweller, Alex Thompson, Srdjan Vucetic, Lisa Wedeen, Colin Wight, and Rafi Youatt.

Many of these colleagues are or were at The Ohio State University, which I joined in 2004 and where most of the work on this book was done. The Political Science department and the Mershon Center for International Security Studies provided tolerant intellectual environments for their new colleague and his strange ideas, and – under Rick Herrmann's leadership in both cases – also remarkable institutional patience as my required yearly productivity reports continually promised that the manuscript would be done "in a year or two."

Among my Ohio State colleagues Allan Silverman deserves signal recognition for commenting in detail on large parts of the manuscript, and for being a good friend and constant source of intellectual guidance – and humor – throughout my time here.

Jerome Busemeyer, Doug Porpora, and Abe Roth were exceptionally kind to take several days out of their lives first to read and then discuss the whole manuscript at a Mershon workshop in 2013. Their thorough and probing comments clarified many smaller points and also affected the framing of the book as a whole.

Zac Karabatak undertook the daunting task of reconciling the footnotes against the bibliography, which saved me an immense amount of time and would have driven me crazy in the process.

Ann Powers was the first person to agree unambiguously with my thesis and provided essential moral and administrative support throughout its realization. I especially appreciated her irreverent attitude toward academics, which reminded me that just because the Experts say something is true doesn't mean that it is.

Turning finally to family, Chris Wendt, my brother and a physicist by training, patiently answered my many naïve questions over the years about quantum mechanics, which was an invaluable resource that saved me from numerous elementary mistakes (though he bears no responsibility for those that undoubtedly remain).

My father, Hans Wendt, taught me that the way forward in science is to look for anomalies, not just patterns, which as you will see found its way into the structure of many parts of the book.

I owe a special gratitude to Jennifer Mitzen, who has been my companion the whole way, which means that more than anyone she had to deal with all the anxiety and stress that accompanied my undertaking. Despite her own many responsibilities and hesitation in the face of quantum theory she was always willing to drop everything to talk about my ideas. Jennifer also gave me extensive comments on several chapters, both line-by-line ones that sharpened many points and structural ones that forced me to recast whole sections. The final product would have been much the worse without her.

Quantum Mind and Social Science is dedicated to Emma and Otto, who have never known a day of their lives when I was not working on it, and who may live long enough to find out whether Daddy's conjecture is right.

1 Preface to a quantum social science

Why are we here?

Almost from its inception as an academic discipline in 1919, International Relations (IR) has featured "Great Debates" about what we today would call the relationship between ideas and material conditions, human agency and social structures, and naturalist and anti-naturalist modes of inquiry. While often disparaged as mere "meta-theory," at least implicit positions on these essentially philosophical questions play an important role in the field. Intellectually, they structure our substantive theorizing, methods, empirical findings, and ultimately the normative and policy implications we draw from our research; and sociologically, they affect who we hire (and sometimes, fire), where we publish, and how we train our graduate students. Unfortunately, despite considerable disciplinary investment in meta-theory since the 1980s, from my own vantage point, as someone who has been involved in these debates for 25 years, I see no progress toward ending them. IR scholars have a better sense today of what the issues are and how, why, and when they matter, but the debates remain as intractable as ever. When it comes to the ontological and epistemological foundations of IR scholarship, we are in a "Land of Confusion"[1] from which escape is nowhere on the horizon.

Of course, the confusion is not IR's alone, but the social sciences' as a whole. Although over the years sociologists, economists, political scientists and others have acquired better data and statistical techniques that have significantly improved empirical understanding of trends and relationships in society, social scientists' ability to cumulate deeper, theoretical knowledge has lagged seriously behind. This is true even in economics, where despite greater theoretical homogeneity, vigorous heterodoxies survive. In contrast to physical sciences like chemistry or geology, where there is broad agreement on the nature of reality and how we should study it, in the social sciences there is no such

---

I am very grateful to Colin Wight for exceptionally detailed comments on a draft of this chapter, especially since he disagrees with the whole idea.

[1] If you'll pardon the reference to the 1986 hit by *Genesis*; cf. *Disturbed*'s 2005 cover.

consensus. As a result social scientific theories rarely die, and if they do, like zombies they inevitably come back to life later.

As I argue below, the reason for this state of affairs is that social phenomena are mind-dependent in a way that chemical elements and rocks are not, and as such do not present themselves directly to the senses. Thus, before social scientists can even "see" what they are studying they must make a number of philosophical assumptions about the mind that are easily contested by those who would make different ones.

In philosophy there is a long-standing suggestion[2] that when debates persist for many years with no discernible progress, this is because all sides are making an assumption that is in fact mistaken. If such an assumption could be identified in the philosophy of social science, then that might enable IR scholars and social scientists more generally to find the Undiscovered Country of philosophical clarity that has eluded us for so long. But what might it be?

My own "aha!" moment came in 2001 after reading Danah Zohar and Ian Marshall's book *The Quantum Society*, which I had picked up almost randomly at the University of Chicago bookstore.[3] Zohar and Marshall were writing for a general audience, so I did not find the discussion of social and political theory entirely satisfying. However, their basic idea – that the mind and social life are macroscopic quantum mechanical phenomena – hit me as just the kind of thesis that could help move philosophical debates in the social sciences forward. That is because it calls into question a foundational assumption taken for granted by all sides – namely that social life is governed by the laws of classical physics. I don't know if the conjecture is right, but I felt it deserved a more systematic treatment that could be subjected to serious academic scrutiny. That is what I have tried to do in this book. Doing so took much more space (and time!) than I expected, and so unlike my first book,[4] which was half philosophy and half IR, this one is all philosophy. So for my colleagues in IR, all I can offer here is the promise of a more IR-focused "volume 2" down the road. In the meantime, I hope they will find something of value in a book addressed to all social scientists.

Introduction

The advent of quantum theory in the early twentieth century revolutionized physicists' description of reality. Exactly what conclusions should be drawn from that description of reality is still being debated today, but the theory is extraordinarily well confirmed and all sides agree on its basic findings. In particular, whereas mathematical symbols in classical physics correspond to the properties of real material objects and forces, in quantum physics they

---

[2] Due, I believe, to Frank Ramsey in the 1920s.   [3] See Zohar and Marshall (1994).
[4] See Wendt (1999).

represent only the probabilities of finding certain properties when they are measured. Moreover, these quantum probabilities, which are expressed by "wave functions," are completely unlike classical probabilities. Whereas the latter denote our ignorance about what is actually the case and as such are incomplete descriptions of reality, the former denote all that could even in principle be known about quantum systems. Despite its probabilistic character, in other words, the wave function is a *complete* description of a quantum system, until its measurement, at which point it "collapses" and just one, classical outcome is observed. So, unlike in classical physics, where we can safely assume that objects have, for example, a momentum or position even when we are not observing them, in quantum physics we have no basis for such an assumption. Wave functions are potential realities, not actual ones.[5]

Understanding how the indeterminate quantum world results in the determinate classical world – a process known as "decoherence" – is one of the deep mysteries of quantum theory. However, its immediate significance in the present context is that, although quantum mechanics subsumes classical physics, its practical applicability is generally thought to be confined to sub-atomic particles. Above that level, it has long been assumed that quantum effects wash out statistically, leaving the decohered world described by classical physics as an adequate approximation of macroscopic reality. That includes social life, the contemporary study of which, I argue below, is all based at least implicitly on the worldview of classical physics.

In this book I explore the possibility that this foundational assumption of social science is a mistake, by re-reading social science "through the quantum." More specifically, I argue that human beings and therefore social life exhibit quantum coherence – in effect, that we are walking wave functions. I intend the argument not as an analogy or metaphor, but as a realist claim about what people really are. Scholars have long pointed to a number of strong analogies between human and quantum processes: between free will and wave function collapse, the holism of meaning and non-locality, observer effects in psychological experiments and quantum measurement, and even double-entry accounting and quantum information.[6] These and other analogies are sufficiently suggestive that one might apply quantum thinking to social life simply on that basis.

While one could read this book entirely in that way, as an interesting analogy, my personal belief is that human beings *really are* quantum systems. I defend that belief explicitly only in the Conclusion, but the book as a whole

---

[5] While there is debate about the ontological status of the wave function, no one argues that it is real in the same sense as classical objects.

[6] See Brandt (1973), Rosenblum and Kuttner (1999), Bitbol (2002), Heelan (2004), Pylkkänen (2004), Filk and Müller (2009), Grandy (2010), Kuttner (2011) and – since you're probably wondering about the case of accounting – Fellingham and Schroeder (2006).

is written with a view toward showing how this hypothesis could possibly be true. This realist stance will take me into controversial, speculative and frankly dangerous territory that could be avoided by an analogical road to "quantum social science." However, it would also come at a cost, which is that it would make quantum theory just another tool for social scientists to pick up – or not – as they see fit, and bracket some of the theory's most profound potential implications. In contrast, if human beings really are quantum, then classical social science is founded on a mistake, and social life will therefore *require* a quantum framework for its proper understanding.

This is not the first call for a quantum social science. Already in 1927 – just weeks after the Solvay conference marking the culmination of the quantum revolution – the President of the American Political Science Association, William Bennett Munro, challenged social scientists to come to grips with the new physics.[7] Philip Mirowski argues that to a limited extent they did, in that its probabilistic "spirit" facilitated social scientists' embrace of statistical methods in the 1930s.[8] But until recently there has been almost no reflection on the significance of quantum theory itself for the social sciences. As if to drive home this neglect, the methods embraced in the 1930s were based on classical probability theory – which came from the *previous*, Newtonian revolution in physics – not quantum probability theory.

While the social sciences have prospered in the ensuing years, there is today a good reason to re-open the quantum question: growing experimental evidence that long-standing anomalies of human behavior can be predicted by "quantum decision theory." This is a quantized version of expected utility theory, which replaces the latter's either/or Boolean logic with the both/and logic of quantum probability theory.[9] Quantum decision theory predicts most[10] of the deviations from rational behavior found by Daniel Kahneman, Amos Tversky and others using expected utility theory as a baseline – order effects, preference reversals, the conjunction fallacy, the disjunction fallacy, and so on. Psychologists have devoted enormous energy to trying to explain these anomalies, but the results have been partial and theoretically ad hoc. In contrast, with a single axiomatic framework, quantum decision theory shows they are not anomalies at all, but precisely what we should expect. Prestigious journals like *Journal of Mathematical Psychology* (2009), *Behavioral and Brain Sciences* (Pothos and Busemeyer, 2013), and *Topics in Cognitive Science* (2014) have taken notice and devoted substantial space to this unfamiliar approach. While the theory is new and its larger reception remains to be seen, its findings are

---

[7] See Munro (1928).   [8] See Mirowski (1989).
[9] See especially Busemeyer and Bruza (2012), which includes an accessible introduction to quantum theory, probability and logic.
[10] My sense is that this qualification is necessary only because the literature is so young that it has not been able to take up all the relevant anomalies; see Chapter 8.

extraordinary. Rarely in the social sciences has one theory explained so much that was so puzzling before.[11] Quantum decision theory seems as clear a case as one could hope for of progress in social science, not just within a research program, but from one research program to the next.[12]

But that's only the half of it. Quantum decision theorists have been cautious in speculating about the philosophical implications of their work, focusing instead on just proving that it predicts previously anomalous behavior. In doing so they have embraced what is known as "generalized" or "weak" quantum theory, which applies the quantum formalism to phenomena beyond the domain of physics – like social life – while remaining agnostic about what is going on underneath.[13] While this "as if" strategy has pragmatic attractions, it overlooks the fact that quantum decision theory's success at the behavioral level fulfills a key prediction of a controversial hypothesis about what is happening deep inside the brain: quantum consciousness theory, according to which *consciousness* is a macroscopic quantum phenomenon.[14] That could help solve one of the deepest mysteries of modern science: the mind–body problem, or how to explain consciousness in scientific terms.

Since the Enlightenment it has been assumed that to explain consciousness scientifically means showing how it is compatible with the worldview of classical physics. Classical physics implies a materialist ontology in which reality is ultimately made up of just matter and energy. It is therefore ironic that quantum wave functions are not *material* at all, at least not in any ordinary sense. This has led some philosophers of physics to argue that, far from materialism, quantum theory actually implies a panpsychist ontology: that consciousness goes "all the way down" to the sub-atomic level. Exploiting this possibility, quantum consciousness theorists have identified mechanisms in the brain that might allow this sub-atomic proto-consciousness to be amplified to the macroscopic level. Modern neuroscience can't test this claim yet, but one of its implications is that human behavior should have quantum characteristics, which quantum decision theory bears out. From this standpoint, in short, there is the possibility not only of a progressive problem shift in behavioral social science, but of a paradigmatic change in the modern scientific worldview.

Social scientists might reasonably doubt that a hoary philosophical controversy like the mind–body problem could be relevant to their work. Yet we have hoary controversies of our own. In social epistemology there is the

---

[11] Something similar may be starting to happen in the biological sciences with the emergence of "quantum biology," which I discuss in Chapter 7.
[12] See Lakatos (1970).
[13] See Atmanspacher et al. (2002) and Walach and von Stillfried (2011). Because it uses the formalism to make quantitative predictions I would say quantum decision theory goes beyond a purely analogical approach.
[14] See Chapter 7 and Atmanspacher (2011) for a recent overview.

"Explanation vs. Understanding" debate between naturalists or positivists,[15] who think there is no essential difference between physical and social science, and anti-naturalists or interpretivists who think there is because people act on meanings that must be interpreted.[16] In social ontology there is the "Agent–Structure" debate, between individualists who think that social structures can be reduced to the properties and interactions of individual agents, and holists who think they can't.[17] And then there is perhaps the biggest debate of all, between materialists who think social life ultimately can be explained by material conditions and idealists (or idea-ists) who think that ideas play an autonomous or even decisive role. This latter debate arguably subsumes the other two, since without ideas in play there would be no meanings to interpret or social structures to reduce. Moreover, this debate is not merely like the mind–body problem in seeming intractable, but of a piece with it substantively, because ideas are dependent on consciousness. Which is to say: some of the deepest philosophical controversies in the social sciences are just local manifestations of the mind–body problem. So if the theory of quantum consciousness can solve that problem then it may solve fundamental problems of social science as well.

I have put a lot of balls in the air and will not try to catch them all. First, except in Chapter 8, I will not deal extensively with quantum decision theory. Work in this vein is in full swing, and now spreading from psychology to the social sciences at large,[18] and with no formal training myself, I am in no position to contribute to it. My focus instead will be on its philosophical implications, which have been neglected so far. Second, only in the Conclusion will I take up the Explanation–Understanding debate. One reason is frankly practical; this book is so long already that to finish it I need to focus its argument as much as possible. Another is that pioneering contributions in this area have already been made by scholars such as Karen Barad, Michel Bitbol, Patrick Heelan, and Arkady Plotnitsky – although they are by no means all in agreement.[19] But most importantly, in my view we will not make clear progress on the epistemology of a quantum social science until we have a firm basis in its ontology, where little work has been done. That leaves just one – albeit still very large – ball to catch, the nature of ideas and consciousness, and its implications for the agent–structure problem.

---

[15] I will use these terms interchangeably, giving 'positivism' a broader meaning than it carries in much social scientific discourse, where it is often juxtaposed to scientific or critical realism. Realists are naturalists and thus positivists in my sense.

[16] See Apel (1984) and Hollis and Smith (1990) for introductions to this debate.

[17] See for example Wendt (1987), and Wight (2006) and Elder-Vass (2010a) for the state of the agent–structure art in IR and social theory respectively.

[18] See, for example, Haven and Khrennikov (2013) and Khrennikova et al. (2014).

[19] See Barad (2007), Bitbol (2002; 2011), Heelan (1995; 2009), and Plotnitsky (1994; 2010).

Since the start-up costs for thinking in quantum terms are high, my goal in this "preface" is motivational: to explain why it is necessary to turn to such an exotic theory to solve basic problems of social ontology. In particular, I show that the agent–structure problem stems from the fact that the ways in which social scientists have dealt with an essential feature of the human experience – namely experience itself – originate in classical assumptions about the mind–body problem. The chapter ends with an overview of the book's positive argument.

The causal closure of physics

There are at least two long-standing anomalies in social ontology: the existence of subjectivity, specifically its conscious aspect; and the unobservability of social structures. The two are related through the agent–structure problem, of which they are in effect opposite sides, and in the end I argue that the second is a function of the first. However, they involve distinct issues and literatures, and as such are treated separately below.

In social theory, subjectivity and unobservable social structures are usually referred to as "problems" rather than "anomalies," but this understates their significance. By calling them anomalies I mean that, given a classical worldview, they simply should not be there any more than the anomalies in physics which sparked the quantum revolution should have been there. To be sure, subjectivity and social structures cannot be seen with the naked eye or recorded on instruments, and as we will see this has prompted some philosophers to argue that they are illusions and thus *aren't* there. However, most social scientists, I suspect, think they are, so before we give in to philosophers of illusion it makes sense to explore all possible means to justify this belief.

But first, I need to do some work on the other side to convince credulous social scientists that subjectivity and social structures are anomalies at all. To do that, in this section I begin with a foundational principle to which all social scientists should agree, the "causal closure [or completeness] of physics" or "CCP."[20]

The CCP means that the social (and all other) sciences are subject to a physics constraint: no entities, relationships, or processes posited in their inquiries should be inconsistent with the laws of physics. The idea is that because physics deals with the elementary constituents of reality, of which macroscopic phenomena are composed, everything in nature[21] is ultimately just physics. This

---

[20] With apologies to the Chinese Communist Party; for good introductions to the CCP and its rationale, see Papineau (2001) and Vicente (2006; 2011).

[21] Or at least everything with causal powers in the temporal world; the CCP does not rule out the existence of God or other spiritual phenomena as long as they mind their own business; see Papineau (2001).

gives physics a foundational role with respect to other sciences, which today are often collectively called the "special" sciences to signify their subordinate status.[22]

At a working level the CCP is almost universally accepted today in the physical and biological sciences. The situation may seem less clear in the social sciences, where even positivists may be skeptical of "social physics," and interpretivists reject naturalistic approaches to social inquiry altogether. Nevertheless, I argue in a moment that the CCP is almost universally accepted in the social sciences as well. But before defending that perhaps provocative claim let me prepare the ground by first emphasizing two things that the CCP does *not* commit us to.

First, epistemologically speaking, the causal closure of physics does not mean social scientific theories must be *reducible* to physics, in the sense of being able to replace their laws with laws of physics without loss of explanatory content. Such reductions have proven elusive even in the physical and biological sciences, the objects of which are often closer to physics in scale and complexity than human beings are. If chemistry is not reducible to physics, then all the more reason to think that social science is not either. Our knowledge of the world is "dappled," in Nancy Cartwright's suggestive image, disparate and fragmented rather than integrated and uniform.[23]

However, as Lawrence Sklar has argued in response to Cartwright, we should not confuse the epistemological point that our knowledge is currently fragmented with the ontological point that the laws of physics *do not apply* to everything in the world.[24] All objects and forces are made up of the phenomena described by fundamental physics,[25] and thus "the laws of the fundamental theory are *as true* of these objects as they are of the carefully isolated systems of small numbers of particles constructed in the laboratory."[26] In other words, whatever law-like processes exist in social life, they cannot force the elementary constituents of nature to violate *their* laws. So while the CCP does not imply reductionism, it does limit ontologically what can exist and happen at the macro-level.

The other thing that the CCP does not commit us to is the philosophical doctrine of *physicalism*,[27] according to which everything in the world is ultimately physical. That may sound counter-intuitive, since 'physical' is usually defined by "whatever physics says there is," so how could the causal closure of

---

[22] See Fodor (1974).   [23] Cartwright (1999); also see Dupré (1993) and Ziman (2003).
[24] Sklar (2003); also see Pettit (1993b) and Hoefer (2003).
[25] Today taken to be quantum field theory.
[26] See Sklar (2003: 433), emphasis in the original; also see Ladyman (2008: 745–746), "[s]pecial science hypotheses that conflict with fundamental physics... should be rejected for that reason alone."
[27] At least as it is currently understood; see below.

physics not imply physical*ism*? And indeed the two are often conflated in the literature.[28] In my view this conflation is a mistake, and since this will form a crucial wedge in my argument it is important to see why.

Physicalism is the modern descendant of classical materialism. Materialists held that reality is ultimately purely material, understood as the little bits of matter and (later) energy described by classical physics. Importantly, these bits of matter were assumed to lack any trace of consciousness within them. With this claim materialists opposed not just theism, which gave God a temporal role, but also all doctrines that gave consciousness or mind a fundamental status, like idealism, dualism, and panpsychism. For materialists, at the end of the day everything is just mindless matter in motion. However, with the quantum revolution materialists were betrayed by their physicist allies, who found that the classical idea of matter broke down at the sub-atomic level. In effect, quantum physics falsified classical materialism.[29] Rather than abandon materialism, however, materialists morphed into physicalists. In doing so they retained their opposition to theism and all doctrines that give mind a fundamental status, but now deferred to the ongoing inquiry of physics to tell us what precisely the fundamental level looks like.

The problem with this is not only that physicalism lacks a stable meaning of 'physical,' which has worried some physicalists themselves.[30] The problem, as Barbara Montero points out,[31] is that unlike classical physics, quantum physics does not rule out the possibility that mind *is* an elementary feature of reality (see Chapter 4). So in the quantum world, 'physical' does not necessarily mean 'material,' and as such, physical*ism* (or more precisely "*physics*-calism") does not entail and might even end up contradicting materialism. Conflating physicalism with the CCP begs the question against non-materialist "physicalisms," in other words, making it non-falsifiable and thereby trivially true.

Faced with this ambiguity we have two options. One is to go with the open-ended definition of 'physicalism' implied by deference to physics, and give up any inherent connection to old-fashioned materialism. That would be in the spirit of the discursive change to 'physicalism,' and of my own argument below, which is physicalist in this broad sense. However, it would be against how physicalism is usually understood today (i.e. as twenty-first-century materialism) and thus potentially confusing. Instead I shall follow Montero and others who argue that physicalism should be defined separately from the CCP as the doctrine of "No Fundamental Mentality," which a future physics might

---

[28] See, for example, Kim (1998: 147), Papineau (2001), and Vicente (2006: 168, note 5).
[29] See Montero (2001: 63; 2009).
[30] A problem known as "Hempel's Dilemma," for a good discussion of which see Crook and Gillett (2001); see Poland (1994) for a comprehensive introduction to physicalism.
[31] Montero (1999; 2001; 2009); also see Crane and Mellor (1990) and Davies (2014).

or might not confirm.³² That preserves the historical continuity of 'materialism' with 'physicalism,' and also makes clearer what I am arguing against. Unless otherwise noted, I will use the two terms interchangeably below.

So accepting the CCP commits us neither to reductionism nor to a materialist physicalism – all we have to accept is that everything that exists and occurs in nature, including social life, is constrained by the laws of physics. It seems hard to disagree with that, since consider the alternative: things happen to which the laws of physics do not apply. But in that case, what – or *where* – are their extra-physical causes? One possibility is God, though in that case we are in the realm of faith and engaged in an altogether different enterprise. The other main historical answer was Descartes' substance dualism, according to which mind is its own reality entirely separate from matter, but still part of nature. But substance dualism is no longer widely seen as credible,³³ and it seems a second-best solution in any case, to be embraced only if a comprehensive physicalism (now in the broad sense) proves impossible to articulate. Since I do not think that this has yet been proven, insofar as we are committed to social *science*, I take it that the laws of physics constitute a basic constraint on what social objects can be and do.

I cannot think of any social scientist who does not accept the CCP. For positivists it is constitutive of the very idea of science, so this case is clear. However, it might not seem so for interpretivists. Interpretivists explicitly reject naturalistic approaches to social science on the grounds that intentional phenomena – mental states such as beliefs, desires, and meanings – play a central role in human life, and do not seem to be anything like physical objects or causes. Thus, if we want to capture the specificity of social life – what makes it essentially different than geology or chemistry – then looking to physics will at least be no help, and might positively hinder our understanding.

Still, I know of no interpretivist, post-modernist, or other critic of naturalistic social science who says that social phenomena can *violate* the laws of physics. To be sure, the people interpretivists study might believe in things that violate the laws of physics, like a God with powers to intervene in the physical world, and on that basis create institutions that have real effects. However, whatever their personal views about God, in their scholarship interpretivists would not

---

³² See Montero (2003), Wilson (2006), Brown and Ladyman (2009), and Göcke (2009); for skepticism about the No Fundamental Mentality constraint on physicalism see Judisch (2008) and Dorsey (2011).
³³ Though see Göcke, ed. (2012) and Swinburne (2013) for recent exceptions, and Stapp (2005) and Barrett (2006) for arguments that dualism is implied by quantum mechanics. The skepticism toward substance dualism does not extend to *property* dualism, according to which complex forms of matter can give rise to irreducible mentality; see for example Koons and Bealer, eds. (2010).

treat such beliefs as true.[34] Interpretivists agree with positivists on the principle of "methodological atheism," which brackets the question of God's reality and temporal role.[35] As Jürgen Habermas puts it, "a philosophy that oversteps the bounds of methodological atheism loses its philosophical seriousness."[36] Similarly, in their work no interpretivist embraces the claims of astrology, divination, or other pseudo-sciences that contradict the laws of physics – or, for that matter, even of ESP, the reality of which is at least open to scientific debate.[37]

Notwithstanding their explicit anti-naturalism, in other words, implicitly interpretivists too seem to accept that social life is physically constrained and constituted. And why shouldn't they? Human beings have material bodies that think and interact with each other through thought, voice, sound, sight and touch, all of which seem indisputably subject to the laws of physics. Intentional phenomena might not be reducible to those laws, but they are still subject to them. From this perspective, therefore, interpretivists are not anti-naturalists but more like naturalists-*plus* – accepting the CCP at the level of fundamental ontology, while offering other, mostly epistemological arguments about what makes the social sciences special.

Classical social science

But the causal closure of *which* physics? Today the "P" in 'CCP' refers to quantum physics, which is universally acknowledged to be more fundamental than classical physics. However, quantum phenomena are also widely believed to wash out above the sub-atomic level, and so it might be thought that, for all practical purposes, the relevant principles of causal closure in social science are classical (call this the CCCP).[38] In this section I argue that this is indeed how social scientists have understood the constraints of the CCP with respect to their work.

Making this argument is complicated in two ways. First, few social scientists have written on the CCP. This is not for want of philosophical reflection more generally, since almost from the start issues of ontology and epistemology have been deeply contested in the social sciences. But since the turn of the twentieth century there has been almost no discussion of our relation to physics specifically,[39] so I am forced to infer implicit views about which CCP social

---

[34] Thanks to Ted Hopf for sharpening this point for me.
[35] For two provocative exceptions see Porpora (2006) and Gregory (2008).
[36] Habermas (2002: 160).   [37] See Jahn and Dunne (2005).
[38] With apologies now to the former Soviet Union; these authoritarian associations of 'CCP' and 'CCCP' are of course entirely accidental...
[39] Though there has been some on our relationship to physical*ism*; see Neurath (1932/1959), Papineau (2009), and for a critical response to Papineau, Shulman and Shapiro (2009).

scientists see as the relevant constraint. Second, as we shall see, what social scientists might *say* about the CCP if asked and what we *do* in our research may be different things. Nevertheless, the simple answer to whether social scientists feel bound by a CCCP is yes. Both historical and substantive considerations point to this conclusion.

On the historical side, substantial scholarship has been done showing that from their origins in the seventeenth century to their consolidation in the late nineteenth the social sciences were deeply influenced by (classical) physics, the most successful and prestigious science of the day.[40] For both intellectual and political reasons, our Founders – Hobbes, Hume, Smith, Comte, Jevons, Walras, Marshall, Pareto, and others – borrowed frequently from physics in their thinking about society. Bernard Cohen shows that this took various forms – analogies, metaphors, homologies, and identities – and argues that efforts to establish homologies and identities usually failed, leaving the overt impact of physics on the social sciences mostly on the analogical and metaphorical level.[41] But even if classical physics was not fruitful for substantive theorizing about social life, at a deeper level its impact was profound. By the early twentieth century the metaphysical assumptions of the classical worldview – materialism, determinism, locality, and so on – were deeply ingrained in the minds of social scientists. These assumptions were taken to be true of reality as a whole, and thus fundamental constraints on social scientific inquiry.

That this history is still with us is suggested by what happened next – or didn't. At the same time that the quantum revolution was transforming physics in the early twentieth century, in the social sciences "physics envy" went out of fashion. Whether because borrowing from physics yielded few insights, because social scientists had become more self-confident, or because they thought quantum effects washed out at the macroscopic level, the effect was that until quantum decision theory came along social scientists had almost never considered the significance of quantum physics for their inquiry.[42] Thus, by default if nothing else, I think social scientists today would appeal to the CCCP as the relevant constraint on their work.

On the substantive side, in turn, classical thinking permeates the ontology of positivist social science. Since it would require a long detour to show this, consider instead the following classical assumptions about social life: 1) mental

---

[40] See for example Mirowski (1988), Cohen (1994), and Redman (1997) with reference mostly to economics, and Gantt and Williams (2014) on psychology. Note that the moniker 'classical' was only added after the emergence of quantum physics.

[41] See Cohen (1994).

[42] There are exceptions – including Matson (1964), Brandt (1973), Weisskopf (1979), Schubert (1983), Karsten (1990), Becker, ed. (1991), and Peterman (1994) – but they were not cumulative and are little known today.

states are set by our neural constitution; 2) neural states are physically well defined; 3) human behavior obeys the laws of classical probability theory; 4) consciousness is epiphenomenal and thus not relevant to explaining human behavior; 5) the mind is a computer; 6) reasons are efficient causes; 7) there is no action at a distance; 8) social structures are reducible to the properties and interactions of individuals; 9) time and space are objective background conditions for action; and 10) in principle we can observe social life without interfering with it. I think positivists would accept most if not all of these principles unhesitatingly, as simple common sense. As evidence, consider the methods training that graduate students across the social sciences are given in formal theory and statistics. It's all based on classical logic and probability theory, which assumes that the world our students will be studying in their careers is a classical one, and in my experience is so taken for granted that the question never even comes up.

On the other hand, interpretivists would reject many if not most of these assumptions.[43] But if social life is not subject to the CCCP then are interpretivists saying it is to the CCQP? Certainly not explicitly, since the question has almost never been raised;[44] yet as I suggested above, neither have interpretivists said social life can violate the laws of physics. Instead, they have opted for epistemological arguments that while the assumptions and methods of the physical sciences may be useful for studying rocks and glaciers, they are not appropriate for studying the intentional phenomena that constitute society. This "Two-Sciences Settlement"[45] makes sense as a pragmatic defense of the autonomy of the social sciences, but it seems to concede that, ontologically, at the end of the day the social world is all just matter and energy. And if that's right, then why are intentional phenomena not amenable to the methods of the physical sciences? In short, what are intentional phenomena *physically*? The problem here is that, like positivists, interpretivists have implicitly equated naturalism with classical naturalism, and so their rejection of naturalism is framed by the classical worldview as well. This is not to say that interpretivism is classical in the same way as positivism, since as we will see its focus on meaning is hard to square with a materialist ontology.[46] Indeed, an important goal of this book is to legitimate that focus, which in the social sciences is

---

[43] As would critical realists like Roy Bhaskar (1979; 1986) and his followers, who espouse a kind of hybrid interpretivist naturalism (also see Wendt, 1999: Chapter 2). The interpretivist aspect of critical realism is subject to the question raised here about interpretivism more generally, whereas its naturalist aspect, especially its concern with identifying unobservable deep structures, is dealt with below.

[44] See Apel (1984) for an exception, though his discussion is focused more on epistemology than ontology.

[45] See Ephraim (2013).

[46] For that matter, as I suggest below, in practice even much positivist social science does not observe a classical physics constraint.

intellectually marginalized. But to do that, intentional phenomena need to be made consistent with the CCP.

The anomaly of consciousness

Even if it is accepted that social life is governed by the laws of classical physics, it may be objected that the constraint is so loose that it is irrelevant to either the content or practice of social science. Positivists are interested in the behavior of people, who are subject to different laws than matter and energy in their simplest forms. And even assuming that interpretivists grudgingly conceded that human action is constrained by the laws of classical physics, so what? It still doesn't tell us anything about meaning, discourse, and other intentional phenomena.

As a social scientist myself I find such skepticism understandable – after all, what do physicists know about social science? So the burden of proof is on me to show that physics could matter to social science in any interesting way. As a first step, in this section I argue, by way of the mind–body problem, that if we approach social science under a classical physics constraint, then intentional phenomena have no place in our work. I develop this suggestion in three stages. I first define the mind–body problem, and more specifically the problem of consciousness, and show how it constitutes an anomaly for the classical worldview. Second, although social scientists might not care about consciousness, I argue it is presupposed by intentional phenomena, which we routinely invoke in our theories. Third – and here's the kicker – if consciousness cannot be reconciled with the classical worldview, then intentional phenomena no more belong in a classically conceived social science than vitalism's *élan vital* belongs in a classical biology.

The mind–body problem

In its most general form, the mind–body problem is how to understand the relationship between mental states, which are subjective, and brain states, which are objective. However, this general formulation has traditionally been twisted by an assumption that brain states must be understood in classical and therefore materialist terms.[47] According to materialism, the elementary constituents of all macroscopic objects are purely material. "The problem" is thereby recast in narrower terms as how to explain mental states by reference to brain states, *the bases of which contain no trace of mentality.*

---

[47] This made sense when the modern mind–body problem was framed by Descartes in the seventeenth century, but such is the grip of materialism that even after quantum physics it continues to be a mostly unquestioned assumption today.

The mind–body problem in this narrower sense is really several problems, which differ in their tractability for a materialist approach. Due to David Chalmers it has become customary to group them into two categories, the "easy problems" and "the hard problem," which deal with different aspects of the mind.[48] Difficult as they are, what makes the easy problems at least easier is that they concern the functional aspects of the mind, or what it *does* – information processing, pattern recognition, and so on – which there is little reason to think cannot be explained by purely material forces. After all, computers process information and recognize patterns, and no one thinks they are not material. So insofar as the mind is like a (classical) computer, as the computational theory of mind has it, we can expect future neuroscience to unravel its functional mysteries.[49] The hard problem, in contrast, is explaining consciousness. The definition of 'consciousness' is highly contested, and for some it encompasses even the functional aspects of the mind.[50] For that reason, following Chalmers, I shall define it as the *experiential* aspect of mind, the feeling, in Thomas Nagel's famous words, that there is "something it is like" to be conscious.[51] Thus, henceforth I will treat 'consciousness' and 'experience' as synonyms.

Especially for social scientists, who mostly study adult human beings, it is important to note that consciousness-as-experience does not imply *self-*consciousness, or consciousness *that* one is conscious.[52] Self-consciousness may be necessary for the kind of social life that humans have constructed, but that is not what is at stake in the hard problem: it is the kind of raw, pre-linguistic experience one might expect a dog, bat, or newborn child to have. Of course, it is open to skeptics to argue that dogs, bats, and newborns aren't conscious, but that seems implausible, since they clearly experience pain. Self-consciousness is not reducible to consciousness in this more primitive sense, but it is dependent on it, so if we cannot explain the latter then there is no hope of explaining the former. The hard problem is not about the reflexive awareness that underlies social institutions, in short, but the simple experience of a subjective point of view.

Explaining consciousness is "hard" for the classical worldview because it is unclear how a purely material world could ever give rise to it. As Joseph Levine has put it, there is an "explanatory gap" between the objective physical descriptions of neuroscience and the subjective experience of those descriptions.[53]

---

[48] See especially Chalmers (1995; 1996); for an excellent review of contemporary mainstream positions on the mind–body problem see van Gulick (2001).
[49] And indeed social scientists themselves are increasingly getting in on this action, as exemplified by the emergence of a new discipline of "social neuroscience."
[50] See Güzeldere (1997); Ram (2009) identifies forty different definitions in the literature.
[51] Nagel (1974); also see Siewert (1998) and Horgan and Kriegel (2008).
[52] See Kriegel (2004) for a good discussion of the distinction.
[53] Levine (1983; 2001); also see Gantt and Williams (2014), who argue that the gap is also ontological.

A thought experiment by Frank Jackson illustrates the problem compellingly.[54] "Mary" has lived her entire life in a black and white room and as such never seen colors, but she is also a brilliant neurophysiologist who knows everything that can be known scientifically about the physics of light and vision. If one day she were freed from her room and could see red for the first time, would she learn anything new? Although philosophers still debate the point,[55] the reason Jackson's argument became a classic is that, intuitively, the answer seems to be yes. She would know what red was *like*, which all the science in the world alone would not have told her. As Chalmers puts it, the hard problem is that "[e]ven if we knew every last detail about the physics of the universe . . . that information would not lead us to postulate the reality of conscious experience."[56]

When confronted with this, social scientists may respond that consciousness must be an "emergent" phenomenon from the immense complexity of the brain, which we are only just beginning to understand. So whatever the physical details, since we know humans are conscious, we can get on with our work, which mostly uses folk psychological rather than physical concepts anyway. But while the idea of emergence has its advocates, as a solution to the hard problem it has not been an easy sell to philosophers of mind.

The basic idea of emergence is that qualitative novelty can appear when parts which themselves lack that quality are organized in a certain way, like unsolid molecules becoming a solid rock. Although this kind of emergence is contested even in the physical sciences,[57] let us grant that emergence in rocks, and other purely material phenomena, is possible. The novelty involved in the hard problem is qualitatively different, so to speak: emergence must explain something that seems more than purely material – namely Mary's *experience* of the material. What has to emerge is subjectivity from objectivity, feeling from no feeling, almost life from death, for which no plausible account has been provided. Thus, critics have argued that whatever the status of emergence in chemistry, when it comes to explaining consciousness it boils down to " . . . and then a miracle happens."[58] If most philosophers of mind are not sold on the emergence solution to the mind–body problem, then it will not get social scientists off the hook.

In sum, while lately there has been progress on the easy problems of mind, after centuries of hard work there seems to have been none on the hard problem.

---

[54] See Jackson (1982: 130; 1986). [55] See for example Cummins et al. (2014).
[56] Chalmers (1996: 101).
[57] See Clayton (2006), Kim (2006), Wimsatt (2006), Corradini and O'Connor, eds. (2010), and O'Connor and Wong (2012) for good introductions to the debate.
[58] As Bedau (1997: 377) puts it, emergence is "uncomfortably like magic." See van Gulick (2001) on emergence and the mind–body problem, Megill (2013) for a defense of emergentism in this context, and Strawson (2006) and Lewtas (2013b) for critiques.

Or at least that is what I take away from philosopher of mind Jerry Fodor's sobering assessment of his field: "[n]obody has the slightest idea how anything material could be conscious. Nobody even knows what it would be like to have the slightest idea about how anything material could be conscious. So much for the philosophy of consciousness."[59] To which he might have added, "and so much for neuroscience too," since the problem Fodor is pointing to is not scientific, as if our modern theory of the brain is on the right track but just not there yet in terms of explaining consciousness.[60] The problem is philosophical. As long as the brain is assumed to be a classical system, there is no reason to think even *future* neuroscience will give us "the slightest idea how anything material could be conscious."

Faced with such a long-standing anomaly, contemporary materialists are in disarray, and some are even beginning to think that something has gone wrong in their framing of the problem.[61] But what? One suggestion, in a Wittgensteinian spirit, is that the hard problem is a pseudo-problem caused by philosophical confusion. David Papineau, for example, argues that the supposed "explanatory gap" is due to our inability to stop thinking in dualistic terms; if we could get over dualism then the gap would disappear.[62] Perhaps, but this is a minority view even among materialists, who one might expect to be receptive to it. Another materialist, Colin McGinn, thinks the problem is genuine but argues that beings with our limited brains are "cognitively closed" to ever solving it.[63] That too might be right, but it's a fishy argument that says materialism can't explain consciousness yet leaves materialism intact as our ontology,[64] so before we go down this road we should be sure that all others have been tried. And then there is the most recent materialist re-think, which is that consciousness (and for good measure, free will too) is actually an *illusion*.[65] I discuss this view in Chapter 9, so here let me just say that it seems a singularly unattractive answer. First, denying the reality of experience is highly counter-intuitive; as

---

[59] Quoted in Kirk (1997: 249). See Levine (2001), Bitbol (2008), Majorek (2012), Nagel (2012), and Lewtas (2014) for good overviews of the challenges facing a materialist solution to the mind–body problem.

[60] My sense is that many philosophers share Noë and Thompson's (2004) doubts that neuroscience alone will solve the problem. However, neuroscientists continue to try; see for example Feinberg (2012).

[61] Another fresh approach is offered by "New Materialists" coming mostly out of the humanities, who are trying to bring a kind of materialism into fields long dominated by social constructivism; for a good overview see Coole and Frost, eds., (2010b). By virtue of its re-thinking of matter, New Materialism has some affinities with my own argument that I address in Chapter 7, but since its proponents are not engaged with the philosophy of mind literature I will bracket it here.

[62] See Papineau (2011), and for more explicitly Wittgensteinian approaches to similar effect see Bennett and Hacker (2003), Overgaard (2004), and Read (2008).

[63] See McGinn (1989; 1999).

[64] Lewtas (2014: 337) likens it to theists' response when confronted with the problem of evil.

[65] See Noë, ed. (2002), Wegner (2002), and Sytsma (2009).

one critic puts it, "believe it if you can."[66] Second, rejecting the explanans (consciousness) rather than the explanandum (classical brain states) is in effect to reject uncomfortable data, which is not rational and makes it unclear how materialism could ever be falsified.[67] Illusionism about consciousness seems moved more by blind faith in materialism than anything else.[68]

Yet illusionism, I take it, is an advance in our understanding, since it seems the logical culmination of the materialist approach to the mind–body problem: given the problem's persistence, *if* materialism must stay then consciousness must go, because like the soul there is no place for it in nature. Unfortunately, given that consciousness is widely seen as essential to the human condition, that means there is no place for *us* in nature either – that we are not "at home in the universe." Thus, many materialists still hope a materialist way to explain consciousness without denying its reality will be found. And perhaps it will.[69] In the meantime, however, the failure to make progress on this issue suggests a worldview in deep paradigmatic crisis.[70] It's not for nothing that consciousness is considered one of the deepest mysteries facing the modern mind.

Intentionality and consciousness

But is it a mystery that should concern social scientists? Judging from our practice the answer might at first seem to be no. Although the elementary objects of social science, human beings, (I think most of us would agree) are conscious, social scientists mostly take that for granted, such that the term 'consciousness' is largely absent from our discourse.

On the positivist side, the ambition is to make social science as much like physical science as possible, generalizable and objective. Since consciousness is idiosyncratic and inaccessible to third-person observation, it is best left aside. Thus, while most positivists routinely attribute intentional states to human beings, the fact that these states are conscious is rarely considered, except perhaps as a methodological barrier to objectivity.[71]

On the interpretive side, matters are less clear, but there is definitely a reluctance to thematize consciousness. Interpretivists mostly focus on what

---

[66] O'Connor and Wong (2005: 674).
[67] See Lewtas (2014) on the irrationality of materialism at this point in history.
[68] On commitment to materialism as a faith see Montero (2001: 69), Velmans (2002: 79) and Strawson (2006: 5).
[69] Though in the absence of reasons to expect such a breakthrough it is unclear why we should still hold out hope for it; see Lewtas (2014: 329).
[70] See Nagel (2012) for a particularly eloquent statement of the crisis, ranging well beyond the mind–body problem to evolutionary considerations and more.
[71] Though if human beings were not conscious then requiring scholars doing research on human subjects to get their work approved by an "Institutional Review Board" would presumably not be necessary.

is public and *shared*, like language and norms, not on what is experienced by individuals. To be sure, many are interested in subjectivity, a concept closely related to consciousness. Yet outside phenomenology, psychoanalysis, and feminist theory, which have always taken experience seriously, the experiential aspect of subjectivity is mostly extruded in interpretivist work, in favor of distinct concepts like *inter*subjectivity, the *discursive production* of subjectivity, and subject-*positions* that do not foreground experience per se.[72] Although I lack the space for exegesis here, consider three giants of interpretivist philosophy: Wittgenstein, Foucault, and Habermas – each in different ways trying to *get away* from a "philosophy of the subject" that they associate with a bankrupt Cartesianism.[73] So despite engagement with the problematic of subjectivity (sic), interpretivists exhibit at least a serious ambivalence about what makes it *subject*-ivity in the first place, namely its conscious aspect. In short, in most of contemporary social science there seems to be a "taboo" on subjectivity.[74]

However, while most social scientists neglect consciousness, we do care about intentional phenomena, which I shall now claim presuppose it. If that is right, then our work makes at least implicit assumptions about consciousness and its place in nature.

Intentionality refers to the fact that mental states like beliefs, desires, and meanings are intrinsically "about" or directed toward things beyond themselves, whether real objects in the world, fictional objects in one's own mind, or the minds of other people.[75] This is in contrast to the states of objects that lack minds, like rocks and glaciers, which are not "about" anything. Although social scientists rarely cite this technical, about-ness meaning of intentionality, it pervades the purposive, folk psychological discourse that we use throughout our work. This is not to say that explaining intentional action is always the goal of social science, much of which studies *un*intended consequences – but those are only meaningful in relation to what was *in*tended. Even self-consciously non-intentional approaches like structural and evolutionary social theories assume purposive action at the micro-level, and insofar as institutions are *collective* intentions, intentionality is present at the macro-level as well.[76] This reliance of social science on intentional discourse is hardly surprising, since in everyday life we routinely attribute intentional states to other people.

---

[72] See Scott (1991) for a particularly sophisticated discussion of experience that I think illustrates this point.
[73] The neglect of subjectivity in modern social theory has led to a number of recent efforts to "bring the subject back in," for which the argument of this book may be seen as providing a physical basis. See for example Frank (2002), Freundlieb (2000; 2002), Henrich (2003), Ankersmit (2005), Ortner (2005), Archer (2007), and Heelan (2009).
[74] See Wallace (2000).
[75] See Jacob (2014) for a good introduction to the philosophical literature on intentionality.
[76] See especially Gilbert (1989) and Searle (1995).

A social science that could not accommodate this fundamental fact would be an impoverished social science indeed.

The relationship between intentionality and consciousness has long been debated by philosophers. Some think that consciousness is dependent upon intentionality, others that intentionality is dependent upon consciousness, and many just ignore one while focusing on the other. However, in recent years the balance of opinion seems to have shifted toward the view that "consciousness is the irreplaceable source of intentionality and meaning."[77] As John Searle puts it,

> I now want to make a very strong claim... The claim is this: Only a being that could have conscious intentional states could have intentional states at all, and every unconscious intentional state is at least potentially conscious. This thesis... has the consequence that a complete theory of intentionality requires an account of consciousness.[78]

To me at least, intuitively this makes sense. Could a machine have genuine intentional states – i.e. of its own rather than ones attributed by us – if it did not also have consciousness?[79] We can program a machine to act *as if* it had intentional states – in the way that a thermostat may be said to be "goal-directed" – but the real intentionality resides in the designer, who is conscious, not the thermostat. Nonetheless, Searle's "very strong claim" is still very much contested.[80] This poses a threat to my narrative if you too have doubts, since if Searle cannot convince his fellow philosophers, then even if I went through all his arguments I should not convince you either.

In the interest of pressing my attack, therefore, I am going to expose a flank here by resorting to stipulation: intentionality depends ontologically on consciousness.[81] Note, with Searle, that this is not to deny the existence of unconscious intentions, as long as they could in principle be made conscious. And nor is it to deny the existence of collective intentions, which are grounded in individual intentions and as such derivatively dependent on consciousness. It is to affirm only that where there is no consciousness there is no intentionality, and so by attributing intentionality to human beings social scientists are also attributing to them consciousness.

---

[77] Siewert (2011: 17).
[78] See Searle (1992: 132), quoted in Kriegel (2003: 273); also see McGinn (1999) and Strawson (2004).
[79] On really having vs. merely ascribing intentional states see Dennett (1971: 91), and Gamez (2008) and Gök and Sayan (2012) for contrasting views on the possibilities for machine consciousness.
[80] See Siewert (2011: 16–19) for an overview of the debate, and Kriegel (2003) for a concise analysis of Searle's (and McGinn's) arguments for the consciousness-first view.
[81] If you'll pardon the military metaphor, blitzkrieg is the only way to wage this campaign, concentrating all of my argument on the weak point in the opposition's lines (the mind–body problem), breaking through, and then bypassing local resistance in the hopes that global success will render it moot.

The threat of vitalism

Nevertheless, the fact that the origin of consciousness and therefore intentionality is a mystery has not stopped social scientists from doing their work, which might suggest that the mind–body problem doesn't matter to us after all. I want to argue now that it does, because the questionable reality of consciousness puts explanations that invoke intentional phenomena into question as well, on the grounds that they are analogous to vitalism.

Vitalism is a theory of what makes life "life," and was widely held in the nineteenth and early twentieth centuries. Against materialists, vitalists argued that the only way to explain life is by reference to an unobservable, non-material *élan vital* or "life force." Materialists were withering in their philosophical criticism of this idea, but what really turned the tide against vitalism were revolutionary scientific advances in biology like genetics, which seemed to eliminate the explanatory need for an *élan vital*. As a result, there are few recently respectable theories that are today as totally discredited as vitalism, which is now considered to be a pre- if not pseudo-scientific doctrine.

What makes vitalism instructive in the present context is that two of the main reasons that scientists and philosophers rejected it apply to explanations that invoke intentional phenomena – and by implication consciousness – as well. First, just as we have no public evidence for the *élan vital*, which vitalists claimed was inherently unobservable, we have no public evidence for consciousness either, only our own experience. Second, as an extra-material force the *élan vital* conflicts with the CCP – or more precisely, the CC*C*P. It's not just that we can't see it because it's unobservable; the *élan vital* can't *be* there because classical physics tells us that no such thing exists. By the same token, if consciousness cannot be reconciled with the CC*C*P then it cannot *be* there either (whence illusionism).

These similarities suggest that a strong analogy exists between the status of consciousness in modern science and the debate a century ago over the *élan vital*. Indeed, Daniel Dennett uses this analogy to criticize those such as Chalmers who think materialism can't explain consciousness. He argues that if that were right then vitalism could be true as well – and since "we all know" that vitalism is false, there must be a material basis for consciousness.[82] Chalmers tries to deflect the criticism by rejecting the analogy, arguing that the vitalists sought to explain only the form and functioning of organisms – akin to the easy problems of mind – which we have since learned can probably be explained by material forces alone; as such, there simply is no "hard problem" of life analogous to consciousness. But that is not so clear. Brian Garrett shows that, historically, some vitalists were concerned with more than

---

[82] See Dennett (1996).

just form and functioning, but the nature of life itself.[83] And as we will see in Chapter 7, notwithstanding the advances in biology that seemed to make vitalism redundant, there is still no consensus on what life *is*, suggesting that there is indeed a hard problem of life. As such, I take Dennett to be right that the two debates are related.

If so, then presupposing consciousness in explanations of human action is like positing an *élan vital* to explain life. We see this conclusion in at least two materialist critiques of intentionalist social science. One is behaviorism, which eschews reference to intentional objects because it invokes causes that cannot be known scientifically. Like Dennett, B. F. Skinner made an explicit analogy to vitalism, arguing that "mentalism" is to psychology what vitalism is to biology (and for good measure, what animism is to physics).[84] The other hardline approach is "eliminative materialism," which foresees the eventual replacement of intentionalist, folk psychological theories with materialist explanations that refer to brain states alone.[85] On both views, social "explanations" that invoke intentional phenomena are at best pre-scientific placeholders until real science comes along, after which they will be seen as no more legitimate than vitalism today.

In sum, there is a tension between social scientists' commitment to a materialist, classical physics constraint on their work, and the routine use of intentional states to explain human action, which are inconsistent with such a constraint. This tension belies any simple claim that "social science is classical," since much of our practice is not. But if we want to keep intentional phenomena in our accounts, and if that presupposes consciousness, then it raises the threat of a "vitalist" social science with no physical foundation.[86] Giving up the quest for such a foundation might be welcomed by some interpretivists, yet even they do not seem prepared to argue that social life could violate the laws of physics. If social scientists want to avoid the charge of pseudo-science, in short, then we will need to rethink what the CCP means for our work.

The anomaly of social structure

Up to this point I have focused on the agent side of the agent–structure problem. Starting there made sense because consciousness poses such a clear problem for the belief that social life is subject to the worldview of classical physics.

---

[83] See Garrett (2006).
[84] See Ringen (1999: 168–169). See Moore (2013) for a recent behaviorist critique of mentalism, and Foxwall (2007; 2008) for a sympathetic treatment of behaviorism and its limits.
[85] See especially Churchland (1988), and more recently Irvine (2012), who argues that the concept of consciousness has no place in science.
[86] In Chapter 8 I shall defuse this threat not by abandoning vitalism but by giving it an alternative, quantum foundation.

However, many social scientists are not interested in what goes on inside actors' heads, but in the structures that constitute macro-level social systems like capitalism, the state, and international system. A good reason for this is that patterns of behavior exist at the macro-level that do not depend on people holding specific desires or beliefs – in the jargon, macro-level patterns are "multiply realizable" at the level of agents.[87] For example, people may follow the law because they think it is legitimate and therefore the right thing to do, because they think it is in their self-interest to obey, or because they are forced to comply. Since motives vary, if the goal is to explain the survival of a state, say, then it makes sense to do so by reference to its structure rather than individual intentions or consciousness.

On this much many social scientists will agree. However, there has long been controversy about how to understand the nature of social structures and their relationship to agents. I address this debate in Part V; here I argue that *any* conception of social structure that depends on intentional phenomena and therefore consciousness will be an anomaly for a classical social science. The main symptom of this anomaly is a feature of social structures that we mostly take for granted: they are invisible. I show how this symptom points to the mind–body problem, and suggest that it poses a "threat of reification."

Where is the state?

Imagine if extra-terrestrials came to the solar system and started surveying Earth from the sky, with sophisticated equipment that enabled them to track the movements of billions of individuals, but not what we were thinking or saying. Would the ETs see any social structures?

Take the state, the ontology of which is typically understood in three different ways, though all of which are ultimately structural.[88] In everyday life and international politics itself the state is usually treated as an agent or "person." These persons pervade the media, our history books, and IR scholarship too, most of which assumes that states are agents with interests, beliefs, rationality, and a capacity for purposive action. Yet ETs would not see any such agents, because notwithstanding clever arguments to the contrary,[89] states do not have material bodies and thus can't *really* be people too. If the real people who act in the name of a state act "as if" they are a unitary agent, then it is only in virtue of a social structure that binds them together.

---

[87] See Wendt (1999: 152–156) for a discussion of multiple realizability and citations.
[88] See the "Forum on the State as Person" in *Review of International Studies* (2004), and also Wight (2006: 215–225) for an account that nicely integrates all three perspectives.
[89] See Wendt (2004).

Students of domestic politics are more likely to conceptualize the state explicitly as a social structure, as a set of institutions that enable collective action on behalf of the common good. Yet our ET friends would not find any states in this sense either. Not just because states are really big and thus difficult to see in their totality, but because institutions are no more material objects than states as agents are.[90] It might be objected that modern states do have well-defined boundaries marked with fences and barbed wire. But how could the ETs distinguish these boundaries visually from the fences and barbed wire which surround cattle ranches or gated communities? Perhaps by plotting the movements of millions of individuals, but in a globalizing world those patterns might be as likely to cross territorial boundaries as to coalesce within them.

Finally, the state may also be seen as a practice. Here it is not being an agent or a structure that constitutes the state, but the material practices of policemen pulling over speeding drivers, diplomats talking to other diplomats, and soldiers shooting enemy soldiers. The ETs might do better seeing states in this sense, since their cameras would at least pick up *something* that is literally the state – its individual agents. But how could they know who those people are without first knowing the (invisible) social structure that constitutes their identities as members of a state?

So where then *is* the state, physically in space? If the question seems strange it is because normally we do not think of the state as something that should have a location or be visible in the first place, like a car or cat. Instead, it is a collective intention, an object of *thought* to which our beliefs and desires may be directed, but is not in itself a material object.[91] The state is a state of mind, in other words, before it is an agent, structure, or practice. Nor is it unique in this regard. The Catholic Church, capital markets, and universities are all collective intentions that can only be "seen" if you already know they are there. Whereas with material objects seeing is believing, with social structures believing is seeing.

Of course, some material objects cannot be seen with the naked eye either, like viruses, distant galaxies, and infrared light. But in these cases, there is no question of their being directly observable at least *in principle*, as the inventions of the microscope, telescope, and infrared glasses attest. That is because they are classical material phenomena and as such mind-independent. In contrast, social structures are mind-dependent, and so no as yet un-invented technology

---

[90] Also see Coulter (2001: 33–34).
[91] For a contrary perspective readers should consult Paul Sheehy's (2006: 97–130) systematic defense of the idea that groups actually are material objects, by virtue of consisting of individuals organized in relations that create causal powers. Overall I am in strong agreement with Sheehy's holist theory of groups (see Chapter 12), but the relations that constitute groups are ultimately mind-dependent and as such in my view cannot be reconciled with a classical understanding of physicality *as materiality*.

will enable ETs to see them. Indeed, even if ETs could scan our brains they would not see them, since social structures are not "in" our brains either, but in our minds.[92] This is not to say that, through careful study of our behavior and perhaps extrapolation from their own experience, ETs could not infer the presence of states. But that would mean coming to see them as we do, by learning to read our minds. Short of that, the ETs would have to report back home that while Earth was teeming with life, perhaps even intelligent life, nowhere were there any states.

The threat of reification

Practically speaking the "location problem"[93] in social ontology is not difficult to solve, since most of us can find states and other socially structured systems when we want to. Moreover, critical and scientific realists have built an entire philosophy of science around the idea that we can know unobservable entities at least theoretically. Nevertheless, given their mind-dependence, it is not clear how social structures are consistent with a materialist ontology. If reality really is nothing but classical matter and energy, then unobservable social structures should not be there, any more than consciousness should be there. So if the latter is ultimately an illusion, then *social structures must be illusions too*.

This poses a threat of reification to those who insist on positing social structures anyway. By 'reification' I mean "the socially induced illusion of thinglikeness."[94] The idea here is that although we often treat social structures as objects out in the world, from a classical standpoint there can be no such objects. Of course, the shared belief that social structures exist makes them real *for us*, since we will act upon those beliefs much like believing in witches will induce people to act as if there are witches. Moreover, social theorists routinely warn against reifying social structures into things in a material sense. But that is precisely the point, for if social structures are not things in a material sense then in a classical world in which everything *is* material, what could they be *if not* illusions? In other words, if we accept a classical physics constraint, then to posit the existence of unobservable social structures is necessarily to reify them.

In saying this I am not trying to suggest that social structures do not, in some sense, exist and have causal powers. Rather, it is that the CCCP provides no grounds for such a claim, and so if we want to retain them in our ontology then it will have to be on a quantum basis instead.

---

[92] As McGinn (1995) points out there is a similar problem in locating consciousness.
[93] The phrase is Hindriks' (2013); cf. Sheehy (2006: 104–107).
[94] Hull (2013: 54); also see Maynard and Wilson (1980). Although originally a Marxist idea, the concept of reification has since been appropriated by other social theories; see Hull (2013) for an excellent, theory-neutral conceptualization.

As if explanation and unscientific fictions

It might be objected that this argument presupposes a realist epistemology. For realists the purpose of science is to disclose the world as it really is and so to invoke intentional phenomena – whether at the agent or structural levels – in social explanations is to posit them as at least provisionally real. As the philosopher of mind Jaegwon Kim puts it, referring to the related issue of mental causation in psychology,

> [t]he possibility of psychology as a theoretical science capable of generating law-based explanations of human behavior depends on the reality of mental causation: mental phenomena must be capable of functioning as indispensable links in causal chains leading to physical behavior. A science that invokes mental phenomena in its explanations is presumptively committed to their causal efficacy; for any phenomenon to have an explanatory role, its presence or absence in a given situation must make a difference – a *causal difference*.[95]

However, many social scientists today subscribe to non/anti-realist epistemologies, from post-structuralism on the one end to empiricism and pragmatism on the other. From their perspective, it might appear that explanations invoking consciousness or social structures only lead to the threats of vitalism and reification if we insist on treating them as real – which there is no need to do.

Consider, for example, how an empiricist or pragmatist might think about explanations that invoke intentional states. On their view, theory should be judged not by how well it discloses the world as it really is – which ultimately cannot be known – but by how well it enables us to predict, solve problems, or otherwise get by in the world.[96] Theory is a tool or instrument, not something to be taken as literally true. Since the assumption that people act *as if* they have intentional states helps us explain their behavior, then even if they are ultimately illusions, it would be a significant loss of knowledge if such states were excluded from our theories a priori while we wait for proper materialist accounts to come along.[97] Thus, whatever the problems that consciousness and social structures might pose for realists, from an instrumentalist or "as if" perspective we should not let ourselves be bullied by philosophers into giving up our best tools, threats of vitalism and reification or not.

Indeed, one might press this objection further against the whole idea of a physics constraint on social science, by pointing out that even in the physical sciences it is common to make assumptions, such as ideal gases and frictionless

---

[95] See Kim (1998: 31), emphasis in the original; also see Maul (2013).
[96] Friedman (1953) is perhaps the most well-known exposition of such a view, but in different forms it is widely held across the social sciences.
[97] See also Dennett (1971; 1987) on the "intentional stance."

planes,[98] which are explicitly fictional. If fictions are essential to the practice of all science, as Hans Vaihinger argued in his "philosophy of the As If,"[99] then why should social scientists eschew reference to consciousness or social structures just because philosophers cannot explain them? As long as they advance knowledge, they should be countenanced just like any other scientific fiction.

Yet, while advocating a liberal attitude toward fictions in science, empiricists and pragmatists nevertheless want to hold the line against fictions that cannot possibly count as scientific,[100] such as God or ghosts, and here the threats of vitalism and reification still have force. What defines an "unscientific" fiction? This has received surprisingly little attention in the literature,[101] perhaps because as methodological atheists modern scientists are not generally inclined to bring supernatural forces into their theories in the first place. But why not? Some Christians think that evil behavior is caused by the Devil. This explanation is coherent, parsimonious, and even supplies a causal mechanism. Yet I suspect most social scientists would reject it a priori. Or take the *élan vital*. It might not explain organisms' functioning, but it does purport to explain the nature of life, the riddle of which materialists themselves have not solved. Yet in today's debate about life, no one considers the *élan vital* a valid construct, *even* as a convenient fiction.

The implicit reason for these exclusions seems to be that in modern science, fictions are legitimate if they refer to something that at least in principle could fall within the CCP, which is to say, is *physical*.[102] As Peter Godfrey-Smith describes fictionalized models, "each model system itself is something that would be concrete if real; it would be an arrangement of physical entities."[103] That makes sense, and as such I too would reject the Devil as a legitimate scientific fiction. However, there is still the question of what 'physical' means, which turns on which physics we are talking about. In quantum physics physicality can encompass mentality, which opens the door to intentional states (and, I will argue, the *élan vital* as well). In classical physics, physicality means materiality, and there seems little prospect that materialism will ever be able

---

[98] Godfrey-Smith (2009: 101). For recent discussions of the value of false models in the social sciences see Rogeberg and Nordberg (2005) and Hindriks (2008).
[99] Vaihinger (1924); see Fine (1993) for a contemporary revival of Vaihinger's ideas, and Contessa (2010) for recent discussion in light of Fine's article.
[100] So much so that Giere (2009) worries that in today's "cultural climate," embracing too eagerly the idea that scientific models are fictions may provide succor to creationists and others who threaten to break the distinction down. Also see Sklar (2003: 438).
[101] Though see Janzen (2012), who draws the line at ghosts.
[102] Such an assumption seems to be implicit in Bokulich's (2012) sophisticated defense of the view that fictions can be explanatory, for example, as it is also in Schindler's (2014) critique of Bokulich.
[103] Godfrey-Smith (2009: 104).

to explain consciousness. Even from an "as if" perspective, in other words, as long as social science is thought to be constrained by a classical CCP, intentional states and social structures have no more place in our work than the Devil or *élan vital*.

My central question, and answer in brief

My point in raising the threats of vitalism and reification is not to suggest that social scientists should abandon intentional phenomena in our explanations, as behaviorists and eliminative materialists would have us do. First, theories that assume intentionality work much better than the alternatives. Behaviorism told us little, and neuroscience is still young (and even when it matures, what about Mary?), so at this point intentional phenomena are the only explanatory game in town. A second reason to keep intentional phenomena in the mix is ethical.[104] It is through attributions of intentionality that our subjects – conscious individuals – make an appearance in our work. Insofar as social science is addressed to those subjects, in the form of normative implications for their behavior, it is important that their subjectivity not be written out altogether – for otherwise who is our addressee?[105] Part of the point of social science I take it is to give meaning to events, by taking what seems inexplicable and fitting it into a pattern that is relevant to people's lives, which will be hard to do if we deny the subjectivity of our audience.

Yet, given a classical CCP, the result is then a de facto dualism between mental and material phenomena.[106] In ontology we face "two incompatible ontologies... the ontology of subjectivity and free agency, on the one hand, and that of things or objects and their relations in the external world, on the other."[107] And in epistemology the best we can do is a Westphalian Settlement, in which positivists and interpretivists live and let live with irreconcilable differences. Since there is only one reality, from a naturalist perspective such dualisms should be accepted only if we have no choice, and indeed going back to the behavioral revolution social scientists have long called for transcending them.[108] However, if the argument above is correct, such efforts are doomed to failure as long as we retain a classical framing of the mind–body problem.

---

[104] Also see Wight (2006: 211–212).
[105] See Frank (2002: 391). For a sampling of views on the link between consciousness and moral cognition see the special issue of *Review of Philosophy and Psychology* edited by Phelan and Waytz (2012).
[106] See for example Wendt's (2006) trenchant critique of Wendt (1999).
[107] Freundlieb (2000: 238).
[108] See Jackson (2008) for a particularly sophisticated recent effort.

Nothing short of a root-and-branch approach will overcome the "bifurcation of nature" that is materialism's legacy.[109]

Hence the central question(s) of this book: (a) how might a quantum theoretic approach explain consciousness and by extension intentional phenomena, and thereby unify physical and social ontology, and (b) what are some implications of the result for contemporary debates in social theory? While obviously philosophical, these questions have practical implications far up the chain of social science, from methods training, to concept formation, theory-building, and empirical research. So the audience to which the book is addressed is all social scientists, not just those interested in the philosophical foundations of their work.

When philosophical debates persist for a long time with no apparent progress, one way to gain traction is to look at what all sides have in common. In the mind–body problem a key, generally unstated assumption is that the nature of the body is clear, and as such "the problem" is with the mind. Specifically, it has been assumed that the matter of which bodies are composed is completely and only material in the traditional sense. This assumption stems both from our experience of ordinary physical objects as things with mass and extension but no subjective inside, and from the development 350 years ago of classical physics, which described a universe of such objects with great success.

Yet since the quantum revolution we have known that at the sub-atomic level matter in the classical materialist sense breaks down into wave functions. Indeed, it is not just that which breaks down, but the whole classical worldview, which is also atomist, determinist, mechanist, and objectivist (Chapters 2 and 3). However, that is not to say that, individually, all of these assumptions are wrong, because it is not clear what precisely quantum physics is telling us about reality. This has been the subject of intense debate since the 1930s between advocates of at least a dozen "Interpretations" of quantum mechanics (Chapter 4). One of the big issues in this debate is which (if any) classical assumptions can be salvaged, and at what metaphysical price. For example, despite the breakdown of the classical view of matter as tiny little objects, quantum theory does not necessarily show that mind is present at the sub-atomic level, which would preserve the core principle of materialism in a broader sense – No Fundamental Mentality. But it does force those who would keep materialism to accept some very radical consequences indeed. And so it goes for every other reading of quantum theory, each of which makes different trade-offs but is equally counter-intuitive.

Regardless of which interpretation of quantum theory one prefers, the existence of such a debate shows that the nature of matter is no less mysterious

---

[109] The phrase is Whitehead's; see Jones (2014), and more generally Barham (2008).

than the nature of mind. In Montero's words again, in the mind–body problem we have not only a mind problem but also a "body problem."[110] Materialists will object that the problem of mind takes place on a macroscopic scale far above the quantum level, where matter has its familiar corpuscular/energetic properties and classical physics is for all practical purposes a valid description. That is not to say that quantum theory does not apply to the macro level, since it applies everywhere; the whole universe is quantum. But except in very special conditions, wave functions collapse or decohere into particles as soon as they interact. This is why the macroscopic world appears to us as it does – classical. So, materialists can say that even if matter in the ultimate, quantum sense is not good old fashioned matter anymore, the kind of matter relevant to the mind–body problem still is.

But what if the orthodoxy is wrong? What if the physics of the mind is not classical, but quantum – not in the trivial sense that all of reality is quantum, but in the substantive sense that *consciousness itself* is quantum mechanical?[111] That is the radical hypothesis of "quantum consciousness theory" (Part II). The theory has two parts: quantum brain theory and panpsychism. The latter does the crucial work in solving the hard problem, but the former plays a key role in overcoming long-standing objections to panpsychism. From these elements I then go further than most advocates of the theory, arguing that it implies a new, quantum form of vitalism.

Quantum brain theory hypothesizes that the brain is able to sustain quantum coherence – a wave function – at the macro, whole-organism level (Chapter 5). How the brain might do this is not agreed on by the theory's advocates, who have explored the possibility from different angles. The pioneering and most well-known approach is due to Stuart Hameroff and Roger Penrose,[112] but as we will see there are other approaches as well. Notwithstanding their differences, however, their conclusion is the same – that the brain is a quantum computer.

Whether quantum brain theory is true is speculative and deeply controversial, but it has attracted a growing number of advocates. Two facts may account for this interest. First, we understand very little about the brain below the neural level, so things that today we "know" as false might well turn out to be true. Second, there is the zero progress from a classical standpoint on the problem of consciousness. Thus, to skeptics who say there is no way a quantum theory

---

[110] See Montero (1999).
[111] That it has to be one or the other stems from the fact that classical and quantum physics are the only two physics we have (relativity theory is part of classical physics). Contemporary physics is admittedly incomplete, but one would expect a future physics to subsume quantum physics in the same way that the latter did classical.
[112] See Hameroff (1994) and Penrose (1994) for their seminal statements.

of consciousness could be true, one could easily retort there is no way a classical one could be either!

Quantum brain theory takes known effects at the sub-atomic level and scales them upward to the macroscopic level of the brain. However, by itself this would not explain consciousness, since it does not tell us why any physical system, even one as mind-bogglingly complex as a quantum computer, would be conscious. This question is addressed by the ontology of panpsychism (Chapter 6).

Panpsychism takes a known effect at the macroscopic level – that we are conscious – and scales it downward to the sub-atomic level, meaning that matter is intrinsically mind*ed*. With this principle of Fundamental Mentality panpsychism opposes not only materialism but also idealism and dualism. Against idealists who privilege the mind panpsychists see mind as only an aspect of matter, not something to which matter can be reduced. By the same token, while panpsychists agree with dualists that mind and matter are distinct, against dualists they do not think that matter is purely material and thus that mind is a substance over and above it. Mind and matter constitute a duality, not a dualism, one that I will argue emerges from an underlying reality that is neither mental nor material (a view known as neutral monism).

Panpsychism can be traced to the ancient Greeks, but it also finds expression in the great modern philosophical systems of Spinoza, Leibniz, Schopenhauer, Whitehead, and others. However, like vitalism, after the 1940s panpsychism became an object of ridicule in Western philosophy, and as such for decades was ignored in the literature on the mind–body problem. It is therefore perhaps symptomatic of the contemporary crisis of materialism that since the 1990s there has been a strong resurgence of the idea within philosophy of mind and – interestingly – the philosophy of physics.[113] For unlike classical physics, there is a clear place in quantum mechanics for mind – the collapse of the wave function. As physicist Freeman Dyson put it, "mind is already inherent in every electron, and the processes of human consciousness differ only in degree but not kind from the processes of choice between quantum states which we call 'chance' when they are made by electrons."[114] To be clear, quantum theory does not *imply* Fundamental Mentality, but it *allows* for it physically, and results in quite an elegant interpretation of the theory. As such, quantum consciousness theory suggests that two of the deepest mysteries confronting modern science – how to interpret quantum theory and how to explain consciousness – are two sides of the same coin. Although I will not be primarily concerned with the former here, I argue that putting them into conversation enables us to bootstrap a solution to the latter.

---

[113] See Malin (2001), Primas (2003), Pylkkänen (2007), and others cited in Chapter 6.
[114] Quoted in Skrbina (2005: 199).

This puts the threat of vitalism in a new light (Chapter 7). Earlier I used this threat to develop a *reductio* of intentional explanation: since "we all know" there is no such thing as the *élan vital*, intentional explanations are no more scientific than vitalism. Now it appears that the vitalists were right all along. Extrapolating from the rapidly growing literature in quantum biology, I argue that there *is* an irreducible "life force," quantum coherence, which can only be known from the inside, through experience. In this "quantum vitalism"[115] we end up with an updated version of the *Lebensphilosophie* of Goethe and the nineteenth-century Romantics, and later Schopenhauer, Nietzsche, Merleau-Ponty, and others. Such a perspective calls into question a basic metaphysical assumption of modern science, that all ultimate explanatory principles must be "dead."[116] By suggesting that experience goes all the way down, quantum consciousness challenges this philosophy of death, echoing the great physicist Eugene Wigner's intuition that it is through biology that the deepest problems of physics will eventually be solved.[117] Far from excluding intentional states from social science, quantum vitalism would be their very basis.

If quantum consciousness theory is true then the physics constraint to which human beings and society are subject is quantum rather than classical. This matters, because in a quantum world lots of things are possible that aren't in a classical one, and so a quantum perspective presents an opportunity not only to overcome dualism in social science, but to expand our conception of social reality altogether.

The basic directive of a quantum social science, its positive heuristic if you will, is to re-think human behavior through the lens of quantum theory. To this end, in the second half of the book I explore some implications of quantum consciousness theory for social ontology, and specifically the agent–structure problem. In doing so I cannot hope to engage properly the vast literature on that problem, which has been framed to date by implicit classical premises. Instead, my goal is to theorize the agent–structure problem through the quantum, as if we were theorizing it for the first time.

In Part III I focus on human agents in isolation from their social context, in order to unpack what individual quantum minds bring to the social table. I devote a chapter each to three mental faculties – Cognition, Will, and Experience. In Chapter 8 I summarize quantum cognition, decision, and game theory, where the evidence for a quantum model of man (sic) is strongest. The upshot is that, in contrast to the classical view that people have a portfolio of actual mental states in their heads upon which they then act, these states exist only as

---

[115] The phrase is Hameroff's (1997).
[116] See Montero (2001: 71); as Schopenhauer put it, materialism "carries death in its heart even at its birth" (quoted by Hannan [2009: 11]).
[117] See Wigner (1970), and also Matsuno (1993).

"superpositions," or wave functions of potential states, until they are elicited in interaction. I link this to the performative view of agency developed by post-structural theorists, arguing that quantum mind is its physical basis. In Chapter 9 I take up Will, arguing that a quantum model supports two claims that comport with common sense but are anomalous from a classical standpoint: that Will is inherently free, and that its causal power is teleological rather than mechanical. Finally, in Chapter 10 I address Experience, in particular our experience of time, which I argue exhibits temporal non-locality. This suggests that, in certain respects, it is possible to change the past, not just narratively, but literally.

In Parts IV and V I turn to the nature of social structure. Notwithstanding my treatment of agents in Part III in isolation from each other, the key idea running through these chapters is that, by virtue of our entanglement from birth in social structures, human minds are not fully separable. Non-separability refers to the fact that the states of quantum systems can only be defined in relation to a larger whole. It is the basis of non-local causation in quantum mechanics, and what makes quantum phenomena irreducibly holistic.

In Part IV I focus on the special case of language, which is the medium of all other social structures. Here I draw on the rapidly growing literature on quantum semantics, which shows that concepts exhibit "semantic non-locality." I draw two major implications from this work. First, against the dominant view that linguistic meaning is compositional, built up out of separable elementary units, semantic non-locality implies that meaning is irreducibly contextual (Chapter 11). Second, this in turn provides a new perspective on the Problem of Other Minds, or how human beings can know each other's thoughts. Building on an analogy to light, which enables non-local "direct perception" of visual objects, I argue that language enables us to do the same with other minds – that language is like light (Chapter 12).

Finally, in Part V I address the agent–structure problem more directly, challenging both the emergentist ontology associated with critical realism and the reductionist ontology associated with rational choice theory. Against the former, I argue in Chapter 13 that social structures are not actual realities existing somewhere above us in space, but potential realities constituted by inherently non-local shared wave functions. In this way, quantum theory underwrites a "flat" rather than stratified social ontology, in which individuals are the only real realities. While that might seem to vindicate individualism, the holism and non-locality of quantum theory belies that conclusion. The key here is the unique character of emergence in quantum contexts. When applied to social life, quantum emergence leads to a solution to the agent–structure problem not unlike the recent "practice turn" in social theory, according to which agents and structures are both emergent effects of practices. In Chapter 14 I suggest that all this points toward a vitalist sociology. Taking the state as an example, I argue that the state is a holographic organism endowed with collective consciousness.

Re-inventing the wheel?

The social ontology developed in Parts III–V recapitulates many ideas that are already held in the social sciences, in some cases widely (such as intentional explanations being legitimate). This is not surprising. Despite my suggestion above that, if asked, most social scientists would say their work is ultimately grounded in classical mechanics, they don't think very often or explicitly about that constraint. Instead, they have pressed ahead with trying to make sense of social life with whatever tools seem to work best, most of which originate in folk psychology rather than physics. Folk psychology relies heavily on intentional phenomena in its accounts, and since I have argued that such phenomena cannot be reconciled with the classical worldview, a good deal of extant social science must at least implicitly have a quantum aspect.

However, that then raises the question of whether taking an explicitly quantum approach to social science will just re-invent the wheel. By the end of this book I hope you will be convinced that the answer is no, that there is real value added in such an exercise, so for now let me just highlight six contributions that I think it can make.

First, by providing a naturalistic basis for consciousness and intentional phenomena, the argument seeks to unify physical and social ontology. If it is correct, that would not only justify theoretical practices that social scientists often take for granted but are illegitimate from a classical point of view. It would also point well beyond the social sciences to philosophy and more, by creating the possibility of giving the human experience a home in the universe.

Second, even when it affirms extant theoretical practices a quantum approach may force a re-thinking of how they are understood. For example, intentional explanations will not be seen as a mechanical unfolding of preexisting mental states, nor will unobservable social structures be seen as really real.

Third, by virtue of these changes phenomena that are currently considered anomalous would be explained. The clearest example to date is the success of quantum decision theory in explaining the Kahneman-Tversky effects, but as we will see there are many other anomalies for classical social science that are predicted from a quantum perspective.

Fourth, and more prospectively, the conceptual, logical, and methodological tools of quantum theory offer the potential for revealing new social phenomena. Consider structural power, a concept often invoked by critical theorists but which from a classical perspective is impossible to see as anything other than a concatenation of local power relations – and thus can only be illusory. Conceptualizing structural power as a form of non-local causation suggests that it is indeed quite real, at least in the quantum sense of the term.

Fifth, if quantum consciousness theory is taken as an explanation for consciousness, then the concept of complementarity could resolve the controversy between positivists and interpretivists. Although I will not address epistemology much in this book, given how bitter and intractable the dispute has been, this may be one of the most important pay-offs of a quantum social science.

Finally, there would be significant normative implications as well. Most mainstream normative theorizing about social life today, especially in the liberal tradition, assumes a world of separable, constitutionally pre-social individuals who then struggle to achieve sociability (the state of nature and all that). It is not hard to see the imprint of the classical worldview on this atomistic and competitive picture, but either way, quantum phenomena are marked by their holistic and "cooperative" character. That points toward a more communitarian and relational starting point for normative theory, which suggests that sociability is less a hard-fought achievement than the pre-condition and norm of human existence, and if so, that our obligations to others run correspondingly deeper. As with the epistemology question, I will only gesture in this direction, but if trying to make society a better place is one of the main reasons social scientists do what they do, then it is a thematic that needs to be developed down the road.

In sum, although the idea that social life is quantum mechanical may seem bizarre at first, and some of the arguments I make are indeed radical, I hope to show that it is actually quite intuitive, far more so than treating social life on the classical model of clashing billiard balls. And not only that, as I argue in the Conclusion, it is *too elegant not to be true*. For the price of the two claims of quantum consciousness theory – that the brain is a quantum computer and that consciousness inheres in matter at the fundamental level – we get solutions to a host of intractable problems that have dogged the social sciences from the beginning. These claims are admittedly speculative, but neither is precluded by what we currently know about the brain or quantum physics, and given the classical materialist failure to make progress on the mind–body problem, at this point they look no more speculative than the orthodoxy – and the potential pay-off is huge. If classical social science is in fact founded upon a mistake, then far from re-inventing the wheel a quantum social ontology would give our wheels the right ground on which to roll.

Having said that, however, I should emphasize that this book may be read in an "as if" rather than realist way. My personal belief in the argument certainly helped make it possible to spend ten years of my life working on it. But the test of my narrative is not that you come away thinking that social life really is quantum mechanical. After all, many of the experts upon whose work I draw, like quantum decision theorists, are themselves agnostic about the philosophical implications of their work, much less ready to embrace panpsychism! The test is whether you come away convinced that a quantum perspective offers at least a fruitful heuristic for thinking about long-standing controversies in social theory, and ultimately for doing empirical social science. If so, then I will count my effort as a success, and we can argue about the realism issue later.

Situating your observer

As we will see shortly, observation plays a critical role in quantum phenomena. Although the nature of that role is contested, what is clear is that observation has a very different status in the classical worldview. There, a subject–object dichotomy is assumed such that we can at least aspire to observe in a neutral and passive way, by accurately recording the properties of objects that are assumed to exist independently of us.[118] Injecting an observer's values or interests into this process breaks down the subject–object distinction, resulting in measurement bias relative to the true values. Observing sub-atomic particles is an altogether different matter. In most interpretations of quantum theory particles cannot be said to exist prior to measurement, and in preparing quantum systems for observation an entanglement is created with the observer that affects what is eventually seen. That does not mean that the observer literally creates reality, but it does mean that she participates in what is actually observed, and as such observation cannot even in principle approximate the classical ideal of separation.

This breakdown of the subject–object dichotomy makes it natural to wonder about *my* relationship to the reality I will be observing for you. On one level that reality is the same as yours, namely social life, which I will be arguing is quantum mechanical. However, I will be making that case not directly, through experimental tests of quantum hypotheses, but indirectly through readings of philosophical and scientific discourses in several disciplines to which I am a complete outsider. While I have worked hard to make sense of these discourses, I cannot speak with any real authority about them. The risks of dilettantism and just plain silliness are obvious, as well as of my observations creating a putative reality rather than just measuring it. In short, *caveat emptor*!

Yet there is one useful epistemic role that I think I can credibly play in this book, which is that of an anthropologist.[119] Each of the bodies of scholarship I take up in effect constitutes a distinct culture, with its own assumptions and concerns. As anthropologist I make no pretense of having an insider's knowledge of these cultures, especially since my "fieldwork" consisted of reading

---

[118] See Jackson (2008) for a good discussion.
[119] More skeptical readers might say Marco Polo is more like it, whose tales of wonder were of dubious veracity – but at least he encouraged others to go see for themselves.

their texts rather than talking with the natives.[120] Moreover, I approached my reading selectively, with a view not toward mastering the cultures on their own terms, but toward supporting my own, quantum view of social life (so more like a nineteenth- than twenty-first-century anthropologist perhaps!). In part for this reason I have provided extensive references to work outside the social sciences that relates to my argument so that readers may follow up,[121] but even then I do not aim to provide comprehensive overviews of each literature. Instead, my goal is to show how things look to *me*, from my situated "somewhere" rather than an objective "nowhere."

While this means that everything that follows should be taken with a large grain of salt, an anthropological attitude does have two virtues, not just for my own argument but – if they will bear with an itinerant and iconoclastic guest – even for my "subjects." First, judging from citation patterns, a striking feature of the cultures I look at, even ones that are talking about the same things, is how isolated they are from each other, making them more like a vast archipelago than a system of contiguous states. "Island hopping" will enable me to highlight potential connections, and perhaps start some conversations in the process. Second, as an outsider I have the freedom to say things that cannot or would not be by insiders – like my Emperor Has No Clothes suggestion above that after 350 years of failing to explain consciousness, perhaps materialism is simply wrong. Whether in each case I am playing the fool or stating the obvious will be up to each of you – also situated, participant observers – to decide.[122]

As you do so, however, my hope is that you will approach this book holistically. Whatever plausibility is due to its argument will stem not from a close reading of everything in the bibliography, much less from expert knowledge, but from how the underlying ideas fit together into a coherent whole. Despite the complexity of the narrative, the book's main thesis is quite simple – that human beings are walking wave functions – and not particularly dependent on the details. Like the social reality it purports to describe, it is emergent in a quantum sense.

---

[120] My discussion in Part I did, however, benefit considerably from conversations with and comments from two physicists, Chris Wendt (my brother) and Badredine Arfi, as well as a psychologist who knows the physics, Jerome Busemeyer.

[121] In contrast, I only cite social scientific scholarship as necessary to make my case, since to reference properly everything that social scientists have written on the topics I cover would vastly expand an already long bibliography.

[122] Either way, for me this book has been "Alex's Excellent Adventure."

Part I

Quantum theory and its interpretation

Introduction

Chapter 1 dealt with one of the deepest mysteries in modern science: how to explain consciousness. This Part deals with another mystery: how to make sense of quantum mechanics. Both involve a problem of reconciliation with the classical worldview, which is one hint that the two mysteries might be related. A second is that, unlike classical physics, which makes no reference to or allowance for consciousness, quantum theory raises the issue of consciousness and its relationship to the physical world in a very direct way. These hints do not necessarily mean there *is* a connection between the two mysteries, but by the end of this Part I hope to have shown why it would be natural to look for one.

My more immediate aim here is to give the social scientific reader an introductory understanding of the experimental findings of quantum theory, its key concepts, and the debates about its interpretation. The discussion assumes no prior knowledge of quantum physics and makes no use of mathematics. Quantum theory without equations might seem like an oxymoron, and certainly reading this Part will not enable anyone to *use* quantum theory. But to a perhaps surprising degree it is possible to understand quantum theory without being able to use it. Its findings can be communicated in ordinary language, and the same goes for its main concepts and interpretive debates. One consequence of this has been the emergence of a virtual cottage industry of books on quantum theory for a lay audience. Most are written by physicists, and many are quite good.[1] However, it is not just popular treatments that do without equations: much of the professional philosophical literature does as well. (I draw on both kinds of literature below.) Philosophers of physics are trained to understand the math, of course, which is necessary for a full appreciation of the issues. But the primary questions about quantum theory are about metaphysics,

---

[1] See for example Zukav (1979), Herbert (1985), Friedman (1997), and Rosenblum and Kuttner (2006); also see A. Goff (2006) on "quantum tic-tac-toe," which was written as an aid for teaching students with no background in the area. For braver souls Haven and Khrennikov (2013) provide an excellent technical overview of quantum theory aimed at social scientists.

not physics. Once we understand the basics we should be in a position to have a reasonably well-informed discussion.

That said, "understanding" in this context is something of a misnomer, since no one, not even physicists, *understands* quantum theory, if by that we mean what it is telling us about reality. Richard Feynman has often been quoted as saying that anyone who claims to understand it clearly doesn't know what he is talking about. It would seem, then, that just as it is possible (within limits) to understand quantum theory without being able to use it, it is possible to use it without fully understanding it. As such, my real goal here is to give readers an understanding of why we do not understand quantum mechanics. It will take most of this Part to do that, but let me summarize the problem here in two ways, one from a common-sense perspective and the other more theoretically.

The simplest characterization of the problem is that the reality quantum theory depicts is nothing like the macroscopic material reality described by classical physics. This is not because its constituents are so much smaller than everyday objects; scale is not the issue. It is because the properties of quantum systems seem fundamentally inconsistent with macroscopic reality: material objects dissolve into fields of potentiality, larger objects cannot be reduced to smaller ones, events seem not to have causes, and so on. Put another way, since the classical worldview is the basis of what we today take to be common sense about the world, at a gut level quantum theory just doesn't make sense – indeed, so much so that John Bell, one of the great physicists of the twentieth century, argued that whatever picture of reality eventually emerges from quantum theory will surely "astonish us."[2]

A more precise characterization of the problem is that the predictions of quantum theory are probabilistic, yet the outcomes of experiments on quantum systems are always definite, classical events. Of course, most social scientific theories are probabilistic as well, but *if* the macroscopic world is classical, then there must be an ontologically deterministic process underlying those probabilities about which we are simply ignorant at present. In the quantum world such an assumption is problematic. Quantum probabilities behave quite differently than classical ones (whether objective or subjective),[3] and although there are ways to make quantum theory deterministic, these are contested and come at a high price in other respects. As such, most physicists today believe that quantum theory is "complete," in the sense that there is no deeper, as yet undiscovered, classical theory or "hidden variable" that could explain its predictions deterministically. Hence the mystery: how to explain the transition

---

[2] The quote is in Rosenblum and Kuttner (2002: 1291).

[3] For a good overview of philosophical questions pertaining to quantum probability theory see the special issue on the topic in *Studies in History and Philosophy of Modern Physics*, June 2007.

from the quantum to the classical world? Is it a real process, and, if so, how does it happen? If not, then what is it?

In short, quantum theory does not answer crucial questions about the reality it describes, and strongly suggests that they *cannot* be answered even in principle by science. As Steven French puts it, we face an "underdetermination of metaphysics by physics."[4] Understanding the theory is therefore a philosophical rather than scientific problem, of "interpreting" it to produce a coherent picture of reality. Since the 1930s over a dozen interpretations have been proposed that make strikingly – indeed, wildly – different ontological and epistemological assumptions (Chapter 4), and once proposed they never die because of the under-determination problem. As they are designed to explain the same data, it is difficult if not impossible to discriminate among them empirically.[5] Fortunately, this does not prevent scientists from *using* quantum theory, which means the philosophical debate is actually of little interest to practicing physicists. But in considering its relevance to *social* science the debate is harder to avoid, since some interpretations suggest there are no such implications, while others suggest there are many.

So we are in something of a bind: unable to determine which interpretation of quantum theory is correct, and unable to determine quantum theory's relevance to social science as long as that is the case. This might seem to call into question my decision to set epistemological questions aside in favor of ontology, and to reduce the question of social ontology to a matter of personal metaphysical taste, but I don't think so. The fact that physicists cannot (yet?) adjudicate scientifically among interpretations of quantum theory does not mean there is no other way to do so: the criteria are simply more philosophical. The philosophy of quantum theory is full of arguments for and against each interpretation. Reasonable people disagree about these arguments, but that is true in any area where the data do not yield definitive answers. No doubt you the reader, when encountering some of these interpretations below, will make judgments about their relative plausibility, and even though you could not prove your judgments empirically, you could offer principled considerations on their behalf. The interpretive debate about quantum theory is about how to weigh those considerations, which thanks to the debate are much clearer today than they were eighty years ago. Moreover, we have an ace in the hole: quantum consciousness theory (Part II). As an independent, ontological reason for

---

[4] French (1998: 93).
[5] However, on the plus side this may account for the mutual respect and relative lack of polemic evident in the debate; strong opinions notwithstanding, almost everyone seems to understand that their preferred interpretation is speculative and could be totally wrong. At least in their writings, in my experience philosophers of physics are the most open-minded academics in the world (though admittedly this is a low bar!).

thinking that the two great mysteries of modern science are linked, it suggests what a correct interpretation of quantum theory should look like.

This Part is organized into three chapters. Chapter 2 summarizes three of the major experimental findings of quantum theory and key concepts that have been developed to describe them. In Chapter 3 I consider six challenges that the theory poses for the classical worldview, which might be considered its "negative" implications, or what it tells us the world is *not* like. Finally, in Chapter 4 I address a sample of five positions in the interpretive debate about the theory's "positive" implications, or what the world *is* like. I concentrate on the debate between materialist and idealist interpretations, with a view toward setting up the panpsychist argument of Part II.

2 Three experiments

Quantum theory is a mathematical formalism that allows physicists to predict the probability of observing different outcomes in experiments on sub-atomic systems. It has been tested more rigorously than any theory in science, and has never been wrong. However, strictly speaking it does not "explain" the behavior of sub-atomic systems, since it does not propose a mechanism to account for it.[1] It tells us *that* they will behave in certain ways, but not *why*. Thus, quantum mechanics is not a theory in the sense familiar to social scientists, namely a body of laws that explains some part of reality. The explanatory question is the subject of the interpretive debates that surround the theory, not of the theory per se. That said, the literature routinely refers to quantum mechanics as a theory, and I shall use the terms interchangeably below.

A common way to introduce quantum theory is through some of the key experiments that have confirmed its predictions. In this chapter I describe three of the most well-known: the Two-Slit Experiment, the Bell Experiments, and the Delayed-Choice Experiment.[2] While painting the same overall picture, each reveals distinct features of the quantum realm. While their interpretation is contested, here I will do my best to present just the theory's findings, saving interpretive questions for later.

The Two-Slit Experiment

This experiment actually had its origins long before the quantum revolution. One of the most controversial questions in classical physics was whether light was made of particles or waves. In keeping with his atomistic worldview, Newton favored the particle or "corpuscular" theory of light, which was the majority view among physicists through the eighteenth century. However, in

---

[1] Squires (1994: 3).
[2] Plotnitsky (2010: Chapter 2) provides a more detailed but still highly readable introduction to the Two-Slit and Delayed-Choice Experiments, which I discovered too late to draw upon in this manuscript.

1801 Thomas Young performed an experiment that seemed to prove conclusively that the wave theory was correct.[3]

In his experiment, Young set up a light source behind an opaque screen with one small opening or slit, through which the light passed in a concentrated beam. The light then struck a second screen with two slits side by side, through which it passed on to a third screen with no slits. The slits on the second screen were close enough together that light passing through them would illuminate a partly overlapping area on the third screen. If light were made of particles, then we would expect the overlapping area to be more brightly lit than the non-overlapping regions, since it was being struck by particles from both slits. But this is not what Young found. Instead, the light hitting the third screen formed an "interference pattern" of alternating bright and dark bands. This is what we would expect if light were made of waves. When one wave crest meets another crest they amplify each other (the bright bands on the screen), whereas when a crest meets a wave trough they cancel each other out (the dark bands). Young's demonstration of this interference pattern seemed to settle the debate in favor of the wave theory, and as such was important in the subsequent development of energy physics, which culminated in Maxwell's theory of electro-magnetic waves in 1864.

But despite Young's achievement, by the end of the nineteenth century two anomalies for the wave theory had emerged, the black box radiation problem and the photoelectric effect. The former was concerned with the question of why objects glow when heated. Classical physics predicted that energy in the form of radiation will be emitted from a heated object in a continuous stream, with the level of radiation increasing with higher frequencies of light. But experiments had shown that the relationship was in fact curvilinear, with radiation levels low at both low and high light frequencies, and peaking in the middle. Even more disturbing for classical theory at the time was the fact that when all the frequencies were added together the result was an *infinite* amount of radiation, which did not make sense. In what he later called an "act of desperation," Max Planck solved these problems in 1900 by assuming that energy does not flow continuously, as posited by the wave theory, but in discrete particles or "quanta." With this as his premise, he introduced a mathematical constant to predict the pattern of radiation at different frequencies, which it did successfully. "Planck's constant" became the fundamental building block of quantum mechanics, and its discovery now marks the beginning of the quantum revolution.

The importance of Planck's discovery became clear in 1905, when Einstein published a paper on the photoelectric effect. It was known that, under certain conditions, shining a light on a piece of metal causes the metal to emit electrons. The wave theory predicted that as the intensity of light at a frequency is

---

[3] The following discussion draws on Zukav (1979: 83–86), Friedman (1997: 51–54), and Malin (2001: 27–29).

increased the energy of electrons emitted should go up proportionally as well; it also predicted that the photoelectric effect should occur at all frequencies as long as the intensity is sufficiently high. But experiments had shown that this was wrong – the intensity increased the number of electrons that were emitted but had no effect on their energy, and the photoelectric effect disappears when the wavelength exceeds a cutoff. Why? Einstein showed that this could be explained by Planck's model. Energy varies according to wavelength and not intensity, and at short wavelengths the energy of particles of light (photons) is higher, enabling them to dislodge electrons in the metal, whereas at long wavelengths the energy of the photons is too weak to produce this effect.

In short, Planck and Einstein's findings implied light was not a wave but a "shower of particles,"[4] suggesting that Newton's corpuscular theory had been right all along. However here's the rub: Young's results still stood. Planck and Einstein had shown that in *their* experiments, which were designed to answer different questions than Young's, light behaved as if it were a particle, not that Young was wrong about *his* experiment, where light still behaved as if it were a wave. Thus, rather than vindicating Newton, the implication of Planck and Einstein's work seemed to be that light was *both* wave *and* particle, which makes no sense in the "either-or" world of classical physics.

If light or energy can behave like a particle, or matter, it soon turned out that the reverse was also true, that matter can behave like a wave. This was predicted theoretically in 1924 by Louis de Broglie in his doctoral dissertation, and confirmed experimentally for electrons two years later. In contrast to the long-accepted view that electrons were tiny objects, it now appeared that they could also be "standing waves."[5] Normally we do not see this wave-aspect of matter because the "matter waves" of ordinary objects are so small relative to the size of the objects that their effect is negligible, but at the sub-atomic level the waves are sufficiently large to have a measurable effect.[6]

That *all* matter-energy can behave like both waves and particles has since been proven in modern, quantum versions of the Two-Slit Experiment. In these experiments a particle "gun" shoots a stream of electrons (or any other type of particle) toward a screen with two slits. They pass through the slits and the location of their hits is recorded on a photographic screen.[7] If we first close one slit, then the distribution of hits is concentrated directly across from the open slit, with a small tail on either side. If we then close the open slit and open the other one, we get a similar result across from the second slit. These results are what we would expect if electrons were particles. That might suggest that if we leave both slits open then the result should be a simple sum of the two

---

[4] See Herbert (1985: 57–58).
[5] Zukav (1979: 122); the phrase is Schrödinger's.   [6] Zukav (1979: 119).
[7] The following discussion draws on Albert (1992: 12–14), Friedman (1997: 53–54), and Nadeau and Kafatos (1999: 46–51).

distributions, with two bells separated by connected tails. However, that is not what we observe: with both slits open, in the region of overlapping tails we get the characteristic interference pattern associated with waves.

To rule out the possibility that the electrons are interfering with each other before they pass through the slits, physicists have recently devised a way for the gun to fire one electron at a time. Yet, even in this case with both slits open, a series of shots *still* generates an interference pattern, which suggests that each electron passes through both slits and then interferes with itself![8] Since an electron cannot be divided, that seems impossible; common sense tells us that a particle must go through one slit or the other. We can test that belief by putting "detectors" on each slit. Sure enough, the detectors show that each electron passes through only one slit, which seems to support the idea that electrons are particles. Yet, adding this extra measurement has an unintended effect: it destroys the wave interference pattern on the photographic plate. This result reveals several other paradoxical features of the quantum world.

Measurement is creative

The build-up of an interference pattern in a two-slit experiment without detectors indicates that each electron goes through both slits, and therefore behaves like a wave before it hits the photographic plate. The disappearance of this pattern when detectors are put on the slits challenges that conclusion, since each electron then goes through only one slit, suggesting that electrons are particles the whole time. But that can't be right, since it does not account for the interference pattern when the detectors are removed. Thus, the correct conclusion is that *as long as the electron is not being observed it behaves as if it is a wave, and as soon as it is observed it behaves as if it is a particle*. Measurement is somehow intrinsically connected to a change in our description of the electron. The implication is that in the quantum world, observer and observed form a *single system*, rather than being separable as in the classical world. Shimon Malin likens this to the "refrigerator door effect" on a small child: whenever she opens the door the light is on, and so she assumes that it must be on when the door is closed.[9] But eventually she learns that opening the door (measurement) is what turns the light on. Although it is less clear that electrons "really" change, something analogous is going on here. Whether in an ontological or merely epistemological sense, measurement is "creative."

Collapse of the wave function

At this point it is natural to ask what kind of ghostly phenomena are these waves, which seem to exist only as long as we do not observe them? This is

---

[8] For discussion see Malin (2001: 45–46).   [9] Malin (2001: 48).

Three experiments

one of the fundamental interpretive questions in quantum mechanics that I am deferring in this chapter; here I just want to discuss how waves are defined within the formalism itself (about which all sides agree), and relate them to the measurement process.

The dynamics of waves in quantum mechanics are described by a mathematical equation discovered in 1925 by Erwin Schrödinger ("Schrödinger's equation"). Since waves vary, the content of wave functions will vary as well, but their definition is always the same: a wave function represents the potential for all outcomes – the location of particle hits – that might be observed when we perform a measurement. Importantly, therefore, a wave function consists only of *possibilities*, and as such the wave it describes is not in any actual or definite state like a classical wave. Instead, in quantum mechanics all of the wave's possible states are said to have potential to exist, in a mathematical sense, simultaneously in "superposition."[10] We might think of a wave function as a "field of potentialities."[11] The *probability* of any of these potentials being actualized is determined by squaring the amplitude of the wave function at each point. These probabilities tell us the likelihood of finding a particle at any given location, but the meaning here is not like our conventional, classical idea of probability. In classical physics, probability refers to our degree of uncertainty about some actual state of affairs, for example how many red balls there are in an urn of red and blue balls. In the quantum case, we cannot say that there is an "actual state of affairs" with respect to the question we want to ask, and so the probabilities refer only to possible observations. Put another way, some kind of reality is out there which gives answers to our questions, but the answers are not out there until we ask them. More than this we cannot know, even in principle. Prior to its measurement, the wave function constitutes a complete description of a quantum system; there is no definite reality hiding behind the wave function about which we could obtain further knowledge if only we had the means.

Wave functions are dynamic and evolve deterministically. This means that, as long as we do not perform a measurement, we can predict exactly how the *probabilities* of particle hits will change over time (though not where actual hits will be observed). It is important to keep this in mind given the popular association of quantum mechanics with *in*determinism; at least with respect to the evolution of the wave function, quantum mechanics is deterministic.

What is *not* deterministic is the process by which waves "change" (sic) into particles. This process goes by various names in the literature, depending on interpretive sensibilities. Those who think it describes a change in the real world usually call it "collapse of the wave function," whereas those who

---

[10] For an excellent introduction to the idea, and puzzle, of superposition states see Albert (1992: 1–16).

[11] Malin (2001: 47).

think it merely describes a change in our knowledge usually prefer the less ontological sounding phrase "state reduction" (for "quantum state"). Either way, all agree that when a measurement is performed, the probability of all the possible outcomes that are not actually observed goes to zero, and that which is observed goes to one. This result is immediate, occurring in "no time"; it is indeterminate, there being no way to predict exactly what outcome will be observed; and it lacks any apparent causal mechanism. So what determines where the particle hits? According to the formalism itself, the answer is purely unpredictable (though this is not to say that all outcomes are equally likely). The nature of wave function collapse is one of the most puzzling issues in quantum mechanics, and we will return to it below.

Complementarity

The fact that wave functions describe quantum systems only up to the instant at which we perform a measurement means that descriptions of these systems as waves and as particles are mutually exclusive but jointly necessary. Depending on the experiment, we always see only one aspect or the other: put detectors on the slits and we get particles, take them off and we get waves; close one slit and we get particles, open both and we get waves. Yet, the two descriptions are also jointly necessary in the sense that a complete description of the system requires both; each by itself is only a partial representation. A particle model cannot account for the interference pattern we observe on the photographic screen, while a wave model cannot account for the hits we observe individual electrons making. Since quantum systems seem to exhibit both wave and particle aspects, this has been dubbed the "wave-particle duality." This duality poses a significant challenge to a classical understanding, since it requires a conceptual framework with mutually incompatible elements.

In 1927 the Danish physicist Niels Bohr came up with such a framework, which he called the principle of "complementarity."[12] On the surface, the principle seems straightforward enough, since in daily life we often think in such terms. Descriptions of a house from different sides, for example, are complementary: they are mutually exclusive but jointly necessary for a total picture. However, in quantum theory the concept is more subtle. When we look at a house from multiple vantage points, there is no suggestion that our descriptions are incompatible, or that when we are looking at one side of the house the other sides do not exist. We *know* the other sides of the house are out there even though we can't see them, and a simple change of view will prove it. This is exactly what we do *not* know in quantum mechanics. If we design an

---

[12] See Bohr (1937). He apparently got the idea from a book by William James. The concept has since taken on several meanings, a good survey of which may be found in Hinterberger and von Stillfried (2013).

experiment to demonstrate the velocity of an electron, then we no longer have any basis for saying that a particle is located out there (and vice-versa), since there is no way to acquire information about its location without destroying the interference pattern across positions from which we infer the existence of velocity. Indeed, "[t]he absence of any such information is *the essential criterion* for quantum interference to appear."[13] So the mutual exclusion here is deeper than the merely perspectival problem in the classical case: knowledge of one precludes knowledge of the other, and to that extent they are inconsistent.

The wave-particle duality illustrates one way to think about complementarity, where a trade-off exists between knowledge of an unmeasured quantum system and that of a measured one: as soon as we perform a measurement (which "creates" particle behavior) the wave function is no longer an appropriate description of the system. Another way into complementarity is through perhaps the most widely known idea from quantum mechanics, the "Uncertainty Principle."[14] In 1925 Werner Heisenberg showed that it is impossible to measure the exact location and momentum of a particle at the same time and indeed that there was a direct trade-off between the two: the more precise our knowledge of a particle's location, the less precise our knowledge of its momentum, and vice-versa. So the particle has either a well-defined position or well-defined momentum; it cannot have both. We can only "know" both if we are willing to settle for approximate information about each, since the experiments necessary to measure position and momentum precisely are mutually exclusive.

The combination of wave-particle duality and the Uncertainty Principle suggests that complementarity is a general feature of the quantum world, and in fact Bohr thought that it might apply at the macroscopic, biological level as well – a hunch I pursue in subsequent chapters. However, as we will see below, it is not clear what inference should be drawn from the fact of complementarity, and in particular whether the problem is only an epistemological one of inherent limits to knowledge, or whether it also has ontological implications. Even Bohr himself took several years after first enunciating the principle to settle on one view.[15] But the point for now is that quantum systems cannot be completely described by either a particle or a wave model alone, and thus the quest for a unified or coherent description in the classical sense must be given up.[16] The best we can hope to do is to bolt mutually exclusive descriptions together within the framework of complementarity.

---

[13] Zeilinger (1999: S289), emphasis in the original.
[14] See Malin (2001: 32–35) for a good discussion of the Uncertainty Principle.
[15] See Bohr (1937; 1948), and Held (1994) for a good discussion of Bohr's evolution, and the problems of interpreting complementarity more generally.
[16] Malin (2001: 37). I say "in the classical sense" here because a description in terms of superposition is in a sense "unified."

The Bell Experiments

In the years following the consolidation of quantum mechanics in 1927 Einstein and Bohr engaged in an intense debate about whether its apparent conclusion that reality was non-deterministic could be correct. Einstein was skeptical, based on the conviction that "God does not play dice."[17] Over the next three years he came up with several seemingly decisive objections to quantum theory, each one of which Bohr was able to parry. After one last, dramatic but failed attempt in 1930, Einstein was forced to concede that at least quantum theory's *description* of the world was accurate. However, still convinced that reality could not be non-deterministic, he then took another tack: instead of challenging the correctness of quantum mechanics, he would challenge its completeness. In 1935 Einstein and his Princeton colleagues Boris Podolsky and Nathan Rosen published a landmark paper (known ever since as "EPR") that seemed to prove that, if we assume that quantum mechanics is correct then it *had* to be incomplete. Guided by an analogy to statistical mechanics in classical physics, in which probabilistic patterns at the macro-level are explained by deterministic causes at the micro-level, EPR argued that there must be a more fundamental, *sub*-quantum theory that can explain, deterministically, the probabilistic character of quantum outcomes.[18]

Their argument was based on an ingenious thought experiment.[19] To prove that quantum mechanics is incomplete EPR had to demonstrate that there are "elements of reality" which cannot be accounted for within the framework. Since what constitutes an "element of reality" had been thrown into doubt by quantum theory, in order not to beg the question they chose as their criterion of reality the following condition: if one can predict with certainty the value of a physical quantity, *without in any way disturbing (i.e. measuring) it*, then it is "real." They then came up with what they thought was a proof that a sub-atomic particle can have a well-defined position and a well-defined momentum simultaneously, which according to quantum mechanics is impossible.

The proof begins with a system of two particles, A and B, moving in opposite directions after interaction. Quantum theory allows us to know the total momentum of the system (the sum of the momentum of each particle), and the distance between them. EPR further assume that each particle has an equal momentum, which means that the total momentum of the system is zero (since they are moving in opposite directions). The question then is: can we know the position of B without disturbing it? The answer is yes: by measuring A we acquire knowledge of A's position, and since we know the distance to B, we can then calculate B's position. Therefore, B must be an "element of reality," even though we have not measured it. Moreover, we can perform the same trick

---

[17] Malin (2001: 63). [18] See Einstein, Podolsky, and Rosen (1935).
[19] The following discussion draws on Malin (2001: 63–66).

with momentum: by measuring A's momentum, we can calculate B's, since we know the total momentum of the system is zero. The position and momentum of B are therefore real, since they are both well defined in the absence of measurement. Since quantum mechanics cannot describe such a particle, it must be incomplete.

EPR's thought experiment hit Bohr like a "bolt from the blue,"[20] and it took him some time to formulate a response. However, three months later he published a reply, in which he argued: (1) that A and B are "entangled" by virtue of their relationship and thus should not be considered fully separable particles, and (2) measurements of position and momentum are still complementary, since only one can be measured at a time. Measuring the position of A and thus knowing the position of B does not imply that B has a momentum sitting out there waiting to be recorded, because momentum was not measured for A. Momentum does not exist until it is created by measurement.[21] These two arguments challenged EPR's conclusion that particles were "elements of reality."

This response had considerable force, but so did EPR, and thus, unlike Bohr's previous debate with Einstein about the correctness of quantum mechanics, it was less clear who had won this one about its completeness. Most physicists sided with Bohr's "Copenhagen Interpretation," though for reasons having as much to do with the sociology of knowledge in physics as the substantive merits of his argument.[22] But EPR had their supporters as well, and so the question of completeness was left unresolved. One reason for this was that at the time both sides thought that it was impossible to test experimentally which view was right. After all, all agreed that the quantum mechanical description was correct, and both views were consistent with the available evidence. EPR had raised a fundamental question about how this evidence should be interpreted, but the question seemed to be only a philosophical one with no empirical implications.

All that changed in 1964 when an Irish physicist, John Bell, figured out a way to test which side was right. EPR had based their argument on two ontological assumptions, both reflecting the classical worldview. The first was realism, the principle that the world exists independent of the human mind. This assumption is implicit in EPR's definition of an "element of reality" as something that can be known to exist without disturbance by human beings, and was the main focus of the Bohr-EPR debate. The second assumption, less remarked upon at the time, was locality, which is the principle that no causal influence can propagate faster than the speed of light. This was meant to rule out what Einstein called "spooky action at a distance." (The combination of these two assumptions is called "local realism.") By modifying EPR's original

---

[20] Malin (2001: 66).   [21] Thanks to Jerome Busemeyer for clarifying this point to me.
[22] For a thorough study of how Bohr's view became dominant, see Cushing (1994).

model, Bell was able to derive a theorem that stipulated some inequalities regarding expected correlations between test results that must be satisfied if local realism was true.[23] It turned out that these inequalities are *not* satisfied when calculated via quantum mechanics, which meant that either the latter was wrong (if the inequalities were satisfied in experiments) or that local realism was wrong (if they were not satisfied).

Bell's Theorem had another consequence. Bell proved that EPR was wrong *if* quantum mechanics is correct (which EPR had accepted). But there was some lingering doubt about that assumption, and as such it was still possible to argue in the face of Bell's Theorem that perhaps it was quantum mechanics that is wrong, rather than EPR. After all, local realism is the foundation not only of classical physics but of all our experience of material objects in daily life; to reject it would be to call everything we think we know about the universe into question. The second result of Bell's paper was that it suggested ways to test local realism – and specifically the locality assumption – experimentally. Beginning in 1972, these experiments culminated in a decisive test by Alain Aspect and his colleagues at the University of Paris in 1981. The "Bell Experiments" proved that quantum theory was indeed correct, and that local reality was therefore *not* a basic feature of the universe.

The set-up of the Bell Experiments is similar to EPR's thought experiment.[24] It starts with a light source that emits two identical photons, A and B, in opposite directions. One of the properties of photons is their polarization, which refers to their "spin" or angular momentum. Like any quantum property, polarization is subject to different descriptions (i.e. there are multiple "bases" for looking at it), but for the sake of simplicity these can be reduced to two: horizontal and vertical. Although mutually exclusive, these descriptions are in superposition, which means that in its wave aspect the photon's polarization does not have a definite value.

The photons are then sent toward two polarizers, behind which are measurement apparatuses that record the hits made by the photons, as in the Two-Slit Experiment. A polarizer is a device that allows photons having the same polarization as itself to pass through, but not photons with a different polarization (as in sunglasses, for example). Thus, if the polarizer is polarized along the vertical axis, then photons polarized vertically will get through, and be recorded as hits on the measurement device standing behind, whereas photons polarized horizontally will not pass and be recorded. (The opposite pattern would be

---

[23] The details of Bell's Theorem are too complicated to review here, but are nicely summarized in Albert (1992: 66–70).
[24] The following discussion draws on Zukav (1979: 307–312), Herbert (1985: 215–227), Nadeau and Kafatos (1999: 70–80), and Rosenblum and Kuttner (2006: 142–152).

observed for a polarizer polarized along the horizontal axis). These effects are what allow us to determine the actual state of the photon.

Now, despite the fact that in their unmeasured, wave state the polarization of each photon is a superposition of vertical and horizontal, quantum theory predicts that when the photons are measured – when their wave functions collapse into particles – whatever the polarization of A, it will be perfectly correlated with the polarization of B. And this is in fact what physicists observed in the initial tests of Bell's inequalities in the 1970s. If the axes of both polarizers are arranged to be parallel, so that each one lets through the same kind of photon, then the measured polarization, whether horizontal or vertical, of each photon will be the same. Every time a hit is recorded for A (because its polarization is on the same axis as the polarizer), a hit is also recorded for B; and every time no hit is recorded for A (because its polarization is on a different axis than the polarizer) no hit is recorded for B. Moreover, if we rotate one polarizer so that its axis is perpendicular to the other, so that each now allows photons through with opposite polarizations, the measured states are still correlated, only negatively so: every time a hit is recorded for A, no hit is recorded for B, and vice-versa. Finally, if we rotate the polarizers so that they are neither parallel nor perpendicular to each other, then for every angle in between the measured states will *still* be correlated. The correlations produced by these experiments violated Bell's inequalities, and as such strongly supported the correctness of quantum mechanics.

However, by virtue of how the initial experiments were designed, there was still the logical possibility that each photon somehow "knew" the state of the two polarizers in advance, either through the initial set-up of the experiment or through the communication of a signal between the two photons after it began. Admittedly, it was not clear how such a possibility could happen or affect the outcome, but since it would save the assumption of locality the skeptics wanted more proof. Aspect and his colleagues provided this proof in 1981 in a decisive final experiment, in which the orientation of the two polarizers was randomly determined while the photons were in flight. There was therefore no way the photons could "know" the final state of the polarizers, and given that no signal can travel faster than the speed of light, nor was there any possibility of communication between the photons. And yet upon measurement their polarizations were *still* correlated. Aspect's results were confirmed in 1982 by Nicholus Gisin at the University of Geneva, who further showed that the correlations did not degrade over longer distances. In the Gisin variant the photons were not measured until they were eleven kilometers apart, yet even then their measured states were perfectly correlated. Given the enormous distance represented by eleven kilometers relative to the sub-atomic scale, this indicated that the polarization of entangled photons would be correlated *no matter how far apart* they were, even across the universe. And, since what goes for photons goes for other particles, and since all particles in the universe have at some time or other been entangled, the upshot is that everything in reality is correlated. The universe, in short, is one big quantum system.[25]

What is going on here? The crucial issue at stake in the Bell Experiments was the classical assumption of local reality: the reality principle and the principle that no signal can be transmitted faster than light. In effect, the conjunction of reality and locality is ruled out; we can retain locality and reject reality or retain reality and reject locality. Bell's experiments showed that even in the absence of a faster-than-light signal one could still get interference across separated particles, which suggests that reality is fundamentally non-local. To see the significance of this, it is important to keep in mind the difference between the polarization state of the photons before and after measurement. Before measurement each is in a superposition of both vertical and horizontal polarization. Thus, the problem is not one of our ignorance: there simply is no "fact of the matter" about a photon's polarization before measurement. Yet, when we measure them correlation is what we find. When the actual polarization of one is determined, the value assigned to the twin now interferes with any other tests in other directions of polarization that can be performed on the twin – and this happens instantaneously, no matter where in the universe the twin is. It is almost as if each photon "knows" what is happening to the other, without any signal being transferred between them ("spooky action at a distance").[26] So what is going on here is not superluminal communication, but something more like "communing." In this regard a distinction is sometimes drawn between "influencing" and "signaling." Unlike local causation, which depends on the transmission of a signal, in "non-local causation" there is mutual influence, but no signaling.[27] Making sense of non-local causation is another interpretive challenge of quantum theory, and it will also play a key role in the social ontology I develop in Part V below.

The Delayed-Choice Experiment

The non-locality exhibited in the Bell Experiments is spatial: instantaneous correlations can occur between events that are widely separated in space. However, in quantum mechanics non-locality can also be temporal: correlations can occur between events separated in time as well.[28] This contradicts the

---

[25] Nadeau and Kafatos (1999: 81).
[26] See Hardy (1998) for a good, if somewhat technical, introduction to this idea.
[27] See Berkovitz (1998 and 2014) for thorough analyses of these issues.
[28] See Filk (2013) for a good overview of temporal non-locality.

Three experiments 55

classical assumption that time is a linear succession of "Nows," and – even more provocatively – suggests that events in the present can, in a certain sense, affect the past. The idea of temporal non-locality began with a thought experiment by the eminent physicist John Wheeler, and has since been confirmed in "Delayed-Choice Experiments."[29]

In these experiments a beam of light enters the apparatus from the left and is split into two by a half-silvered mirror, which "is a piece of glass coated with a layer of silver so thin that it reflects half the light and transmits half the light, rendering the glass half transparent and half reflective."[30] Half of the light beam is deflected away from the mirror along path A, while the other half passes straight through the mirror along path B. The light on A then encounters two regular mirrors set at angles such that it is eventually diverted back toward B. At the point where the two paths rejoin stands another half-silvered mirror that splits each beam into two, so that each half of each beam is now traveling in unison with a half from the other beam. The traveling pairs are then in superposition, which means they should exhibit an interference pattern on detectors. Long before the quantum revolution, physicists had verified that we do in fact observe such an interference pattern.

So far the experiment is operating at an aggregate level, with beams of light consisting of trillions of photons. In this set-up, it might still be reasonable to say – in a classical vein – that some photons travel along A and some along B, even though all together they produce an interference pattern. But consider what happens if we narrow the light down to a series of individual photons. If classical physics is right, then each photon must choose a path. Which path will it take, and when will it choose? Quantum mechanics tells us that we cannot know which path it will take for sure, but one might at least think the choice is made when the photon encounters the first half-silvered mirror at the start of the experiment, since that is where the two paths initially diverge. But that is not the case. (For reasons I do not understand) none of the mirrors performs a measurement and as such there is no collapse of the photon's wave function. It remains in a state of superposition, traveling through both paths simultaneously so to speak. No "choice" is made at all.

Now for the "delayed" part. We first need to make one adjustment to the experiment, mounting one of the corner mirrors on path A on a spring so sensitive it will cause the mirror to vibrate if it is struck by even a single photon. The spring can be set to "on" or "off." When on, its vibration will perform a measurement, collapsing any passing wave function into a particle, in the same

---

[29] See Wheeler (1978). The following discussion draws on Herbert (1985: 164–167), Nadeau and Kafatos (1999: 48–50; 186–189), and Malin (2001: 180–183). For a comprehensive but technical overview see Bahrami and Shafiee (2010).
[30] Malin (2001: 129). I draw heavily on Malin (2001: 128–131 and 180–183) in what follows.

way that the detectors on the slits in the Two-Slit Experiment induce collapse. With the spring on, in other words, the photon will come out of superposition and make a "choice": either a hit will be recorded on the spring-mounted mirror, indicating that the photon took path A, or none will be recorded, indicating that it took path B. When is this choice made? Classical thinking tells us it must have been when the two paths initially diverged at the first, half-silvered mirror. But the facts are otherwise. To see this, we manipulate the experiment in the same way we did in the Bell Experiments. First we run the experiment with the spring off. In this case we observe a wave interference pattern, indicating that no choice of path was made. Then we run the experiment again, only this time we quickly turn the spring on *after* the photon has passed the presumed point of choice at the first, half-silvered mirror. In this case we observe a particle hit (or not), indicating that the wave function collapsed and the photon took one path (or not). However, since the change in the spring's state occurred while the photon was in flight, after the choice point, what this experiment shows is that what we do in the present is instantly correlated with what happened *in the past* (i.e. temporal non-locality). In other words, by our intervention the photon's choice is "delayed" until after the point at which it could have chosen in a classical sense.

Wheeler highlights the far-reaching implications of this finding with a thought experiment on a cosmological scale.[31] He begins with Einstein's discovery that gravity is a curvature in space-time that bends light. Gravity is associated with objects, and the larger the object, the greater the curvature. Now consider a quasar, a phenomenon billions of light years away that is an intense source of light. The fact that it takes light that long to reach us means the information it carries reflects what was happening on its source back then, not today; for all we know the quasar could have long since disappeared. So in looking at quasars we are quite literally seeing into the very distant past. Next, assume that a galaxy stands somewhere precisely between the quasar and ourselves. This galaxy exerts a massive gravitational pull that causes the light coming from the quasar to curve around the galaxy to both its left and its right. After passing the galaxy the curved beams of light then meet up again and cross before continuing in separate directions. The question is: will a photon from the quasar pass to the left or the right of the galaxy?

Intuitively one would think that the answer must have been determined long ago, when the photon encountered the galaxy. Yet, quantum theory shows that depending on where we measure the photon we can get different descriptions. If we measure it at the point at which the left and right beams cross, then we get a wave interference pattern, suggesting that the photon went both ways (as in the Two-Slit case with both slits open and no detectors). However, if we measure

---

[31] See Wheeler (1994: 124–125).

it after the beams diverge – i.e. on one of the two, post-galaxy trajectories – then we get a particle hit rather than an interference pattern, indicating that the photon took one of the two paths (as in the Two-Slit case with detectors on the slits). So despite our powerful intuitive assumption, it turns out that our choice of measurement procedure today affects how we should describe what happened in the past.

Does this mean that we can literally change the past? Well, in a sense – although here we begin to get into interpretive issues.[32] Recall that as a wave, before measurement, a photon is in a superposition of possible states. This means that relative to a classical frame of reference it exists as a potential, not an actuality. Thus, when we measure it, what we are changing is not its actual properties before we measured it but its potential to take on one actual property rather than another. In other words, once a measurement is done, the photon's actual past or "history" is determined.[33] However, the process by which that history was made does change the photon's *potential* past, or what could have happened. That might not sound counter-intuitive; after all, in everyday life we routinely assume that the future is "open," which implies that what actually does happen – history – is contingent and thus could have turned out differently. But as we will see below, this way of talking is loose and inconsistent with the classical worldview. Classical reality is ontologically deterministic; in such a world, there is no meaningful sense in which what could happen can be different than what does happen. By contrast, in the quantum world what could happen includes events that might not actually happen, and it is the act of bringing actual events into being that constitutes what could have happened in the past. In short, if we are going to argue that the future is open and history is contingent, it will have to be on quantum grounds – but, as I suggest below, we then have to accept the possibility that the past is open as well.

---

[32] The following draws on Herbert (1985: 167), Malin (2001: 180–183), and Grove (2002). See Chapter 10 for further discussion in the context of the human sciences.

[33] Surprisingly, however, this does not mean it cannot be undone. Another bizarre (and experimentally verified) prediction of quantum theory, known as "quantum erasure," allows one to return a collapsed state to its superposition state; see Aharonov and Zubairy (2005).

3 Six challenges

Despite scholarly consensus on the empirical facts, ever since its inception in the 1920s there has been deep disagreement about how to interpret quantum theory. "To interpret a physical theory is to say what the world would be like, if the theory were true."[1] All scientific theories require interpretation, since strictly speaking what they describe are our experiences of the world rather than the world itself. Thus, whether we are trying to explain or merely describe the world, we are always engaged in inference.[2] To make inferences we need a context for interpretation. For any given claim this context is provided locally by related theories, more broadly by disciplinary paradigms, and ultimately by worldview claims about the nature of reality. The challenge in interpreting scientific theories is to integrate them into this hierarchy of knowledge. Usually this is not too difficult. New theories often require adjustments to nearby theories, but in doing so we can rely on paradigms to make them cohere. More rarely paradigmatic assumptions are themselves challenged, but in that case scientists can fall back on their worldview to make sense of the needed changes. By implication, the most difficult interpretive problems arise when our worldview is called into question, which leaves us without any frame of reference on which to fall back.

Quantum theory poses a worldview problem. Prior to the quantum revolution the classical worldview could not be tested scientifically and so it was essentially metaphysical. Even though it corresponded to our experience of material objects, we lacked proof that it adequately represented the deep structure of reality. By giving us access to that deep structure for the first time, quantum mechanics made it possible to do such a test, or "experimental metaphysics."[3] Unfortunately, the results seemed to contradict classical assumptions, at least on the sub-atomic level. This led to a widely acknowledged "crisis" in the classical worldview, and to the suggestion that we are on the "edge of a paradigm shift."[4] But shift to what? Without a still higher-order framework of knowledge

---

[1] Ruetsche (2002: 199). [2] King, Keohane, and Verba (1994).
[3] Esfeld (2001: 225). [4] See Peacock (1998).

on which to rely, it is very difficult to make sense of what is going on.[5] In short, while quantum theory tells us definitively *that* the world behaves in certain non-classical ways, it does not tell us *why*.

In the rest of Part I I address the problem of interpreting quantum theory, which I have split into two. Chapter 3 pits the experimental results against six assumptions of the classical worldview: materialism, atomism, determinism, mechanism, absolute space and time, and the subject–object distinction. Call these the negative implications of quantum theory, about which there is considerable agreement. In each case I begin with a brief review of the classical assumption, show how it is challenged by quantum theory, and conclude with a few comments about where contemporary thinking is at. (This format will create some redundancy with Chapter 2, but on this unfamiliar terrain that is probably not a bad thing). In Chapter 4 I turn to the more contested positive implications of the theory.

Two caveats are in order before proceeding. First, the language one should use to describe the negative implications of quantum theory depends partly on whether it is interpreted "realistically" (as referring to reality) or "instrumentally" (as just a tool for making predictions). This is itself a matter of dispute and so even at the stage of just describing the challenges posed by quantum theory it is hard to avoid interpretation. Since it poses the challenges most clearly in this chapter I shall adopt a realist discourse, but will note as needed how an instrumentalist might see things differently.[6] Second, even on a realist construal, quantum theory does not necessarily mean the classical worldview is completely wrong (though there are significant constraints on how *right* it can be).[7] Although the classical worldview is a coherent whole, it may be possible to save some of its assumptions if others are jettisoned, and much of the interpretive debate over quantum theory is about just that. But even if not all of the classical worldview is wrong, what is clear is that the overall picture of reality emerging from quantum theory will not be classical.

The challenge to materialism

Materialism is the view that at base reality is composed exclusively of matter, within which there is no trace of mentality. The elementary constituents of classical matter are very small, but their properties are understood by reference to the macroscopic matter with which we are all familiar. Classical material particles are objects, things, or substances; their existence is context-

---

[5] Squires (1990: 177).
[6] Those uncomfortable with this binary choice might consider Richard Healey's (2012) pragmatist approach to quantum theory. I thank Stefano Guzzini for raising this point.
[7] See Ferrero et al. (2013).

and mind-independent; they have hardness or mass; they have definite location and extension in space; and they are dead. Macroscopic phenomena that seem not to have these properties, like life and consciousness, are to be explained in terms of more microscopic phenomena that do.

Quantum mechanics casts doubt on this ontology. Sub-atomic particles are not objects or things in the materialist sense. We have no basis for saying they exist until we measure them, and so they are not context-independent.[8] Prior to measurement, what "exists" (using that term loosely) at the quantum level is the wave function, which does not have hardness or mass. And given the Uncertainty Principle, we cannot even say that quantum systems have definite locations in space. About the only materialist assumption that is not immediately threatened by quantum mechanics is that matter is dead, though I argue later that even this is open to doubt. In short, until we measure it, in the quantum world there seems to be "no there there." The bedrock material objects of the classical worldview, upon which it was thought we could anchor our knowledge of macroscopic reality, seem to have vanished into thin air.

What conclusion should we draw from this? Contemporary physicalists try to save materialism by broadening our definition of 'material' to incorporate the findings of quantum physics. However, as we saw in Chapter 1, it is problematic to move the goal posts in this way, since the likelihood of future discoveries in physics means that the content of materialism could change yet again. Moreover, what if future physicists conclude (as some already have) that mind is an elementary property of reality? Given that it is precisely *against* such an idealist view that materialism has always been defined, should we then redefine materialism as idealism? That would make it non-falsifiable and trivially true. For materialism to be interesting it must at the very least mean No Fundamental Mentality. Beyond that constraint, however, quantum theory does not tell us what ontology to replace classical materialism with. Various non-mental candidates for fundamental status have been proposed – "information," "dispositions," "processes," "events" – none of which squares easily with classical materialism's thing-ontology. Bottom line: as a result of quantum theory the nature of matter is just as problematic as the nature of mind.

The challenge to atomism

Atomism makes three claims: 1) large objects are reducible to the properties and interactions of smaller ones; 2) objects have definite properties; and 3) objects are fully "separable," meaning that their identity is constituted solely by their internal structure and spatio-temporal location rather than by their

---

[8] Paty (1999: 374).

relationships to other objects.⁹ These claims generate a methodological injunction for scientists to try to explain phenomena by decomposing wholes into parts. In the social sciences this demand for "micro-foundations" leads to methodological individualism, the doctrine that explanations are not complete until we have shown how social outcomes are produced by the properties and interactions of independently existing individuals.

All three of these atomist claims are problematic today. The first, reductionism, has been challenged even on classical grounds. Wholes often seem to be "more than the sum of their parts," and as a result the anti-reductionist idea of emergence has gained increasing acceptance.[10] Defenders of atomism now typically defend only the weaker claim of "supervenience," according to which macro-level facts are determined or "fixed" by micro-level ones, but not reducible to the latter.[11] However, even supervenience presupposes the other two claims of atomism, both of which are called into question by quantum theory.

The requirement that objects have definite properties is threatened by the Uncertainty Principle, which suggests that quantum systems lack such properties until measured; until then they are superpositions of potential properties. The claim of separability, in turn, is threatened by quantum entanglement and non-locality. When two or more quantum systems are entangled, the parts of the combined system are not fully separable, since their properties depend on their relationship to the whole. The fact that the parts of entangled quantum systems have only *relational* properties upsets the fundamental principle of atomism that the nature of wholes is determined by the attributes of parts, and even raises doubts about whether "parts" exist at all.[12] Paul Teller offers a helpful analogy to clarify the latter problem.[13] If we put two identical coins in a piggy bank on successive days, and then take one of them out, it makes sense to ask whether we drew the first or the second coin; this corresponds to the idea of individuality in classical physics. In contrast, if we deposit a dollar in a checking account on two successive days, and then withdraw one a day later, it makes no sense to ask this question. The dollars in a checking account do not have an individual existence before they are withdrawn ("measured"). Even if we do not go as far as Teller in rejecting the very existence of parts, however, it is widely agreed that at least the *properties* of entangled quantum systems

---

[9] On separability see for example Healey (1991), Esfeld (2001: 207) and Belousek (2003).
[10] See Sawyer (2005) for an excellent overview of emergentism, especially as it pertains to the social sciences, though he remains within a classical frame of reference; cf. Humphreys (1997a).
[11] See Horgan (1993) for a good introduction; these ideas are discussed at length in Chapter 13.
[12] See Teller (1986) and Esfeld (2004). This is a matter of debate. Some argue that even if their properties are entangled, we can still speak of individual particles; others argue the whole idea of individuality must go. See for example Castellani, ed. (1998), French (1998), Esfeld (2001: Chapter 8), French and Ladyman (2003), Arenhart (2013), and Dorato and Morganti (2013).
[13] Teller (1998: 114–115 and passim).

are dependent on the state of the combined system, which therefore forms an indivisible whole.[14] And since entanglement is a universal phenomenon, this suggests that at the quantum level the assumption of separability is nowhere correct.

If not atomism, then what? In contrast to the uncertainty about what should replace materialism, there seems to be a broad consensus that quantum mechanics implies holism, atomism's traditional rival. How exactly holism should be understood is less clear;[15] but some kind of holism seems essential. Note that this does not necessarily imply holism at the macroscopic level, where separability might still hold for all practical purposes. However, in that case, macroscopic atomism would be a contingent effect of the ongoing process by which quantum reality becomes classical, which strips it of its ontologically privileged status.

The challenge to determinism

Determinism is the view that there is no inherent randomness in nature, that what happens in the present and future is completely fixed by the laws governing the motion of matter in the past.[16] In a sense, the present and the future "have already happened."[17] Of course, this does not mean we can necessarily *know* what will happen in the future. Determinism is an ontological rather than epistemological thesis. Given the complexity of the world and the imperfections of our science, there are many things that we cannot predict and as such appear random. But the fact that an event appears to be random does not mean that it is random. We cannot predict the outcome of the throw of a fair die, but according to determinism it is completely fixed by its properties and the conditions of its fall. That seems intuitively plausible, and indeed like much of the classical worldview determinism conforms to common sense, except in one respect: it conflicts with our experience of free will. Determinists have wrestled with this anomaly for centuries, and as we will see in Part III many have concluded that free will is an illusion.

The probabilistic behavior of quantum systems poses a more serious threat to determinism than does free will because it cannot be dismissed as an illusion. Early in the quantum revolution this led to much hand wringing, especially by convinced determinists like Einstein, who concluded that quantum mechanics could not be a fundamental theory. The problem is rooted in the collapse of

---

[14] For a good discussion see Maudlin (1998).
[15] See Teller (1986), Esfeld (2001; 2004), Seevinck (2004), and Morganti (2009) for good overviews of the options.
[16] Shanks (1993: 21). [17] Malin (2001: 23).

the wave function. Before measurement the wave function evolves deterministically, just like a classical system; the Schrödinger equation yields precise values that enable us to predict its motion over time. However, as soon as we perform a measurement it collapses instantaneously into a particle whose location cannot be predicted in advance; all we can know is the probability that it will be in one place or another.

The implications of this indeterminism are not clear. Some interpreters resist drawing any ontological implications, and treat it as an epistemological problem only. Others would rescue determinism by assuming that hidden within the wave function is a real particle whose behavior is "piloted" by the wave to its final location. On this view, the stochastic aspect of quantum theory is "superimposed" on an underlying deterministic process.[18] However, many physicists today accept that indeterminism is a fact of nature.[19] Another question is whether quantum indeterminism extends to macroscopic reality. Given that the statistical significance of quantum phenomena quickly washes out above the molecular level, many would probably agree with John Searle in rejecting such an extension. In his view, quantum theory does not imply "any indeterminacy at the level of objects that matter to us."[20] On the other hand, Bruce Glymour and his co-authors have argued that the price of this conclusion is that we have to give up the physicalist view that macro-states supervene on micro. They believe this is a price not worth paying, and as such conclude that quantum indeterminacy must "percolate upward" to the macro level.[21] This relates to a final question of whether quantum theory provides any support for free will. From the start of the quantum revolution some have argued that it does.[22] But even if we grant upward percolation, most have thought that any linkage to free will founders on the fact that a random process is inconsistent with the idea of "will" or "choice." While unpredictability to an outside observer may be a necessary condition for free will, it is not sufficient. I will come back to this question in Chapter 9.

The challenge to mechanism

The classical worldview assumes that all causation is mechanical and local. Although in the social sciences the term 'mechanical' has narrow, machine-like connotations, in physics it means nothing more than that causation involves the transmission of force or energy from one object to another. The "object"

---

[18] This is the Bohmian or "pilot-wave" interpretation.
[19] For a different line of criticism of indeterminism as a "philosophical sham," this one driven by the Bell Experiments, see Shanks (1993).
[20] The quote is from Griffin (1998: 168).   [21] See Glymour et al. (2001).
[22] E.g. Eddington (1928).

assumption is crucial here; causation can take place between material phenomena only, which precludes genuine mental causation. Other, non-mechanical, ways in which causation has been understood, like Aristotle's categories of material, formal, and final causation, are also ruled out, for being either not really causal, or, in the case of final causality (teleology), impossible. 'Locality,' in turn, means that no causal influence can propagate faster than light. This implies a temporal separation of cause and effect, which rules out instantaneous causation or "action at a distance." Mechanism and locality presuppose materialism, atomism, and determinism, and especially atomism or separability.[23] To a significant extent, therefore, the fate of mechanical and local causation is bound up with the fate of those assumptions.

As if reinforcing its challenge to the latter, however, quantum theory provides independent evidence for the inadequacy of the classical view of causation.[24] This comes in two main forms: (1) the collapse of the wave function, which occurs instantaneously upon measurement with no apparent cause, and (2) the Bell Experiments, which proved the existence of non-local correlations between entangled quantum systems. These findings do not mean there is no such thing as mechanical causation at the macro-level. But they undermine the view that mechanical causation *exhausts* how things happen in the world, and raise a hard question of what conception of causality, if any, should be put in its place.

Three main alternatives have been proposed, none obviously satisfactory. The first is to abandon the language of causality at the micro-physical level and replace it with another framework, such as "complementarity."[25] However, that leaves unanswered the question of how wave functions collapse and non-local correlations happen. The second alternative is to broaden our view of causality to include non-local causation. The challenge here is doing that without simply restating the fact that we have no idea what "causes" non-local causation. One option is to take a page from Aristotle's pluralistic approach and conceptualize non-local causation as "extended" or "structural" causality.[26] Another would be to adopt David Lewis' "counterfactual" model of causation, which already has wide currency at the macro-level. Lewis' model is agnostic about the means by which causal influences are transmitted, which would allow quantum entanglement to count as one means.[27] The third and most radical alternative is to explain wave function collapse and non-local causation by reference to a non-material mental field underlying reality. The problem for all three alternatives is that their causal mechanisms are hard to reconcile with materialism. Whatever

---

[23] On the latter connection see Healey (1994: 346), Esfeld (2001: 221), and Malin (2001: 22).
[24] For a comprehensive introduction to the quantum challenge to causation see Price and Corry, eds. (2007).
[25] Bohr (1937; 1948).   [26] See, for example, d'Espagnat (1995: 414–415).
[27] On this approach see Esfeld (2001: 219–220) and Frisch (2010).

the choice, quantum "mechanics" is something of a misnomer, there being nothing "mechanical" about it at all.[28]

The challenge to absolute space and time

In classical physics space and time are defined in absolute terms, as objective realities independent of other phenomena in the universe.[29] The latter are thought to exist "in" space and time, which were likened to a neutral stage on which events unfolded. In addition, space is assumed to be local, which gives the concept of distance a precise meaning; time is uni-directional, always flowing from the past to the future; and both are assumed to be continuous, infinitely divisible into smaller units.

This absolutist picture was partially overturned by relativity theory, according to which space and time are relative to objects and observers. Relativity shows that space bends or warps as a function of objects' gravitational fields, and that time varies with the speed at which an observer is traveling. Space and time are therefore not completely independent of the content of the universe. However, relativity theory continues to make the classical assumptions that space is local, and that both time and space are continuous.[30] Thus, it would still make sense to conceive of objects and processes as existing "in" space and time, even if that relationship is more complex than in the absolutist view.[31]

Quantum theory overturns these remaining classical assumptions about space and time. The challenge to classical intuitions about space is three-fold. First, in most interpretations of quantum theory particles lack definite locations in space until they are measured.[32] Second, the Bell Experiments break decisively with locality. If entangled quantum systems are correlated non-locally then the concept of distance, and with it the whole idea of space as something "in" which objects are situated, no longer has a precise meaning. Finally, quantum theory shows that space is discrete rather than continuous, which means there is a limit to our ability to divide it into units; at the "Planck scale" (several orders of magnitude smaller than the atomic scale) space is "granular."[33] The challenge to classical intuitions about time, in turn, stems from the Delayed-Choice Experiment, which suggests that the future can in a certain sense affect the past. For different reasons that I address in Chapter 10, Huw Price has argued that quantum mechanics is also compatible with "backwards causation," which would equally imply a reversal in the "arrow of time."[34] Finally, there is the problem that, strictly speaking, wave functions do not evolve "in" time so much

---

[28] See Hiley and Pylkkänen (2001: 127).
[29] For an overview of classical and relativistic conceptions of time see Ehlers (1997).
[30] Monk (1997: 3).   [31] D'Espagnat (1995: 322).
[32] Ibid: 324.   [33] See Monk (1997: 8).
[34] See Aharonov and Vaidman (1990) and Price (1996); cf. Grove (2002).

as collapse "into" time when they are measured. This has led to the suggestion that like space, time is discontinuous: "the past consists of a discontinuous series of events when measurements have taken place, with nothing happening in the intervals between these measurements."[35]

I will not try to characterize the esoteric debate about how space and time should be positively defined, if not in classical or relativistic terms. Nicholas Monk surveys no less than six different proposals for conceptualizing space.[36] These are intimately related to the current effort to develop a theory of quantum gravity, itself seen as the key to the unification of quantum theory and relativity theory – the incompatibility of which stands as one of the biggest unresolved problems in physics today.[37] The debate about the nature of time is no less wide open.[38] The one thing that seems to be clear, however, is that the classical idea that objects and processes exist "in" space and time is now dead, and that space and time should instead be seen as phenomena that somehow "emerge" from relationships. As Lee Smolin puts it with respect to space, the world is *nothing but* an evolving network of relationships that are not themselves "in" space. If there were no relationships there would *be* no space; relationships define space, not vice-versa.[39] His comments could readily be extended to the conceptualization of time as well.[40]

The challenge to the subject–object distinction

Finally, the classical worldview assumes that a categorical distinction exists between subjects and objects. This places human beings in a spectator relationship to nature, as onlookers rather than participants in its workings.[41] Maintaining this separation is considered essential to science, since otherwise there is the danger that knowledge will be contaminated by our choices and subjective beliefs. If all goes well the result will be a 1:1 correspondence between theory and reality. Of course, no one denies that human beings are part of nature, and since we can only acquire knowledge of the world through our minds, the question arises even within the classical worldview whether it is really possible to excise subjectivity from science completely. Schrödinger, for example, argues that although we take for granted that the world exists independent of us, the subject–object distinction is in fact *produced* by acts of "objectivation," since

---

[35] Feinberg et al. (1992: 638).   [36] See Monk (1997); also see Boi (2004).
[37] For a highly readable survey of these efforts see Smolin (2001).
[38] See Price (1996) and Albert (2000) for good introductions to questions of time in the context of quantum mechanics.
[39] See Smolin (2001: 20, 96).
[40] See Wheeler (1988: 124), and the discussion of time-symmetric approaches to quantum theory in Chapter 6 below.
[41] See Matson (1964: Chapter 1) and Schrödinger (1959: 38); cf. Hutto (2004).

it is only by taking the mind *out* of nature that the world can be constituted as an object.[42] This bifurcation of nature suggests a paradox, in that observers are necessary to know reality yet are assumed to be external to it. However, in the classical worldview this paradox is benign because experiments paint a relatively constant picture of reality regardless of who is observing. The question of the observer's status can therefore be bypassed in practice, and subjects treated "as if" they were distinct from objects.[43]

The question is harder to avoid in quantum mechanics, where the observer is no longer clearly extraneous to the system under study. The intrusion of subjectivity into knowledge about the quantum domain is known as the "measurement problem" because it concerns the apparent impossibility of an objective measurement of quantum systems. The problem arises at two distinct points in the measurement process, today sometimes referred to as the "Heisenberg choice" and "Dirac choice."[44] In both cases the difficulty is that measurement is associated with a kind of creativity: before measurement quantum systems are potentialities, yet afterward we get actualities.[45] Since perhaps more than any other this problem sets the stage for the debate about quantum theory's positive implications, let me discuss it in more detail than the others.

The Heisenberg choice refers to the choice of what to measure, such as position or momentum. When implemented in an experiment, it is always associated with a change in the quantum system from a superposition of states to a definite outcome. Thus, the problem is not just that the results of the measurement depend on what we choose to measure, which would be true in any experiment. Rather, it is that before measuring, say, position, a quantum system literally does not *have* a position (as far as can be known), but afterward it does. Moreover, had we performed a different measurement, we would have ended up with a different actuality.

By construing the Heisenberg choice as one of "what question to ask of nature," Wheeler offers a useful analogy to a quantum version of the game of Twenty Questions.[46] In the standard version of the game one player is asked to leave the room while the others decide on a word; the first player then gets twenty questions to guess what it is, to which the others answer only yes or no. There is no measurement problem here, since the chosen word exists independent of the first player. In the quantum version of the game, while the first player is out of the room the group decides *not* to choose a word. Instead, they agree that as the first player goes around the room asking questions, the

---

[42] Schrödinger (1959: 36–51); also see Malin (2001: 233 and passim).
[43] See, for example, Shimony (1963: 755–756), Rosenblum and Kuttner (2002: 1274).
[44] See Stapp (1999: 153), Malin (2001: 113–114), and Rosenblum and Kuttner (2002).
[45] Malin (2001: 49). This apparent creativity is the basis for Diederik Aerts' "Creation-Discovery" Interpretation of quantum theory; see for example his (1998).
[46] See Wheeler (1990: 11) and the discussion in Malin (2001: 213–214).

members of the group will still answer yes or no, but are required in each case to have some word in mind that is consistent with all the other answers that have already been given. Although there are some differences from the quantum case, Wheeler sees intriguing similarities:

> Second, in actuality the information about the word was brought into being step by step through the questions we raised, as the information about the electron is brought into being, step by step, by the experiments that the observer chooses to make. Third, if we had chosen to ask different questions we would have ended up with a different word – as the experimenter would have ended up with a different story for the doings of the electron if he had measured different quantities or the same quantities in a different order.[47]

In short, in contrast to the classical view that objects ("words") exist independent of the choice of what to measure, in quantum theory no such assumption can be made.[48]

The Dirac choice refers to the process by which the wave function collapses into a particle. The problem here stems from the fact that this collapse does not happen, independent of the act of measurement. That seems to implicate the observer as a participant in the transition from quantum potential to classical actuality, meaning that she is entangled with the system under observation – in effect forming a larger quantum system of both observer and observed.[49] The problem here is that the experimental apparatus and the observer herself are in classical states (since both are experienced only as actualities), and the result of the experiment – a particle hit – is also a classical phenomenon. Thus, it seems impossible to dispense with classical concepts altogether.[50] So somewhere in the transition from quantum to classical states a "cut" gets made. The question is, where?

The answer is one of the major sources of disagreement among interpretations of quantum theory, and we will come back to it below. But by way of set-up: the physicist John von Neumann (also of game theory fame) tackled the problem by breaking the collapse process down into a series of steps, known since as the "von Neumann chain."[51] He concluded that the collapse occurs at the very end of the chain when the results of the experiment are registered in the observer's mind, since this was the only point in the process that does not seem to consist of mere molecules in motion.[52] However, von Neumann's argument implied that the observer's body and the measurement apparatus must be in

---

[47] The quote is from Malin (2001: 213).
[48] See Bitbol et al., eds. (2009) for diverse perspectives on the problem of object-constitution in quantum physics.
[49] See Esfeld (2001: 275) and Heelan (2004). [50] Shimony (1963: 768), Esfeld (2001: 274).
[51] See Shimony (1963), Herbert (1985: 147), and Esfeld (2001: 275).
[52] See Shimony (1963: 757), Herbert (1985: 148), and French (2002: 469).

quantum states (i.e. entangled with the system in question), even though we never observe macroscopic systems in superpositions.

Von Neumann's conclusion flies in the face of common sense, and it was in part to demonstrate its absurdity that Schrödinger – who, with Einstein, sought to defend the classical worldview – devised one of the most famous thought experiments in quantum physics.[53] In his scenario a cat is put inside a sealed box with a vial of lethal poison gas, the release of which is determined quantum mechanically, i.e. randomly. We know that if we open the box (perform a measurement) we will observe the cat to be either alive or dead. The question is what is going on inside the box before we open it? According to the classical worldview the cat must be either alive or dead; since objects exist independent of subjects, opening the box merely confirms what has already transpired. However, if we take seriously that the cat is itself in a quantum state by virtue of entanglement with the quantum mechanically determined release of the gas, we must conclude that as long as the box remains sealed the cat is in superposition, which is to say that both alive *and* dead have some potential to be observed at each moment. Since that seems absurd, Schrödinger argued that macroscopic systems could not be quantum mechanical, thereby justifying a subject–object distinction at that level. Yet, the question then arises, if the cut between quantum and classical systems takes place someplace in the von Neumann chain *before* the observer and her measurement device, where exactly is that? There is no obvious place to make it, or, put another way, no obvious limit to the domain of quantum theory.[54] Thus whether or not Schrödinger was right about his cat, the measurement problem is still with us: at *some* point in the chain measurement becomes "creative."

Enormous intellectual effort has been devoted over the past few decades to the measurement problem, and with the advent of several recent proposals some analysts are suggesting that it has finally been solved.[55] However, my own sense is that Bruce Rosenblum and Fred Kuttner are right when they argue that these apparent solutions are only that: apparent – more pragmatic devices that enable physicists to ignore the problem than genuine solutions.[56] Be that as it may, since this issue is central to the debate about the positive implications of quantum mechanics, let us now turn our attention there.

---

[53] See Zukav (1979: 94–96), Herbert (1985: 150–152), Nadeau and Kafatos (1999: 56–57), and Esfeld (2001: 275).
[54] Esfeld (2001: 274).   [55] Rosenblum and Kuttner (2002: 1291).   [56] Ibid.

4 Five interpretations

While there is general agreement that quantum theory poses a fundamental challenge to the classical worldview, there is none on what worldview should replace it. This lack of clear positive implications is the heart of the interpretation problem in quantum theory, or understanding what the world must be like assuming the theory is true. The difficulty is twofold. First, if the theory is correct and complete then there seems to be no way to adjudicate empirically among its interpretations, forcing us to rely on metaphysics instead. Second, the classical metaphysics upon which human beings have relied in the past to interpret new theories is now itself problematic, leaving us without an interpretive backstop to anchor our thinking. Bell is surely right in expecting whatever picture of reality eventually emerges from quantum theory to "astonish us," but by the same token it is very hard to see what that picture should be.

In this chapter I address the debate about the positive worldview implications of quantum theory. Of all the chapters in this book it is the farthest removed from the concerns of social scientists. However, it plays an important dialectical function in my argument, which is to introduce and help justify the panpsychist interpretation of quantum theory that I will embrace and discuss at more length in Chapter 6. By 'justify' I do not mean that I will defend this interpretation as superior to others. My purpose is more modest, which is simply to show why, despite what might seem to be its bizarre and counter-intuitive nature, panpsychism is taken quite seriously by philosophers of physics as a way to interpret quantum theory. That's because (a) there are strong analogies between quantum systems and the human mind, and (b) the alternatives are *equally* bizarre and counter-intuitive. If readers are prepared to accept that panpsychism is a valid approach to the mind–body problem, then this chapter may be skipped – which I have made less costly by deferring until Chapter 6 a discussion of time-symmetric approaches like the Transactional Interpretation, which will play an important role later in the book. But if – like me at first – you have a hard time wrapping your head around the suggestion that mentality could go all the way down to sub-atomic particles, I urge you to read on, to see how such a position could seem reasonable.

I first review some challenges that any interpretation of quantum theory must address, and then turn to the interpretations themselves. With over a dozen in the literature there is no way to be comprehensive.[1] Instead, I have selected five for discussion, guided by a desire to include some of the most well-known and to capture something of the full range of metaphysical options, both materialist and idealist, on the table. My treatments will be more descriptive than evaluative, though with a few words about the recognized strengths and weaknesses of each account.

The problem and a meta-interpretive framework

It would be helpful if physicists agreed on what the specific source of the interpretive problem was, but even that is elusive. As Richard Feynman put it, "I cannot define the real problem, therefore I suspect there's no real problem, but I'm not sure there's no real problem."[2]

Nevertheless, there seems to be some convergence on the view that the fundamental question is how to integrate the two processes by which quantum systems evolve.[3] Before measurement, they evolve according to the Schrödinger equation, which is deterministic and linear; whereas upon measurement wave functions collapse into particles, which is non-deterministic and non-linear. The problem is that these two forms of evolution do not receive a unified treatment within the theory; or, more precisely, that quantum theory is really only about the first, the second being bolted on to the theory ad hoc to account for what we actually observe. Moreover, there is no obvious reason, from the standpoint of quantum theory, that macro-level objects like cats or observers should not also be wave functions and as such have quantum properties – but when we observe them they always have definite properties. So the problem of relating the two types of evolution is in part explaining how and where the "cut" between quantum and classical systems takes place. In short, how can we square classical appearances with quantum reality?

Closely bound up with this question are three others whose answers – were they known – would help us to interpret the theory. The first is the physical significance of the fundamental concept of quantum theory, the wave function.[4] Technically it represents the probabilities of finding a particle at various locations on a measurement device. But since we have no evidence that particles exist before they are measured, what exactly are we measuring? In the classical world probabilities refer to our degree of uncertainty about the properties of a

---

[1] For more thorough reviews see Albert (1992), d'Espagnat (1995), Home (1997: Chapter 2), and Laloe (2001). Maudlin (2003) provides a remarkably succinct introduction to the debate.
[2] Quoted in Bub (2000: 597). [3] See Home (1997: 76) and Laloe (2001: 680).
[4] On the debate about the nature of the wave function, see for example Matzkin (2002), Friederich (2011), Gao (2011), and Ney and Albert, eds. (2013).

real physical thing or process, which, we can safely assume, has definite properties. In the quantum world this assumption is problematic, since the properties of wave functions appear *themselves* to be indefinite. This suggests that the wave function either does not describe a real phenomenon, or, if it does, that it is not material. Both options have problems.

A second question is the nature and cause of wave function collapse. Again we face the issue of whether the collapse is a real process, or merely a mathematical artefact of our description. And if it is real, what brings it about? We know that it is not caused in any classical sense of causation, but then how does it happen? Perhaps it "just happens," randomly. But how can objective chance exist in nature?

Finally, and implicit in the preceding questions, is a basic metaphysical question about the place of mind in nature. In the materialist worldview of classical physics it has no place: one way or another, the mind must be reducible to the brain. Quantum theory does not directly challenge this assumption, and some interpretations seem to justify and/or presuppose it. But the breakdown of the subject–object distinction in quantum measurement does put the question of mind squarely on the table, and raises the possibility that not only is mind not reducible to matter, but is itself a fundamental constituent of reality.[5]

The answers in the interpretive literature to these questions vary along so many dimensions that it is not obvious how they should be categorized. A number of meta-interpretive frameworks have been proposed that group interpretations in different ways, each structured by what its author sees as the central issues at stake. While illuminating, none brings sharply into focus the central issue at stake in this book, namely the mind–body problem. As such, I shall organize the literature in my own way, around two distinctions.

The first, which is standard in the literature, is between realist or ontological interpretations and instrumentalist or epistemological ones. Realists think that quantum theory is telling us something about reality, whereas instrumentalists are agnostic about reality and see the theory as just a device for predicting the outcomes of experiments. The majority of interpretations today are realist, but since the most widely known one, the Copenhagen Interpretation, is instrumentalist, I shall begin there. Within the realist category, in turn, I make a second distinction between interpretations that assume a materialist ontology and those that reflect what I will call an idealist one, addressing two examples of each. Materialists want a quantum theory consistent with the assumption of No Fundamental Mentality, whereas idealists assume that mind plays an irreducible

---

[5] Marin (2009) shows that this question was recognized already in the 1930s, and helped give quantum theory an air of "mysticism." Eighty years on the debate about whether the theory needs to invoke consciousness is no closer to being settled; see for example Yu and Nikolic (2011) and Pradhan (2012).

role in nature. This organizing framework will make for some strange bedfellows, since idealism is often thought to be incompatible with realism, but it makes sense for reasons that emerge below.

Instrumentalism: the Copenhagen Interpretation

The Copenhagen Interpretation is due primarily to Bohr, who spent his career at the University of Copenhagen. It was the dominant reading of quantum theory from the 1930s to the 1970s, and although less popular today among philosophers it is still considered the orthodoxy among working physicists. Without minimizing its intellectual appeal, the reasons for its dominance had more to do with the sociology of the physics profession, and the fact that it justified physicists turning away from the seemingly intractable questions of metaphysical interpretation and toward simply getting on with concrete applications of quantum theory.[6] The Copenhagen Interpretation is also notoriously difficult to define. This is partly because it has many "interpretations" of its own; Bohr's student Heisenberg, for example, is usually seen as a subscriber, but had his own spin that was in some ways antithetical to Bohr's.[7] But the difficulty also stems from the opaque (or subtle, depending on your biases) quality of Bohr's writings. In 1972 Henry Stapp attempted to cut through the confusion with a definitive statement, and his essay is today perhaps the most widely cited secondary treatment.[8] Even so, it gives the approach a pragmatist gloss that Bohr might have resisted.

What all sides seem to agree on, however, is that the Copenhagen Interpretation is epistemological rather than ontological. It brackets as unanswerable the question of what the quantum world is really like, and focuses instead on the more limited question of what kind of knowledge can be obtained about the quantum world.[9] Thus, its concern is not with reality per se but the *description* of reality; physics is not about how nature *is*, but about what we can *say*.[10] This ontological agnosticism is motivated by the breakdown of the subject–object distinction in quantum measurement. In Bohr's view quantum systems can only be known via a description that includes the totality of the experimental situation (i.e. both observer and measurement device), which means that, unlike the case of macroscopic objects, we have no warrant for saying that the attributes of those systems (position, momentum, and so on) inhere in quantum systems themselves.[11] By virtue of this emphasis on the indivisibility of quantum

---

[6] See Cushing (1994) for a comprehensive study of this question.
[7] On the varieties of "Copenhagenism" see Henderson (2010).
[8] Stapp (1972/1997); cf. Healey (2012). For a lengthier reconstruction of Bohr's philosophy see Honner (1987).
[9] Honner (1987: 84).   [10] Shimony (1978: 11).
[11] Herbert (1985: 160–161), d'Espagnat (1995: 223).

systems and the experimental situation the Copenhagen Interpretation is holistic, but in an epistemic rather than ontological sense, since we also have no warrant for saying that measurement creates the attributes of quantum systems.[12]

This epistemological holism creates a problem for the possibility of objective knowledge. This is because in describing quantum systems we are necessarily limited to classical concepts like causality, space, and time, since these are the concepts that pertain to the macroscopic world in which we live – yet they do not apply to quantum phenomena. The inapplicability of our classical conceptual framework is what generates the paradoxes of quantum mechanics described earlier in this chapter, and points to the need to find a new framework to ensure unambiguous communication about quantum phenomena. Bohr's master concept of complementarity was intended to serve this function, enabling us to entertain incompatible classical descriptions without degenerating into paradox. But it does so at the price of eschewing ontological claims about the nature of quantum reality. Thus, in considering the four questions above that drive the interpretive debate, the answers provided by the Copenhagen Interpretation amount to: "we simply cannot know."

Bohr's refusal to engage ontological questions has led some commentators to perceive his approach as anti-realist in the strong sense of denying that there *is* any quantum reality. One of Bohr's former assistants, for example, quotes him as saying "[t]here is no quantum world. There is only an abstract quantum physical description."[13] Statements like this have been taken by some to provide support for the post-modern or relativist view that knowledge of an independently existing reality is impossible, and thus science is not an epistemically privileged discourse.[14] However, that was not Bohr's own view. While agreeing that objective knowledge in the classical sense is impossible in the quantum domain, he was centrally concerned with securing the conditions for unambiguous communication about quantum systems, and as such defended what Bernard d'Espagnat calls "weak" objectivity. This is the kind of objectivity constituted by *intersubjective* agreement, which can be had about the quantum world (i.e. in an experiment all observers can agree that they saw the same thing).[15] As such, Bohr's approach is neither relativist nor subjectivist. Moreover, he remained a realist about the macro-level, since he thought that the material objects of everyday experience can be known in classical terms, and he was even a realist about the sub-atomic level insofar as he accepted that quantum systems exist independent of observers and cause their experiences.[16] In view

---

[12] D'Espagnat (1995: 221), Esfeld (2001: 232–235).
[13] Shimony (1978: 11); also see Herbert (1985: 158).
[14] See for example Plotnitsky (1994), and for a realist response see Norris (1998).
[15] See d'Espagnat (1995: 324, passim).
[16] See Honner (1987), Stapp (1972/1997: 140), and Barad (2007: 125–131).

of these qualifications it is perhaps better to see the Copenhagen Interpretation as "*a*realist" than "anti-realist." It simply says that, at the end of the day, we cannot know what quantum reality is really like.

By the same token, however, if by an interpretation of a physical theory we mean an account of what the world must be like if the theory were true, then Copenhagen is in a sense not an "interpretation" at all.[17] This is taken to its logical conclusion by Christopher Fuchs and Asher Peres, who argue that quantum theory *needs* no interpretation.[18] In their view, while in science we would ideally be able to extract a clear ontology from experimental results, the inability to do so in quantum mechanics is not a problem because it does not prevent us from using the theory. As such, the only "interpretation" we need of quantum theory is that it is a useful algorithm for computing the probabilities of events. Absent new empirical predictions, which most interpretations do not provide, speculating about how the theory relates to reality inevitably leads to paradox and the mere "illusion" of understanding. It is enough that quantum theory can predict, even if it cannot explain.[19]

Notwithstanding its attractions to some, this refusal to deal with ontological issues also underlies the main objection to the Copenhagen approach: that it is essentially incomplete.[20] In particular, it offers no answer to a crucial epistemological question: *why* is our knowledge of the quantum domain so unlike our knowledge of the macro-world?[21] Why the Uncertainty Principle? Why complementarity? Why non-local correlations? The answers cannot be found solely in epistemology, but only by engagement with ontology. Proponents of instrumentalism would say that such questions simply cannot be answered, and that trying to do so anyway will yield only unscientific speculation. As such, we are better off with a "therapeutic" approach to quantum theory, in which we try to cure the desire for answers to ontological questions by clearing up conceptual confusion.[22] However, critics argue that such a conclusion is premature, and that instrumentalism's "there be dragons" mentality about ontology slows the advance of knowledge. Indeed, in the 1930s Einstein worried that if physicists adopted Bohr's approach they would abandon efforts to "complete" quantum theory, and Murray Gell-Mann has argued that the long dominance of Copenhagen delayed needed philosophical reflection on the implications of

---

[17] In fact, Bohr had a rather different view of what an "interpretation" is, one that focuses not on what reality is like but on reconciling the theory with what we observe and can communicate (Omnes, 1995: 607); cf. Ruetsche (2002: 199) and also Friederich's (2011; 2013) "epistemic" conception of quantum states.

[18] See Fuchs and Peres (2000). That said, Fuchs has since developed his approach into a Quantum Bayesianism (or "QBism") that makes an interesting connection to the kind of neutral monism I defend in Chapter 6; see Fuchs and Schack (2014: 104).

[19] Ruetsche (2002: 208).   [20] Squires (1990: 183).

[21] See, for example, Shimony (1978: 12–13), Esfeld (2001: 234–235).

[22] See Friedrich (2011: 150), who invokes the spirit of Wittgenstein's philosophy.

quantum mechanics.[23] There are risks to an ontological turn, since without new predictions our judgments will lack an empirical basis. On the other hand, empirical testing might not be the only way to advance knowledge, and this now seems to be the majority view among philosophers of physics. So let me turn to four realist interpretations, grouped into materialist and idealist categories, which assume that quantum theory has something to tell us about the world.

Realism I: materialist interpretations

Of all the assumptions of the classical worldview, materialism – understood as No Fundamental Mentality – may be the hardest to give up. Although I have not seen any surveys, I would guess the majority of physicists and philosophers of physics are materialists,[24] and most interpretations of quantum theory fall into this category as well. This is not surprising, since the connection between materialism and physics runs deep, as suggested by the contemporary term for materialism – "physicalism." Given that quantum theory forces us to give up *something* from the classical picture, therefore, we should expect materialism to be the last assumption to go. In this section I discuss two prominent materialist approaches to quantum theory.

The GRW Interpretation

One approach has focused on providing an account of wave function collapse that would bridge the gap between the linear evolution of the Schrödinger equation and the non-linear evolution of the collapse. Since orthodox quantum theory does not offer such an account, one possibility is that it is not (quite) correct.[25] In that case the solution would be to postulate new laws that integrate the two kinds of evolution, thereby modifying the theory.

Several interpretations of quantum theory fall under this heading, but the most widely accepted one has been proposed by Giancarlo Ghirardi, Alberto Rimini, and Tullio Weber, and is usually known as "GRW."[26] The key problem as GRW see it is with the Schrödinger equation itself, which allows superpositions of macroscopic states that are never in fact observed. Their solution is to modify the equation by positing that in addition to its deterministic evolution, quantum systems are subject to "spontaneous localizations": for any given time interval there is some probability that the system will randomly collapse from a quantum to a classical state.[27] Building a model of this process is tricky, since it must

---

[23] See, respectively, Ruetsche (2002: 201) and Squires (1990: 180).
[24] Not to mention most neuroscientists, cognitive psychologists, and social scientists.
[25] Squires (1990: 178).
[26] The original proposal is in Ghirardi, Rimini, and Weber (1986); see Ghirardi (2002) for a good overview that responds to a number of criticisms.
[27] Ghirardi (2002: 33).

satisfy two different requirements: respecting predictions for quantum systems considered in isolation, while generating collapse once they become entangled with measurement devices.[28] In other words, too much modification does not preserve what we know about quantum systems, too little does not yield definite macroscopic objects. GRW thread this needle by adding to the Schrödinger equation a probability term for random collapse that is set low enough that for isolated quantum systems there is virtually no chance of collapse, but high enough that as soon as they interact with a measurement device an avalanche of correlations with the trillions of particles in the device is triggered, causing the systems to collapse extremely rapidly.

The GRW approach has several virtues. It solves the key problem of yielding definite states at the macro-level, while preserving superposed states at the micro-level. In the GRW model, Schrödinger's cat is both alive and dead for only a "split second."[29] Importantly, it achieves this result in an integrated fashion, as a consequence of the normal evolution of the wave function rather than by treating wave function collapse as an independent process.[30] Moreover, it does so without attributing any special role to the consciousness of the observer, making it consistent with materialism. Finally, unlike other interpretations, its modification of the Schrödinger equation means it is essentially a new theory, and as such it generates predictions that in some cases differ from those of orthodox quantum theory, making it at least in principle testable (the jury is still out on its empirical success).[31]

Yet, the GRW approach is not without problems. Apart from various technical difficulties that have dogged the program,[32] the key criticism is that it amounts just to an assertion that spontaneous localizations occur, not an explanation.[33] Since we already knew that wave functions behave strangely, it is not clear we know anything more with GRW than we did before. We have a mathematical description of wave function collapse, one widely admired as a technical achievement, but in the end it still leaves us asking *why* it happens, giving it an ad hoc feel. Various answers compatible with GRW's materialism have been offered, but none has been compelling enough to command broad assent.

The Many Worlds Interpretation

A different materialist approach is offered by the Many Worlds Interpretation (MWI), an approach so radical that those encountering it for the first time might wonder how physicists could take it seriously. But they do: Laura Ruetsche calls

---

[28] Ibid: 36. [29] Ibid: 37. [30] Laloe (2001: 684). [31] Ibid: 686.
[32] See Albert (1992: 100–104) and Ruetsche (2002: 210), and for a response Ghirardi (2002).
[33] Squires (1990: 189); Laloe (2001: 685). See Lewis (2005) and Dorato and Esfeld (2010) for recent assessments and interpretive extensions.

it the "preferred framework" of working physicists,[34] although that is probably not true of philosophers of physics, among whom there is much more division. Its origins lie in Hugh Everett's 1957 doctoral thesis, but in fact he never used the phrase "Many Worlds," and what many today understand MWI to be is actually due more to a variant of Everett's theory developed in 1970 by Bryce DeWitt.[35] Indeed, there are *many* Many Worlds Interpretations,[36] particularly if a descendant of MWI, the "Many Minds" Interpretation, is included in this category. Given this confusing situation, I will concentrate on the assumptions of the approach and DeWitt's canonical reading of it, and at the end briefly consider Many Minds.

More than any other interpretation, MWI takes quantum theory at face value.[37] Rather than bring in exogenous considerations to interpret or supplement the theory, it uses the structure of quantum theory itself to guide the interpretation.[38] Namely, since the theory tells us that the fundamental physical entity in the world is the wave function, and all wave functions are entangled, we should assume that the universe is one gigantic wave function. Since the theory says that wave functions evolve solely in accord with the deterministic Schrödinger equation, we should assume that they never actually collapse. Since the theory makes no provision for a cut between quantum and classical systems, we should assume that macroscopic objects are also quantum systems. And since the theory makes no reference to consciousness, we should assume that observers are purely material systems.[39] The difficulty, of course, is that these assumptions conflict with what we observe in experiments, where we get apparent collapses into classical objects. However, if we take the theory literally, then this should not be seen as a problem so much as a challenge of deducing classical appearances *from* the theory.

What follows is a breathtaking claim: when a wave function is measured *all its possibilities are actualized*, but in different "worlds." Each measurement causes the universe to "split" or "branch" into separate universes, one for each possibility in a given wave function. And since acts of measurement are going on everywhere all of the time, this means that the universe (or "Multiverse"?) is constantly splitting into zillions of sub-universes.[40] These many worlds all exist simultaneously, in superposition with each other as part of the Universal wave function. However, the laws of quantum mechanics do not allow us to

---

[34] Ruetsche (2002: 217); Jeffrey Bub (2000: 613) makes the same point about the popularity of MWI in the quantum computation literature.
[35] Lockwood (1996: 168); see DeWitt and Graham, eds. (1973).
[36] D'Espagnat (1995: 247), Barrett (1999: 149).
[37] Laloe (2001: 690–691), Matzkin (2002: 289). The most systematic account of this interpretation of which I am aware is Barrett (1999). D'Espagnat (1995: 247–253) offers a particularly clear short presentation.
[38] Barrett (1999: 64).   [39] D'Espagnat (1995: 247).
[40] D'Espagnat (1995: 247), Butterfield (1995: 132), Barrett (1999: 150).

perceive these other universes,[41] which accounts for the *appearance* of a single, classical universe in our experiments.

Despite its counter-intuitive quality, MWI has several virtues. In particular, if we set aside the matter of proliferating worlds, it is seen as the most parsimonious and aesthetically pleasing of all approaches to quantum theory.[42] By following the theory relentlessly through to its logical conclusion, it accounts for the appearance of a classical world without appealing to extraneous considerations like new dynamics, hidden variables, consciousness, or wave function collapse – none of which, after all, appear in the bare theory. Moreover, even though the appearance of a classical world is now only that – an appearance, behind which stands the reality of the many worlds of the Multiverse – since we have no access to those worlds we can ignore them for all practical purposes. Thus, in a sense, by shoving all the weirdness of quantum theory off onto hidden worlds, MWI actually vindicates the materialism, determinism, and realism of the classical worldview. Indeed, by arguing that quantum systems evolve solely according to the Schrödinger equation, which is classical and local, MWI is even able to eliminate the troubling problem of non-locality.[43]

Yet there remains a sense among many that, in Ruetsche's understated assessment, MWI is philosophically "suspect."[44] In particular, many critics cannot get past the "metaphysical monstrosity" or "ontological extravagance" of claiming that the universe is constantly splitting into zillions of sub-universes.[45] This seems to violate the principle of Occam's Razor – that we should not multiply entities unnecessarily – and suggests that the theoretical parsimony of MWI comes at a high price in ontological profligacy.[46] There are also other issues.[47] One concerns the compatibility of MWI with other physical theories, especially the principle of the conservation of mass. The latter is based on and corroborated by our experience of *this* world, which is difficult to reconcile with the idea of many other worlds that we cannot see. Then there is the question of what exactly constitutes a world in the first place. We experience our world as if it had identity over time, which seems inconsistent with the proposition that it (and also each of us) is constantly splitting into new worlds.[48] "History" thereby seems to disappear. Further questions have been raised about how exactly the world splits, why we do not feel it splitting, and why we cannot see other worlds. Finally, there is MWI's assumption that human beings are purely material entities. As we saw in Part I, it is not clear then why we should experience classical appearances at all.

---

[41] D'Espagnat (1995: 248).
[42] Zukav (1979: 92–93), Squires (1990: 198–199), Laloe (2001: 691).
[43] Squires (1990: 199), Lockwood (1996: 164).   [44] Ruetsche (2002: 217).
[45] Stapp (1972/1997: 133), Barrett (1999: 155), Esfeld (2001: 280).   [46] Barrett (1999: 156).
[47] See Barrett (1999: 154–179) for a comprehensive overview of critiques.
[48] Butterfield (1995: 143).

Objections like these have given rise to a variant of MWI, the Many Minds Interpretation (MMI), which provides a useful bridge to the idealist interpretations considered below.[49] MMI retains the overall interpretive frame of MWI, but proposes a different understanding of the branching process that occurs when all the possibilities in a wave function are actualized during measurement: instead of the world, it is minds that branch. The starting point of this argument is that when an observer measures a quantum system he or she becomes entangled with it, now constituting a larger quantum system with the original object. Although the details vary with different versions of MMI, the basic idea is that for each possibility in the wave function there is a corresponding mental state of the brain.[50] When the results of the experiment come in only one of these "minds" is experienced (i.e. we see a definite particle hit). The others branch off and become inaccessible to us, but are still in some sense real.

MMI has many of the same virtues as MWI without its heaviest metaphysical baggage, and zeroes in on the fact that the real anomaly for quantum theory, and thus what needs to be explained, is that observers have definite *experiences*. However, like all interpretations it comes at a cost: reliance on a debatable and exogenous theory about how experience relates to the physical world.[51] If quantum mechanics is complete and universal, then we should be able to deduce consciousness from the theory (reduce mind to brain), but it is not clear how this can be done with a materialist ontology. There is nothing in quantum theory itself requiring that each possibility in a wave function correspond to a mental state. Moreover, since all physical states (including brain states) are superposed quantum states, their associated mental states should be superposed as well, but this is not what we experience.[52] Why are only some mental states experienced and not others? Acknowledging such difficulties, David Albert and Barry Loewer argue that the mind must be something intrinsically different than the physical reality described by quantum theory, which actually leads them to reject materialism in favor of an explicit mind–body dualism.[53] Michael Lockwood sees this as a "desperate expedient," and has developed an alternative form of MMI that tries to save materialism through the idea of supervenience of the mental on the physical.[54] If my claim in Chapter 1 is correct, however, this approach to consciousness is unlikely to pan out in the end.

---

[49] This is actually closer to Everett's original argument; see d'Espagnat (1995: 251–252). On the Many Minds Interpretation see for example, Lockwood (1996), Home (1997: 92–94), and Barrett (1999).

[50] Butterfield (1995: 148), Ruetsche (2002: 216).   [51] Lockwood (1996: 170).

[52] At least in a given instant; the feeling of ambivalence between two states of mind might be what it is like to be in a superposition over a period of time.

[53] Albert and Loewer (1988); also see Barrett (2006).

[54] See Lockwood (1996: 176 and passim).

Five interpretations 81

In my view two lessons should be taken away from this discussion of MWI/MMI. The first is the powerful grip of materialism on the modern scientific imagination. That physicists could take seriously an ontology as extravagant as MWI, let alone as a preferred framework, shows that they are willing to pay a very high price indeed to save materialism. The second is that, in MMI, the internal logic of MWI has nevertheless forced at least some materialists to come directly to grips with the recalcitrant nature of consciousness. This is a significant development, and suggests that more explicitly idealist interpretations of quantum theory are worth a look in themselves.

Realism II: idealist interpretations

By "idealist" interpretations of quantum theory I mean those that assign an explicit role to consciousness in quantum processes. This may be seen as either a virtue or a vice: in my view it opens the door to the possibility of solving two "hard problems" with one throw; for materialists it is an undesirable and unnecessary expedient.[55] Like the materialist interpretations discussed above, however, these idealist ones are realist in the sense that they ascribe ontological status to quantum systems. This may seem counter-intuitive, since philosophers have often treated idealism as the *opposite* of realism: if realism assumes that science gives us access to the world "out there," then traditional idealism denied the possibility of such access. But that presupposes that realism implies materialism, which is precisely what idealist views of quantum theory reject.[56] They find within quantum theory a basis for arguing that consciousness is objective and real, but not reducible to material stuff.

There are two basic ways in which consciousness might enter into quantum theory: what might be called "exogenously," through the role of the human observer in the measurement process, and "endogenously," through sub-atomic particles themselves possessing a primitive form of mentality.[57] Since historically the former came first let me start there.

The Subjectivist Interpretation

Unlike most interpretations of quantum theory, what Walter von Lucadou[58] has called the Subjectivist Interpretation lacks a systematic, canonical statement,

---

[55] See for example Yu and Nikolic (2011).
[56] The result with respect to the realism question might be what Bernard d'Espagnat (2011) has suggestively called "open realism."
[57] See Ward (2014) for a good overview of different ways to think about the potential connection between quantum mechanics and consciousness.
[58] See von Lucadou (1994).

but it does have a recognizable family of proponents going back to the 1930s.[59] Their basic claim is that it is only as a result of interaction with the mind of the observer, which stands outside the quantum system in question, that the latter's wave function collapses. As such, it treats quantum theory from the start as a theory of mind–matter interaction, as impossible to do without reference to conscious observers who in effect help "create" reality.[60] As Wheeler puts it, "[n]o elementary phenomenon is a phenomenon until it is a registered (observed) phenomenon."[61] This does not mean the approach is anti-realist, since reality is still "real." The point is that we need to rethink our relationship to reality from one that assumes a strict subject–object distinction to a "participatory" view.

The Subjectivist starting point is the von Neumann chain (see above, p. 68). Against the Copenhagen treatment of measurement devices in classical terms, von Neumann argued that any physical device will immediately become entangled with a quantum system it is used to study, creating a larger quantum system that would preclude a definite outcome to an experiment. This problem will propagate on up the chain to the body of the observer, which as a material object will also become entangled with the system in question. However, since in the end we do observe actual outcomes, von Neumann reasoned that the measurement process must have a *non*-material terminus in the observer's mind.[62] In short, quantum systems collapse at the interface of the brain and the mind.[63] On the basis of this argument von Neumann added the "projection postulate" to quantum theory to describe wave function collapse, which enables us to connect the theory to our experience of definite outcomes.

However, in his thought the philosophical significance of the projection postulate remains obscure.[64] Fritz London and Edmond Bauer clarified matters somewhat in an important 1939 paper,[65] but it was not until the work of Eugene Wigner in the early 1960s that the metaphysical implications of this approach began to receive sustained attention.[66] Wigner begins with what has become a well-known thought experiment. Consider the situation if in place of a measurement device we could substitute a friend of the experimenter with similar abilities. From the experimenter's standpoint the friend is a physical system and so like the original device should be in a superposition, unable to register a definite state. However, after doing the experiment, if the experimenter asked her friend whether she observed a definite outcome, she would say yes, since with her capacity for subjective experience she stands in an identical

---

[59] See for example London and Bauer (1939/1983), Wigner (1962; 1964), Stapp (1993; 2001), Wheeler (1990; 1994), and French (2002).
[60] Stapp (2001: 1470), Wigner (1962: 285).   [61] Wheeler (1994: 120).
[62] French (2002: 469).   [63] Butterfield (1995: 130).   [64] French (2002: 469).
[65] See London and Bauer (1939/1983).
[66] See Wigner (1962), and Esfeld (1999) for a good overview of Wigner's work.

relation to the quantum system as the experimenter herself. The "Paradox of Wigner's Friend" supports von Neumann's idea that consciousness must play a different role in quantum theory than purely material measuring devices.[67] But Wigner drew a clearer conclusion: that there are two "kinds of reality," physical reality and consciousness, and since the latter selects the former, consciousness is "primary."[68]

Wigner's view is not widely held today, for two main reasons.[69] First, given the manifest dependence of the mind on the brain, it seems to imply that if the brain is in superposition, then the mind should be as well. Yet, in any given instant we do not experience superpositions of our mental states, but actual ones.[70] One could avoid this problem by adopting an explicit mind–body dualism. However, second, we then have Descartes' problem of explaining mind–matter interaction. Wigner says that the mind causes changes in the physical world, yet he proposes no mechanism by which this could occur. How could a non-physical mind choose physical outcomes if they are different substances? Wigner tried to answer these questions, but most physicists were unconvinced.[71]

Despite this skepticism, in recent years Wheeler and Stapp have advanced new forms of Subjectivism that avoid some of these problems. In contrast to Wigner, Wheeler emphasizes not the role of the observer in inducing wave function collapse (the "Dirac choice"), but her decision of what question to ask nature in the first place (the "Heisenberg choice"), which determines whether a quantum system has, for example, a definite position or a momentum. This leads to Wheeler's idea of "it from bit": "every *it* – every particle, every field of force, even the spacetime continuum itself – derives its function, its meaning, its very existence entirely – even if in some contexts indirectly – from the apparatus-elicited answers to yes-no questions, binary choices, *bits*."[72] Like Wigner, for Wheeler reality is "observer-dependent," but the role of observers is more indirect. Rather than directly causing state reduction, with its dualistic implications, they merely "participate" in the process.[73] Admittedly, this leaves unresolved the cause of wave function collapse. Moreover, from the standpoint of Subjectivism there is an important ambiguity in Wheeler's theory. It seems to presuppose the mind's freedom and irreducibility to the material world, yet nowhere does Wheeler address the ontological status of consciousness,

---

[67] Wigner (1962: 294). [68] See Wigner (1964).
[69] See French (2002) for critical discussion of the ensuing debate, and Butterfield (1995: 130) for a summary of current objections to this approach.
[70] Though see Lehner (1997). I argue later that this is because it is the *un*conscious part of the mind that is in superposition.
[71] Though see Barrett (2006) for an argument using Wigner's approach to defend mind–body dualism.
[72] Wheeler (1990: 5). [73] Wheeler (1988: 113; 1990: 5).

and in contrast to Wigner's positing of mind as the primary reality, he sees "information" as basic. However, it will be argued below that there is an intimate relationship between information and consciousness, and as such it makes sense to treat Wheeler's view as a form of Subjectivism.

Stapp's approach emphasizes both the Heisenberg and Dirac choices, which he associates with the "active" and "passive" roles of mind respectively.[74] As such, it might be seen as combining the approaches of Wheeler and Wigner. On the Heisenberg choice Stapp follows Wheeler, while being more explicit that the posing of questions to nature involves "top-level guidance" by the mind of the brain, or free will.[75] However, Stapp also sees a role for mind in the Dirac choice, though one different than that posited by Wigner. Whereas Wigner argued that consciousness causes collapse, Stapp sees the role of the mind here as more passive, as coming to know the answer nature returns to a question.[76] Importantly, the two roles of the mind both involve the brain/mind complex. In contrast to Cartesian dualism, therefore, Stapp's ontology is more like a psycho-physical *duality* or parallelism, in which every quantum event is actually a pair: a physical event in an entangled brain-world quantum system that reduces the wave function to an outcome compatible with an associated (not causal) psychical event in the mind.[77]

Stapp makes a start connecting his approach to the philosophy of mind,[78] but Steven French's "reinterpretation" of the Subjectivist Interpretation I think best grounds it philosophically.[79] French argues for a "phenomenological" solution to the measurement problem in the spirit of Edmund Husserl.[80] He begins by criticizing both Wigner and his critics for assuming that the mind or ego is some kind of Cartesian substance that floats above experience. The assumption is an easy one to make, since in ordinary language we talk as if the ego is a distinct entity that "has" experiences, but according to Husserl this is wrong. When we look closely for the subject of experience, the "I," we only find the unity of experience itself. There is no further object "experiencing" experience.[81] French then applies this to quantum measurement, which he describes as an ensemble of three systems – the quantum object, the measurement device, and the observer. These systems are entangled – i.e. not fully separable – through a von Neumann chain, and as such described by a global wave function. His next move is the key. From the observer's standpoint the object and the measurement device are part of the external world, but with ourselves we have a more intimate relationship by virtue of the "faculty of introspection," which enables us to gain "immanent knowledge" of her own (brain) state. By acquiring that knowledge

---

[74] See especially his (2001); for earlier treatments see Stapp (1993; 1996).
[75] See Stapp (2001: 1483 and 1488).   [76] Ibid: 1485.
[77] Ibid: 1486; for a more recent discussion along similar lines see Pradhan (2012).
[78] See especially his (1993).
[79] French (2002); von Lucadou's (1994) reinterpretation of Wigner is similar.
[80] Also see Heelan (2004).   [81] See French (2002: 476–479).

through performing a measurement she "separates" herself from the ensemble's wave function, which cuts the von Neumann chain and collapses the wave function into a definite outcome. Thus, what is going on is not a preexisting mind with definite properties interfering with an already fully distinct matter, but the creation through an *act* of reflection of the subjective/objective distinction in the first place.[82]

French's phenomenological approach addresses at least three of the charges that have been directed at the Subjectivist Interpretation. (1) It avoids dualism by its refusal to treat the mind as a distinct substance. As with Stapp, what we have here is more of a duality, in which the brain's introspective capacity is associated with the production of both mind and matter. (2) By the same token, it avoids having to posit a mysterious causal interaction between mind and the physical world: their relationship is not causal but *constitutive*, a "mutual separation" of an ego-pole and an object-pole through an act of reflection.[83] (3) Finally, it explains the contradiction between brains being in superpositions and the fact that we do not experience such superpositions: the "I" that is the "subject" of experience can only be posited *after* the separation of the ego- and object-poles has already taken place.[84]

Nevertheless, there is one important element missing from French's model, which is an explanation of how it is possible for an observer to have "immanent knowledge" of her superposed brain states, and why this should be associated with the experience of consciousness. By his rigorous phenomenological description of the ego French gives us strong reasons for rejecting both dualism and materialism, but this does not yet amount to a positive ontology of consciousness. However, what is clear, as he suggests cryptically at the end of his article, is that this will require "an utterly radical re-conception of the natural world and our place within it."[85]

The Bohm Interpretation

Panpsychist interpretations of quantum theory offer such a radical view. Whereas Subjectivism treats consciousness as a property only of the observer and thus exogenous to the sub-atomic level, panpsychism endogenizes it, giving mind a place within the formalism itself. There have been a number of panpsychist interpretations of quantum theory lately, which I address at more length in Chapter 6.[86] Here I discuss just one of the earliest arguments along these lines: that of David Bohm.

---

[82] Cf. Schneider's (2005) view of quantum measurement as a "speech act."
[83] French (2002: 484). [84] Ibid: 485. [85] Ibid: 489.
[86] See for example Atmanspacher (2003), Nakagomi (2003a, 2003b), Primas (2003), Pylkkänen (2007), Gao (2013), and Seager (2013).

For two decades after the consolidation of the Copenhagen Interpretation in the early 1930s it was thought that no other reading of quantum theory was possible. This reflected a belief that the wave function was an absolutely complete description of quantum systems, and as such there was neither any need nor any way to incorporate "hidden variables" into quantum theory that might explain its predictions.[87] However, in 1951 Bohm developed a hidden variables theory that was observationally equivalent to orthodox quantum mechanics. At the time Bohm's model went almost completely unnoticed, but by the 1970s it was being taken seriously, partly because it has many classical elements, which to some offered the hope of restoring the classical worldview to its rightful central place.

Nevertheless, Bohm himself saw his ontology as non-classical, and nowhere is this more apparent than in his view – emphasized more in his later works – that the wave function has a primitive form of mentality.[88] Curiously, this suggestion is often completely ignored in the interpretive debate *within* the Bohm Interpretation, giving it a more materialist spin than he perhaps intended.[89] Such neglect is possible because although panpsychism is a natural inference from his approach, it is not logically entailed by it and as such has an ad hoc quality.[90] Moreover, Bohm may have added to the confusion by criticizing "exogenous" idealists like Wigner specifically for bringing mind into quantum theory.[91] In any case, given this slippage, I shall discuss Bohm's view in two parts, first just laying out its structure, and then addressing its implications for the idealist/materialist debate.

Bohmian mechanics makes three fundamental claims.[92] First, contrary to every other interpretation of quantum theory, it posits that particles are real material objects with definite positions and trajectories. Because of the Uncertainty Principle the latter cannot be known before measurement and as such are hidden variables, but by using Bohm's rewritten Schrödinger equation they can be calculated after the fact. Quantum theory can then be read as a theory about the evolution of particle positions.[93] Second, the wave aspect of particles describes a real phenomenon – a genuine field, which Bohm calls a "quantum potential" – as opposed to being just a mathematical expression from

---

[87] See Home (1997: 16 and 54).
[88] See especially his (1990); this thesis is less prominent in his classic (1980; though see pp. 207–208), but was already present in his (1951: 168–172).
[89] See Sole (2013) for a recent illustrative overview; also see Albert (1992), Home (1997), and Ruetsche (2002). See Pylkkänen (2007) for a comprehensive study and panpsychist reading of Bohm's philosophy, and Seager (2013) for a diagnosis of its "Janus-faced" reception in similar terms.
[90] Stapp (1993: 137).   [91] Bohm and Hiley (1993: 24).
[92] Bohm and Hiley (1993) is the most systematic treatment of the theory.
[93] Albert (1992: 134).

which the statistical properties of observed phenomena are derived.[94] The job of the quantum potential is literally to guide particles in their movement toward observed outcomes.[95] Although the details differ, this idea is similar in spirit to de Broglie's early "pilot-wave" model, in which the wave function is assumed to refer to a real wave that pilots particles to their destination.[96] Finally, a quantum system is an indivisible union of two separate entities, a particle and a wave. This means that the wave function is not a complete description of a quantum system, since hidden within any given wave is a real particle with a definite position.[97]

Bohm's framework has a number of classical features. (1) It restores the idea of particles as tiny material objects. (2) The quantum potential is treated as a real phenomenon, not unlike a field in classical physics, by which particles are carried and influenced like a cork floating on the sea. (3) It is causal and ontologically deterministic. By enabling us to retrodict particle positions it connects a quantum system's initial state and the outcome of an experiment, while faithfully reproducing the *ex ante* epistemic indeterminism of the orthodox theory.[98] (4) Wave functions no longer mysteriously collapse in reality, even though they appear to when we measure them.[99] (5) Finally, the observer plays no special role in the theory, thereby reinstating a clear subject–object distinction.[100] All the action is taking place endogenously rather than through the intervention of a human being. In effect, Bohm's approach *adds* the quantum wave function to a classical description of the world, rather than forcing us to choose between them. For this reason proponents see it as the "natural embedding of the Schrödinger equation . . . into a physical theory."[101]

Given these classical characteristics it is perhaps not surprising that the secondary literature has neglected the idealist implications that Bohm himself drew from his approach. However, there is that new force that Bohm introduces, the "quantum potential," which Euan Squires calls a "very peculiar object."[102] While akin to a field in classical physics, it has two non-classical features. First, whereas the effects of a classical field depend on both its form and amplitude, the effects of the quantum potential depend only on its form. Thus, in contrast to a cork bobbing on a wave of water, which will move less the further it is from the center of the wave, a particle bobbing on a quantum field will do so at full strength regardless of its distance from the source. As such,

---

[94] Kieseppa (1997: 56).
[95] Albert (1992: 135). This is where Bohm introduces non-locality into this theory.
[96] Home (1997: 37–40); for an overview of the pilot-wave model see Squires (1994: 79–84).
[97] Callender and Weingard (1997: 25).
[98] Albert (1992: 164); Home (1997: 44). For this reason it is also sometimes called a "causal" interpretation of quantum theory.
[99] Albert (1992: 163), Hiley (1997: 39).   [100] Hiley (1997: 39).
[101] Goldstein (1996: 163).   [102] Squires (1990: 195).

even remote features of the environment can affect a particle's movement.[103] Second, whereas the effects of a classical field are transmitted to objects in a push/pull fashion, the quantum potential doesn't work that way. Indeed, how it does work is not entirely clear, which raises the metaphysical question of what it is made of.

The view of Bohm and his followers is that the quantum potential consists in "*active information*."[104] Information here is understood to be objective rather than a measure of our knowledge, which accords with the conventional definition of information due to Shannon.[105] However, unlike Shannon information, which Bohm and his co-author Basil Hiley call "passive" because it cannot do anything without a subject to use it, active information exercises causal agency on its own. The powers of this agency are not mechanical but "informational"[106] – the quantum field informs its associated particle about the environment, giving it a "perspective," to which the particle responds deterministically according to the Schrödinger equation. In this way a quantum field can affect a particle without providing much of the energy needed to move it.[107] As such, it is possible for even a very weak quantum field to move a particle with a great deal of energy, an implication that will prove important later in thinking about the mind–body problem.

What all this amounts to is that information is a fundamental aspect of reality. This suggestion is being heard increasingly today both within quantum physics (by Bohmians and non-Bohmians alike) and outside.[108] But Bohm's idea that information in the quantum world is "active" gives it an added and provocative dimension. In particular, it leads to the view that, by virtue of their indivisible union with quantum fields, particles have an inherent (if primitive) form of mentality.[109] Mind is not exclusive to human beings, or even organisms in general, but goes all the way down in nature. Given that physicalism bases itself on physics, this raises the question of what 'physical' means. Hiley and Paavo Pylkkänen argue that the quantum world is still "physical," but its physicality is "subtle" and thus "mental." As such, they see Bohm's theory as an "objective idealism": idealist because it posits mentality as irreducible and objective because this mentality exists independent of people.[110] The more common name for this position is panpsychism.

In sum, quantum theory calls into question all of the metaphysical assumptions upon which the classical worldview is based: materialism, atomism, determinism, mechanism, the subject–object distinction, and that space and time are

---

[103] The example is from Bohm and Hiley (1993: 31–32). [104] Ibid: 35–36.
[105] See Shannon (1949). [106] Pylkkänen (1995: 340).
[107] This is proposed to come from the quantum vacuum; see Bohm and Hiley (1993: 37).
[108] See, for example, Wheeler (1990) and Chalmers (1996).
[109] Bohm (1990: 281); also see Hiley (1997) and Pylkkänen (2007).
[110] Hiley and Pylkkänen (1997: 76).

absolute. Again, "calling into question" does not necessarily mean they are wrong, but at the level of appearances there is a challenge to every one. Most interpretations of quantum theory save one or more classical assumption, but while the balance of philosophical opinion has swung this way and that, there has been little interpretive progress over time. No interpretations have been definitively eliminated, and all have their problems; the only question is which problems you want to have.[111] About the only thing everyone seems to agree upon is that the eventual solution will astonish us.

One of the deepest divides in this debate is over the role of consciousness. Materialists preserve the classical assumption that everything is ultimately material. Idealists give this assumption up and argue that consciousness plays an irreducible role in nature. Considered in the context of quantum theory alone, there is no definitive choosing between them. What I want to suggest now is that if we link the problem of interpreting quantum theory to the problem of explaining consciousness, it becomes clearer what choice should be made.

---

[111] This reminds me of Kenneth Waltz's (1979: 18) characterization of IR, where "nothing seems to cumulate, not even criticism."

Part II

Quantum consciousness and life

Introduction

We saw in Part I that in addition to explaining consciousness there is another hard problem for the modern scientific worldview – understanding what quantum physics is telling us about reality. And if anything this one is harder. In philosophy of mind there is at least consensus enough for an orthodox position to exist (materialism), and a widespread expectation that it will be vindicated by future neuroscience. By contrast, in philosophy of physics all sides understand that quantum theory is compatible with a variety of metaphysical interpretations, and have little hope that new empirical discoveries will settle the matter. But both problems are "hard" in the sense that, despite tremendous intellectual effort over many decades, no clear progress has been made toward solving them.

The basic idea of quantum consciousness theory is that bringing the two problems into contact may be the key to solving both (though I will be addressing only the mind–body problem in what follows).[1] That this could be a novel idea attests to the fact that their philosophical discussions have evolved quite separately. While philosophers of physics have long been interested in the role of consciousness in the measurement process, they have engaged little with the mind–body problem per se; and philosophers of mind have been almost unanimous in dismissing the suggestion that consciousness has anything interesting to do with quantum physics. Social scientists might echo such skepticism, wondering how adding the problem of interpreting quantum theory to the problem of consciousness will help *them*. Besides piling up jargon and arcane debates, in moving to the sub-atomic level we are now even farther removed from the human world, and seemingly poised to engage in the most vulgar reductionism.

Yet, naïvely one might expect the two problems to be related, for two reasons. The first is the intriguing analogies mentioned in Chapter 1 between the mental and quantum domains. Of course, these might be nothing more than analogies,

---

[1] On what it might mean for the quantum debate see Aerts' (2010) interpretation of quantum particles as "conceptual entities," modeled explicitly on concepts in daily life, and also Hameroff and Penrose (1996) and Manousakis (2006).

with no basis in reality (it would be more telling if they did *not* exist). But they suggest that a similar conceptual architecture might be applicable to both domains, which has since been confirmed by quantum decision theory.[2] As philosopher of physics Michel Bitbol puts it, "certain domains of the human sciences (economy, psychology of perception, rational choice theory, etc.) share *exactly the same* (and not just analogous) characteristics and backbone structure as quantum mechanics."[3]

The second is that the problem in each case is a mirror image of the other, although perhaps not as one might expect. Normally in the mind–body case the mind is seen as the problem, whereas in quantum theory it is matter. But if 'problem' refers to what is anomalous relative to what is already established, then in the mind–body problem what is known first is consciousness ("I think, therefore I am"). If we were not sure about *that* then there would be no more reason to problematize materialists' failure to explain consciousness than their failure to explain ghosts. So what is recalcitrant here is really a kind of *body*,[4] one that is conscious. Similarly, in quantum theory what is known first is the Schrödinger equation, which describes a world of deterministically evolving potentialities. Yet when we measure this world we experience one actual reality that emerges non-deterministically. So what is recalcitrant here is what is present to *consciousness*, or the role of the mind in physics.[5] Thus, there is a complementarity between the two issues that, while not evidence for a link per se, is also suggestive and provides a frame for relating them.

Quantum consciousness theory builds on these intuitions by combining two propositions: (1) the physical claim of quantum brain theory that the brain is capable of sustaining coherent quantum states (Chapter 5), and (2) the metaphysical claim of panpsychism that consciousness inheres in the very structure of matter (Chapter 6). Of these claims it is the second, panpsychism, which does the crucial work in explaining consciousness. However, quantum brain theory offers a solution to what has long been a key objection to panpsychism, the "combination problem" of how the zillions of proto-conscious elements in matter combine into the unitary consciousness of the brain, and as such I will treat it first. Then, after discussing panpsychism, I defend a third claim that I take to be an important implication of quantum consciousness theory: that quantum coherence is the essence of *life* (Chapter 7). This leads to a quantum form of vitalism, in which quantum coherence is the elusive *élan vital*.

---

[2] See Filk and Müller (2009) and Bitbol (2011).
[3] Bitbol (2012: 247), emphasis in the original; also see Filk and Müeller (2009) and Pradhan (2012). Bitbol here is referring specifically to epistemological questions, but his point I believe holds for ontological ones as well.
[4] See Montero (1999) on the "body problem" in the philosophy of mind.
[5] See the discussion of the "von Neumann Chain" in Chapter 3, p. 68.

This argument amounts to an epistemological double movement, taking what is known at each level – the third-person knowledge of quantum theory and the first-person knowledge of consciousness – and projecting it toward the other, scaling all the way up and all the way down respectively. The goal of this maneuver is not to reduce one kind of knowledge to the other but quite the opposite, to keep them separate until they are face to face across the micro-macro spectrum. There they can then be joined in the phenomenon of life, which embodies both the first- and third-person (sic) points of view, understood as complementary in the quantum sense. This might sound like a new form of Cartesianism, but unlike Descartes' ontological dualism, which posited two distinct and unrelated substances, the dualism here is only epistemic (so one might call it a 'dual*ity*' vs. a 'dual*ism*'). With respect to ontology, I argue for a quantum version of "neutral monism," which posits a single underlying reality that is neither mental nor material but from which the distinction itself emerges. With respect to epistemology, it is true that by keeping the argument on parallel tracks a subjective–objective distinction is built in from the start, which might be criticized for privileging subjectivity. However, the "taboo on subjectivity" ultimately stems from the inability of the classical worldview to explain a phenomenon which we all know intuitively, is real – first-person experience – and as such, philosophers must find ways to deny, dismiss, or deconstruct it instead. Since quantum consciousness theory purports to offer such an explanation, there is no reason to embrace that taboo here.

But if consciousness goes all the way down, then what kind of "explanation" of it will this be? To answer that we first have to define what precisely the mind–body "problem" is. In the classical view, it is to explain consciousness by reference to phenomena that are purely material, which is to say contain no trace of consciousness. Such an explanation might take the form of reducing consciousness in a causal, functional, or logical sense to material reality, or, for those favoring a more hierarchical ontology, explaining its emergence at some level of material complexity.[6] Either way, according to this definition of the problem, it will not be solved until the explanandum is wholly material. But that begs a key question, which is whether the ultimate constituents of reality are indeed wholly material. As we have seen, in the quantum world that assumption is debatable: there, physicality is not equivalent to materiality, and as such is compatible in principle with mentality. Changing the physics constraint in this way, from the CCCP to the CCQP, would re-found "the problem" on an altogether different basis, abandoning materialism in favor of a more encompassing naturalism, in which consciousness might itself be part of the foundation.

---

[6] For a good discussion of the traditional options see van Gulick (2001).

But ultimately the argument I make may be less about explaining consciousness than about understanding it. This distinction is familiar in the epistemology of social science, where it has been used to distinguish naturalists who think that social science is not essentially different than physical science from anti-naturalists who think that it is. However, the distinction also crops up in the philosophy of physical science. There, against Hempel's view that the two are synonymous, some philosophers have recently argued that understanding provides an epistemic benefit over and beyond explanation, and may even exist when the latter is absent.[7] This benefit is characterized in various ways – "intelligibility," "pragmatic skill," or "grasping" the phenomenon in question – but what they share is a psychological, user-relative dimension. And that seems quite apropos here, both because my argument puts subjectivity front and center, and because what counts as explaining depends on one's ontology and as such – in the quantum context at least – may be contested. So while I personally think that what follows is an explanation of consciousness, I will be satisfied if you merely think that it helps us understand it.

---

[7] See de Regt and Dieks (2005), Grimm (2006), Lipton (2009), Khalifa (2013), and Van Camp (2014); and for a defense of Hempel's view see Trout (2002). This literature makes curiously little reference to its social scientific cousin, which is seen as pertaining to the "special case" of social cognition only (Khalifa, 2013: 162).

5 Quantum brain theory

In recent decades it has become common throughout the social sciences to think of the human mind simply as a very complex computer. Drawing on an older materialist metaphor of the mind as a machine,[1] the "computational" model of the mind took off after the mid-century invention of the computer and subsequent cognitive revolution in psychology, and has since permeated the social sciences in the form of rational choice and other prominent social theories. The contributions of this model have been huge; but it has always been assumed that the computations going on inside our heads are classical. Quantum brain theory challenges this assumption by proposing that the mind is actually a quantum computer. Classical computers are based on binary digits or "bits" with well-defined values (0 or 1), which are transformed in serial operations by a program into an output. Quantum computers in contrast are based on "qubits" that can be in superpositions of 0 *and* 1 at the same time and also interact non-locally, enabling every qubit to be operated on simultaneously.[2] The idea of a quantum computer was first proposed in the 1980s and is still far from being realized technologically, but it has fired the imagination of scientists by raising the prospect of unimaginable increases in computational power, and other neat tricks besides. If quantum brain theory is true, then our model of the mind would be similarly radically altered, irrespective of the issue of consciousness.

Quantum brain theory hypothesizes that quantum processes at the elementary level are amplified and kept in superposition at the level of the organism,[3] and then, through downward causation constrain what is going on deep within the brain. On this view, information from the environment is continuously transformed from the macro- to the micro-level, and then channeled back

---

[1] On the history of the mind-as-machine metaphor see especially Mirowski (1988), and also Cohen (1994) and Maas (1999).

[2] See Siegfried (2000) for a very accessible, if now somewhat dated, introduction to quantum computation.

[3] The possibility of such amplification has a long pedigree among physicists who think that quantum mechanics and biology might be related, going back to Pascual Jordan in the 1930s. For subsequent work see Elsasser (1951), Platt (1956), and Gabora (2002).

upward into an "internal quantum state,"[4] a decoherence-free sub-space of the brain within which quantum computational processes are performed.[5]

Postulating a non-trivial role for quantum processes in the brain breaks with a foundation of modern neuroscience, the "Neuron Doctrine," according to which neurons are the smallest units of the brain relevant to explaining consciousness. With about 100 billion in the average human brain neurons are already extremely small, but still orders of magnitude larger than sub-atomic particles, and well within the classical domain. As such, the Neuron Doctrine assumes a priori that the physical states of the brain are "collapsed."[6] Proponents of the Neuron Doctrine acknowledge of course that quantum processes take place in the brain, since they take place everywhere. But they argue that in an environment as "warm, wet, and noisy" as the brain, interference from zillions of interactions induces all wave functions to collapse, or "decohere," above the molecular level. Since this is far below the neural scale we do not need quantum theory to explain neural behavior. To take on the Neuron Doctrine, therefore, quantum brain theorists must identify physical structures and processes in the brain that can solve the decoherence problem, the difficulty of which no one denies.

But to keep things in perspective the Neuron Doctrine too faces formidable difficulties. There is the hard problem of consciousness that got us here in the first place, plus at least three hard *sub*-problems: (1) the transition from unconscious to conscious states: most information processing in the brain is unconscious, so why not all of it?; (2) our experience of free will; and (3) the unity of consciousness or "binding problem," the fact that in experience huge numbers of neurons fire at the same time. Various classical hypotheses have been proposed in an attempt to solve these problems, so saying the issues are unresolved does not mean quantum brain theory is true. However, its ability to solve them in a clear and unified way is an important test of how well the argument is doing, and so I will come back to them at various points below.

Your quantum brain

Quantum brain theory is not one theory so much as a family of hypotheses, all of which suggest that the brain is able to "continuously generate quantum coherent processes among particles/waves distributed along its volume."[7] However, different branches of the family work at different levels of analysis, and between them they have proposed a variety of possible supports for

---

[4] See Igamberdiev (2012: 24–28).
[5] See Conrad (1996: 97) and Glymour et al. (2001). [6] See Pereria (2003: 101).
[7] Vannini (2008: 176); for an overview of much of this literature see Tuszynski, ed. (2006).

sustaining quantum coherence in the brain. While these hypotheses seem mostly complementary some might be inconsistent, and/or some might turn out to be true while others are false. Add the fact that the subject matter is neuroscience, and this is a difficult discourse for outsiders to penetrate. Fortunately, most of the technical details are not relevant here, since as long as at least one of the proposals pans out the larger, the macro-level hypothesis would be true. Thus, rather than try to review the literature in detail, my primary objective is just to give social scientists a sense that serious work is being done, organize this work into rough categories, and supply references for those who want to follow up. In the second half of the chapter I then take up some critiques of quantum brain theory and assess where the debate stands now.

In parsing quantum brain theory an initial distinction should be made between two different arguments that are often discussed under this heading. What might be called the "weak" argument hypothesizes that the firing of individual neurons is affected by quantum processes, but it does not posit quantum effects at the level of the whole brain. Friedrich Beck and John Eccles have advanced the most detailed proposal along these lines, which they link to an explicitly dualist solution to the mind–body problem.[8] Because the decoherence problem is much less severe at the neuronal than whole-brain level their model has the advantage of facing lower physical barriers to its realization. However, the weak thesis is not where the action is in quantum brain theory these days, and since it has few discernible implications for social science (no walking quantum computers . . .) I shall set it aside below, while recognizing that it might be as far as the theory can go. That leaves the "strong" argument, which does posit quantum effects at the organism level.

Here, two main branches of the family may be distinguished, originating in articles published independently just one year apart – by Luigi Ricciardi and Hiroomi Umezawa in 1967 and by Herbert Fröhlich in 1968.[9] Although these branches have on occasion intersected, for the most part they constitute separate research programs within the quantum brain "paradigm." Their differences originate in the levels of analysis of their starting points: the Umezawa tradition uses quantum field theory to think about the whole brain and from there scales downward; whereas the Fröhlich tradition is interested in what is going on deep inside individual neurons and from there scales up. (Neither is very concerned with the intermediate, neural scale of the conventional wisdom). Although Umezawa beats Fröhlich by a year, the latter sets up the larger argument better and so I will begin there.

---

[8] See Beck and Eccles (1992; 1998), and Smith (2001) on the Cartesian roots of Eccles' thinking, and Clarke (2014) for a critique. For an unheralded earlier model with some similar features, see Bass (1975).
[9] See Ricciardi and Umezawa (1967) and Fröhlich (1968).

The Fröhlich tradition

Fröhlich's contribution was to show that, at least in theory, a particular kind of quantum coherence, Bose-Einstein condensation, might be possible within individual cells, and from there perhaps extend to the whole brain. Quantum coherence refers to a situation in which the wave functions of two or more particles are entangled, such that they form a superposition that can be described with one equation. Concretely, this means the properties of the system's elements are correlated non-locally, so that a measurement on one instantly tells us something about the others. Ontologically, it means the elements have lost some of their identity, such that the system can no longer be disaggregated ("factorized") into distinct parts. This loss of unit identity need not be complete, however; quantum coherence does not require that particles exhibit identical states, only that they be correlated in superposition.

Bose-Einstein condensates (BECs) exhibit a specific kind of coherence in which the particles *are* in identical states, and stay that way, enabling the wave function and the information it contains to be sustained through time.[10] This feature has been linked to a uniquely quantum form of *emergence*,[11] which I will discuss in Chapter 13 as it pertains to social structures. And by the same token it makes BECs an attractive candidate for the physical basis of consciousness, which – recalling the binding problem – has a unitary quality that requires its physical correlates to move in coordinated fashion over time.[12] The idea is that "excitation" of the condensate constitutes a "reservoir of energy" available for various uses,[13] including – by hypothesis – cognition. In this vein Danah Zohar likens BECs to a "blackboard" to which "writing" (thought) is applied by excitations of the condensate.[14]

Is Bose-Einstein condensation possible in the brain? The main reason to be skeptical is that BECs are normally found only at extremely low temperatures, far below those in brain cells, which are immersed in a "heat bath" formed by the rest of the organism. Heat causes molecules to move about randomly, which leads to *de*coherence of quantum states, not to coherence. What Fröhlich showed was that if enough energy were continuously pumped into a cell it could sustain a BEC even at higher temperatures. Zohar offers a useful classical analogy.[15] Imagine a group of compasses on a table in a room shielded from the earth's magnetic field. Because of the shielding their needles will point every which way, and if the table is jiggled they will move about randomly (jiggling plays the role of the heat bath), so that to describe their motions we have to

---

[10] For overviews of the theory and experiments on BECs see Ketterle (1999), Reimers et al. (2009), and Healey (2011).
[11] See Healey (2011); cf. Humphreys (1997a).
[12] See Marshall (1989), Zohar (1990), Ho (1997: 269), and Worden (1999).
[13] Fröhlich (1968: 648).   [14] Zohar (1990: 86).   [15] Zohar (1990: 82).

write a separate equation for each compass. However, if we keep the shielding but pump electromagnetic energy into the compasses they will begin to exert a pull on each other, and at a certain point when the energy is strong enough their needles will "condense" and point in the same direction ("coherence"). At this point we no longer have to write separate equations, but just one for the whole system.

Tantalizing in theory, evidence for this "Fröhlich effect" has been elusive. A first generation of experimental work in the 1970s and '80s was suggestive but inconclusive, and despite a new wave of experiments in recent years, some again promising, it remains unclear whether the effect can actually exist in living systems.[16] Proponents point out, with some justice, that this may be because the "biological mechanisms developed to isolate and protect quantum coherence mechanisms could also make their detection quite difficult," leaving only classical properties to see.[17] Skeptics of course have a different view, arguing that the effect has not been shown because it cannot exist. And at first glance a recent critique of the Fröhlich hypothesis by Jeffrey Reimers and his colleagues looks definitive – until you notice that its primary target is actually just one particular version of the hypothesis in a model from the late 1970s.[18] The critique might still be definitive, but writing in the 1960s and '70s Fröhlich was necessarily vague about some important details, and today there are several different models of how the effect might be produced, with perhaps more to come in the future. Thus, it seems quite possible that proponents will be able to take the Reimers et al. critique in stride, as helping to rule out certain options and thus narrowing the parameters for their search, rather than as proof that their search is misguided. But that remains to be seen.

One research program coming out of Fröhlich's work and today the dominant branch on this side of the family tree is the one pioneered by Stuart Hameroff and Roger Penrose in the early 1990s, focusing on tiny bits of neurons called microtubules.[19] The Neuron Doctrine tells us to look for the physical basis of consciousness in relations *between* neurons – neural networks – rather than inside them. However, neurons themselves are fabulously complex.[20] Each is a single cell made of cytoplasm confined by a membrane, composed of roughly 70 percent water molecules, 20 percent proteins, and 10 percent other elements.

---

[16] See Clark (2010), Craddock and Tuszynski (2010), Lloyd (2011), Igamberdiev (2012), and Plankar et al. (2013) for sympathetic discussions. The existence of BECs in the non-biological realm is well established.
[17] E.g. Hameroff (2001a: 25); see also Clark (2010: 177).
[18] See Reimers et al. (2009). This article is nevertheless very helpful in distinguishing different kinds of BECs and showing that only the most demanding are at issue.
[19] The seminal statements are Hameroff (1994), Penrose (1994), and Hameroff and Penrose (1996). For a comprehensive recent statement see Hameroff and Penrose (2014a).
[20] For good overviews of the internal structure of neurons see Tuszynski et al. (1997) and Satinover (2001).

The proteins are organized in web-like structures making up the "cytoskeleton," which gives neurons their form and regulates their connections. Microtubules – so named because they look like hollow tubes – are the building blocks of the cytoskeleton, and there are typically thousands in a single neuron. And the complexity does not stop there: the walls of the microtubules consist of thirteen columns of protein "dimers," of which there may be 10 million in a neuron. In short, each individual neuron has literally *billions* of parts.

What is all this complexity for? The conventional wisdom used to be that the cytoskeleton is a mere physical support for the neuron, like bone, and thus passive and inert.[21] However, the more scientists have looked, the less plausible this has become. Microtubules are dynamical systems performing numerous functions, including perhaps computation, given that their internal structure bears a striking resemblance to cellular automata, which are widely thought to engage in computation. However, there is still the question of whether any computation going on within microtubules is quantum or classical. Here the case for a quantum answer begins by pointing out that microtubules are at precisely the right scale to mediate between quantum processes at the sub-atomic level and classical, neural computation. From there the argument highlights a number of intriguing features of microtubules that suggest they can amplify those quantum processes into coherent superpositions at the microtubular level.[22]

Assuming that to be the case, in turn, to scale the argument up two additional problems must be solved: (a) how coherent microtubular states avoid decoherence in the noisy environment within each neuron; and then (b) how they do so in the even more complex environment outside, and not just survive but join other coherent microtubular states to form a global superposition in the whole brain. Addressing (a), several mechanisms have been proposed by which quantum coherence within microtubules might be shielded from the intra-neuronal context, among which is a special kind of water – "ordered water" – that is known to fill and surround microtubules.[23] Addressing (b), along with ordered water a frequently mooted possibility is that quantum "tunneling" is involved.[24] A well-established phenomenon in quantum physics, tunneling refers to the ability of electrons to penetrate barriers that in the classical world should be impenetrable – in this case the "gap junctions" between neurons.[25] There is considerable evidence that neurons connected by gap junctions (versus synapses) fire simultaneously, as if they were a single neuron.

---

[21] Satinover (2001: 163); also see Tuszynski et al. (1997).   [22] Hameroff (2001b: 86).
[23] Ordered water is water in which the nuclei of its constituent molecules have been put into a quantum coherent or "ordered" state by an electromagnetic charge; see Marchettini et al. (2010) and Ho (2012).
[24] To my knowledge this was first proposed by Evan Harris Walker (1970), an early quantum brain theorist working independently of both Ricciardi/Umezawa and Fröhlich.
[25] See Hameroff (1998: 1881–1882), and Hameroff et al. (2002: 162–164).

Quantum tunneling could explain this effect, while providing a mechanism for scaling up quantum coherence all the way to the top.

The Umezawa tradition

As this discussion shows, quantum brain theory faces explanatory challenges on multiple levels of analysis at once. Since most of these are below the cellular level, trying to meet them has fallen largely to scientists working in the bottom-up spirit of Fröhlich. However, there is also the challenge of understanding the structure and functioning of the whole brain as a quantum system. This is the concern of the top-down approach of the Umezawa tradition, which in recent years has been carried on and further developed in the work of Mari Jibu and Kunio Yasue, Giuseppe Vitiello, and others.[26]

Animating Ricciardi and Umezawa's 1967 paper and much of the later work in this vein is the puzzle of memory. Two puzzles really: (1) how can memory recall entrain huge numbers of neurons firing synchronously in highly structured phase and amplitude (this is an aspect of the binding problem above)? And (2) how can memories last a lifetime when we know the molecules making them up live at most a few months? Stuart et al. call these the non-locality and stability problems of memory.[27] Both involve long-range correlations among the brain's elements, the one spatial and the other temporal, the explanation of which is a central concern in this literature and memory science more generally. The orthodox view is that neural networks are the answer, but after decades of intensive research this approach still faces a number of hard questions.[28] This is perhaps not surprising given that memory retrieval is by definition conscious, and so might not be understood until we understand consciousness.

The central claim of the Umezawa tradition is that the solution to the puzzles of memory lies in a branch of quantum physics called quantum field theory (QFT). Traditional quantum mechanics (QM) is centered on the Schrödinger equation, which can solve systems involving only individual or at most a small number of particles, a highly idealized case. Many-particle systems like the brain cannot be solved by QM and so we need a different mathematics. If the system in question is one in which the elements do not exhibit quantum coherence and thus their motions are essentially random, then quantum statistical mechanics (QSM) is called for;[29] if on the other hand they are coherent,

---

[26] See especially Jibu and Yasue (1995) and Vitiello (2001); for a good introduction to this stream of quantum brain theory see Jibu and Yasue (2004).
[27] See Stuart, Takahashi and Umezawa (1978); for a recent quantum model of memory see Brainerd et al. (2013).
[28] See, for example, Arshavsky (2006) and Forsdyke (2009).
[29] See Vitiello (2001), and Svozil and Wright (2005) for an application of QSM to the social sciences.

then we need QFT. The Umezawa thesis is that non-living matter instantiates the first, disordered case and living matter the second, ordered one. I will say more about living matter in Chapter 7; for now what matters is that it changes the physics constraint on memory from a classical to a quantum constraint at the collective level. More specifically, the Umezawa proposal is that suffusing the brain's neural network is a quantum field, the dynamics of which drive its elements into coherent motion (solving the binding problem) and keeping them there (solving the stability problem).[30] As such, it side-steps the problem of unit-level coherence that bedevils the Fröhlich approach, since the properties of this quantum field are multiply realizable.[31]

The implications of the Umezawa hypothesis for what memory is and where it is located are quite interesting in their own right but peripheral in the present context, which I want to wrap up instead by addressing how its top-down logic relates to the bottom-up logic of the Fröhlich tradition. The answer is not clear, for reasons that a social scientific analogy might clarify. Umezawa cuts into the quantum brain from what social scientists would call a "sociological" standpoint that is concerned with the whole almost to the exclusion of its parts, whereas Fröhlich cuts in from a "psychological" (or "social psychological") standpoint that highlights the parts more than the whole. To unify quantum brain theory the two would ideally meet up seamlessly in the middle – and interestingly they do find common ground in the significance both attach to the role of ordered water in the brain. But as social scientists know, in a complex system the relationship between micro and macro can be anything but clear – and we have thought long and hard about the problem, whereas in the quantum brain literature there has been little explicit discussion of it.[32] Where do brain wide quantum fields come from, if not from the interaction of their elements (sic), and if so how? Is the decoherence problem at the microtubular level really *completely* irrelevant to a quantum field theory of the brain? And so on. The two traditions share a common conclusion, speak to each other's silences, and occasionally invoke each other as supporting arguments in their narratives. But beyond that the relationship between them has yet to be worked out.

Assessing the current debate

Despite the growing research on quantum brain theory and now two dedicated journals,[33] the idea has been ignored by most neuroscientists and philosophers

---

[30] See Vitiello (2001: 114). Note that in QFT the concept of "element" or "particle" is problematic, since from this perspective particles appear as properties of the field.

[31] As Vitiello (2001: 52) puts it, living matter exhibits "plasticity" with respect to its underlying constituents. For good overviews see Jibu and Yasue (2004) and Vitiello (2006); cf. John's (2001) non-quantum approach to field theorizing about the brain.

[32] Though there has been considerable discussion of part–whole relations in quantum field theory itself; see, for example, Castellani (2002).

[33] *Neuroquantology* and *Quantum Biosystems*.

of mind. One reason may be that in this area theory is well ahead of evidence, and given Hume's dictum about miracles, that "extraordinary claims require extraordinary evidence," it is easy to see why a claim would be ignored for which the evidence is anything but. To be fair, studying sub-neural processes in living tissue is extraordinarily difficult, and there is the further problem that quantum processes are inherently elusive insofar as measuring them will itself induce decoherence. But whereas in quantum physics it is common for predictions to be taken seriously even when the technology to test them does not exist, quantum brain theory has received no such free pass. Indeed, its reception by defenders of the Neuron Doctrine has been strikingly hostile, perhaps for paradigmatic reasons. While the theory, if true, would not invalidate what we already know about the brain, it would revolutionize how we should think about the brain in the future, a prospect that is probably no more attractive to contemporary neuroscientists than it would be to any other group of scholars who have devoted their careers to a particular paradigm.

Still, if in academic life it is better to be criticized than ignored, then quantum brain theorists can take heart in the fact that whereas their idea was once simply dismissed on the grounds that "it can't be true, therefore it isn't," it is increasingly being subjected to detailed critique.[34] It is probably going too far to say that these critiques have begun a "paradigm war": from the perspective of the orthodoxy they are more like skirmishes in a far off colony to quell a pesky heresy. But the heretics are growing in number and increasingly getting the Empire's attention, in part because of the current difficulty of testing the theory empirically. This has enabled advocates to play defense against frontal assaults by classical neuroscientists trying to prove *theoretically* that the theory is false – and thus, like insurgents anywhere, being able to claim victory merely by staying in the field.

As a social scientist, I am in no position to assess the technical debate and as such cannot hope to persuade you of quantum brain theory on the merits. However, it may still be useful to convey impressions of my own "anthropological" encounter with this community of discourse. In that encounter I am admittedly a biased observer, hoping the critiques are not decisive, but I also have no interest in writing a book based on an idea that is demonstrably false. Thus, in reading the debate I have asked myself, do the critiques actually target what they claim to target? Are they cumulative, with defenders forced to concede ever more ground? Are there escape routes from the critiques? And so on. With this dialectical purpose in view, let me briefly report three "case studies" in the ongoing debate. In doing so I hope to persuade you not that

---

[34] For an exemplary early dismissal see Grush and Churchland (1995). There has also been increased sparring within the fold, which is productive for the theory's development but does not generally address the deeper questions; see Rosa and Faber (2004), Mureika (2007), and Craddock and Tuszynski (2010).

quantum brain theory is true, but that the experts have not yet shown that it is false.

The first is a paper published by physicist Max Tegmark in *Physical Review E* in 2000, which got considerable attention in the science press as the definitive refutation of the quantum consciousness hypothesis (note the conflation with quantum brain theory), and continues to be widely cited as authoritative today.[35] An important virtue of the critique is that it is the first to zero in on the key question of whether quantum coherence can be sustained long enough in the brain to do computational work. Through detailed calculations of decoherence rates Tegmark tries to prove that no such coherence is possible. However, in a spirited rebuttal Hameroff and two co-authors leave the reader with the strong impression that Tegmark has missed his mark.[36] They point out that half the article is devoted to showing that neurons are incapable of being in superposition, but Hameroff never suggested they could be – his claim is about microtubules and then the whole brain, not neurons – so this section may be correct but is beside the point.[37] Then, when Tegmark does turn to microtubules, rather than critique an existing model in the quantum brain literature he invents a hybrid that ignores some key mechanisms that have been hypothesized to substantially lengthen decoherence rates – thereby leaving much of the theory out of his sights. So while Tegmark's attack forced Hameroff et al. to clarify and elaborate their position, it was hardly definitive.

This is evident from a second critique six years later by Abninder Litt and colleagues at the University of Waterloo, which appeared in *Cognitive Science*.[38] The authors make three arguments against the idea that the brain is a quantum computer. Their first, "computational," argument is that the time scales on which quantum events can endure are too short to influence neural firing. Here they rely on Tegmark, while acknowledging (in a footnote) that Hameroff's model is about microtubules, not neurons – to which Litt et al. in effect respond: see our other arguments below. But that begs the question and as such dialectically this first argument does not advance beyond Tegmark's critique.

---

[35] Tegmark (2000a); also see his (2000b). Although Tegmark has become something of a villain for quantum brain theorists, he himself has recently boarded the panpsychism train (2014), arguing that consciousness is a state of matter, which some may see as confirming his reputation in the physics community as "Mad Max" (Turausky, 2014: 233).

[36] See Hagan et al. (2002), and for further discussion Rosa and Faber (2004), Davies (2004), Mavromatos (2011), and Georgiev (2013), who argues that even if Tegmark is right about decoherence times this would not preclude quantum effects in the brain.

[37] Also see Alfinito and Vitiello (2000: 219). However, there are those who think that quantum effects are important for how neurons work; see for example Melkikh (2014).

[38] See Litt et al. (2006); for Hameroff's own response, which emphasizes the scientific details, see his (2007).

Their second, "biological" argument does better and makes several claims – (a) that quantum processes are insufficiently isolated in the brain to prevent rapid decoherence (again citing Tegmark); (b) that quantum theories about microtubules "lack any empirical support"; (c) that microtubules are found throughout the plant and animal kingdoms and so the theory implies that "carrots and rutabagas" engage in quantum computation too; and (d) that quantum computation in organisms has no survival value and so would not be selected for in evolution. Yet, these attacks too can be deflected relatively easily: (a) relies on contested claims about what is possible in theory but no one knows in fact. If the "any" in (b) is meant literally then Litt et al. overreach – what they should have said is "lack *much* empirical support." (c) Is not an argument at all, but an assertion of the authors' belief that plants are incapable of computation – a belief called into question by the recent discovery that quantum processes are involved non-trivially in photosynthesis (see Chapter 7). And on (d) too there is debate, with some scientists arguing that there is survival value in quantum computation, while others that it is not selected for in evolution because it is constitutive of and thus emerges with life itself.[39]

We are left then with Litt et al.'s last, "psychological" argument, which comes down to the explanation of consciousness. Here they use the case of anesthesia to affirm their belief that consciousness will be explained in classical neurocomputational terms. However, not only do they assume we now have a complete understanding of how anesthesia works (ironic, given that Hameroff is an anesthesiologist!), they hold a belief about the physical basis of consciousness for which – given the hard problem – there is arguably no evidence at all.[40] So in the end, taking all three arguments together, we have a critique the conclusion of which might be right, but the basis for which fails to prove beyond a reasonable doubt.

At first glance the subject of my last case study seems more definitive.[41] This is an article published in *Physical Review E* in 2009 by the same group that raised questions about the Fröhlich hypothesis above, only this time led by Laura McKemmish.[42] Its authors begin by attacking the heart of the Penrose-Hameroff model, the claim that tubulin dimers oscillate between two states and thus could be elementary units (qubits) for quantum computation in the brain. Importantly, McKemmish et al.'s criticism here is not just theoretical but empirical, based on new evidence about the structure and function of microtubules which, they argue, points strongly away from the model. They then consider

---

[39] On this discussion see for example McFadden (2001) and Castagnoli (2009; 2010).
[40] Moreover, it has been suggested that if computationalism is true then it cannot avoid panpsychism; see Bartlett (2012) for an overview of this line of argument.
[41] A more recent, somewhat more open-minded critique is offered by Baars and Edelman (2012), with a response by Hameroff (2012b).
[42] See McKemmish et al. (2009).

whether the Penrose-Hameroff approach could be salvaged by modifying it in light of this new evidence. As in other critiques a key issue is whether quantum coherence can be sustained long enough to do any computational work. Invoking Fröhlich's hypothesis as the only systematic suggestion for how this might be done, McKemmish et al. reprise their companion critique. The last sentence of this article sums up what they think they have accomplished:

[t]he basic physical assumptions upon which the [Penrose-Hameroff] model depends simply do not hold either from a structural, dynamic or energetic perspective and we hope that with this work we can finally put to rest this intriguing but fundamentally flawed model of cognitive function.

Have McKemmish et al. succeeded where previous imperial generals have failed, in delivering a death blow to the quantum brain insurgency? Judging from subsequent publications it does not seem so. Using state-of-the-art technology, a research group led by Anirban Bandyopadhyay claims to have found for the first time indirect evidence of quantum vibrations within individual microtubules.[43] That does not in itself speak to quantum coherence in the brain as a whole, though the fact that ordered water plays a key role in this finding is suggestive to that effect. Invoking this research in a recent updating and elaboration of their theory, Hameroff and Penrose are untroubled by the McKemmish/Reimers critique, and go on to offer a lengthy point-by-point rebuttal of it as "largely uninformed and basically incorrect."[44] In short, it appears that quantum brain theory is still in the fight, though whether further experimental work will ultimately confirm the theory is anyone's guess. In the meantime, for those of us on the outside there are four reasons to reserve judgment.

The first concerns burdens of proof. It is essential to preserving the epistemic authority of science that, in its advancement over time, whatever is embraced as new knowledge be thoroughly vetted against the old. The practice of science is therefore generally focused on trying to avoid "Type I" errors, or accepting a claim as true that is in fact false. In that dialectical context the burden of proof is on advocates of a new theory to prove beyond a reasonable doubt that their claims are substantiated; by analogy to the law, the presumption is that a new theory is guilty (false) until proven innocent. However, the question today is not whether quantum brain theory is true and so should be counted as knowledge. Given current technology, even those who believe it cannot claim to *know* it is true. The question is whether it *might* be true and as such warrant further investigation. Here the operative worry is not a Type I error but the usually much less salient Type II, or deciding that a claim is false which is actually

---

[43] See Sahu et al. (2013a; 2013b). See Sahu et al. (2011) for a good overview of the debate about microtubules over the past half century.
[44] See Hameroff and Penrose (2014a: 67–68) and Hameroff and Penrose (2014b: 104) respectively.

true.[45] In that context the burden of proof is reversed: onto critics to prove beyond a reasonable doubt that quantum brain theory is wrong. As we know, "innocent until proven guilty" is a higher bar to meet, and one can admire the willingness and skill of the critics who take on the challenge; even if the theory is true, without able prosecutors probing its weaknesses we will never find out. (In that sense the critics are contributing to the development of the theory themselves). But it remains a higher bar, and as such demands a measure of skepticism toward claims to have "finally put this theory to rest."

Second, even assuming that McKemmish et al. have proven that the Penrose-Hameroff microtubule-based model is wrong, this is not the only possible physical realization of quantum brain theory, such that if it fails the theory is necessarily dead. The most popular and well-specified model perhaps, but not the only possibility, even within a bottom-up approach.[46] Moreover, in this respect it is important to note that all of the critiques of quantum brain theory of which I am aware focus on the Fröhlich tradition; none have engaged the top-down approach of Umezawa. Researchers in that tradition also sometimes invoke microtubules, but in doing so they seem more agnostic on whether they are an essential piece of the story. So even if the Penrose-Hameroff Line is breached, there are redoubts to which the quantum brain forces can fall back.

Third, there is the rapidly growing evidence from quantum decision theory for quantum cognition at the *behavioral* level, which I mentioned in Chapter 1 and will review in detail in Chapter 8. But the evidence is not limited to humans: algae, plants, birds, and other organisms have been shown to make use of quantum effects as well, which is spurring the rise of "quantum biology" (Chapter 7).[47] To be sure, behavioral evidence alone does not tell us what is going on inside organisms, where so far at least it remains theoretically possible that the mechanisms involved are classical. However, from an aesthetic standpoint that is decidedly inelegant, since we then have to explain why at both the sub-atomic and behavioral levels we observe quantum effects, whereas at the meso-level, within organisms, everything is classical. If these behavioral findings hold up, in short, it would open up an entirely new front in the war over quantum brain theory, one that would outflank its skeptics and put them on the defensive.

Finally, if we reject quantum brain theory now, then we are back to square one on explaining consciousness. Recall that the attraction of the theory is not just that the brain might be a quantum computer, marvelous though that would be.

---

[45] On Type I and II errors in science see Lemons et al. (1997).
[46] See McFadden (2007), who is critical of the Penrose-Hameroff approach, and Cooper (2009), who does not even mention it, but nevertheless affirms the possibility of quantum information processing at the elevated temperatures within the body. Other new directions are represented by Romero-Isart et al. (2010) and Igamberdiev (2012).
[47] For a good overview see Abbott et al., eds. (2008).

It is also in offering the possibility of re-founding the mind–body problem on the radically different physical basis of quantum theory, in which, unlike the classical worldview, there is a natural place for consciousness. So as long as the theory is still in the field it seems useful to launch a counter-attack of sorts, by exploring whether assuming it to be true might help solve the mind–body problem. If it can, then that should increase our confidence that the insurgents will eventually win.

6 Panpsychism and neutral monism

Although this book is about the implications for social science of a quantum approach to the mind, much of that potential stems from its ability to integrate consciousness – a particular aspect of the mind – into a naturalistic worldview. And when it comes to the hard problem, quantum brain theory is only a necessary condition for such integration, not sufficient. Necessary, because it would enable the body to exploit for macro-level purposes a physical but non-material space for consciousness at the quantum level – collapse of the wave function – that does not exist in classical physics. Insufficient, however, because quantum brain theory embodies an objective, third-person standpoint and as such can tell us no more than classical brain theory can about why brains have a subjective, first-person point of view in the first place. The explanatory gap is still there, in other words, only now pushed down into quantum theory. To cross this gap, therefore, in addition to a new physics of the brain we will need a new *meta*-physics as well. As I argue in this chapter, it is this metaphysics that does the real work in solving the mind–body problem.

Most social scientists have an instinctive aversion to 'metaphysics,' which in our world connotes wooly-headed and speculative thinking that has no place in science. Yet in moving onto this terrain the situation is both worse and better than that. It is worse because we have no choice. The Neuron Doctrine itself is based on a metaphysic, materialism, which looks increasingly unlikely to ever explain consciousness;[1] and in thinking about the meaning of quantum theory metaphysical debate can hardly be avoided, precisely because physics provides little guidance for choosing among its interpretations.[2] So the only question is whether to change our metaphysics of the brain from classical to quantum. But the situation is also better than one might think, because even if by definition it cannot be scientific, metaphysics can be *rational* – in the sense of being consistent with known facts, logical in argument and coherent in structure.

---

[1] See Bitbol (2008) and Nagel (2012) for excellent overviews of the contemporary case against materialism.
[2] Also see Nakagomi (2003b) on physics and world models.

So while I might not convince you that the ontology put forward below is correct, I do hope to prove that a reasonable argument can be made.

Recalling my epistemological double movement above, the argument here starts like the one for quantum brain theory, with what is taken to be known. Only now the gaze is reversed, and what is taken to be known is not the external, material world of the brain but the interior, subjective world of experience. I take inspiration here from Schopenhauer, for whose system this strategy was a defining feature.[3] In his view, "we ourselves are the thing-in-itself. Consequently, a way from within stands open to us to that real inner nature of things to which we cannot penetrate from without."[4] But it also finds expression in the work of physicists like Diederik Aerts: "what happens in our macro-world, i.e. 'people using concepts and their combinations to communicate,' already took place in the micro-realm, i.e. 'measuring apparatuses, and more generally entities made of ordinary matter, communicating with each other, where the words and sentences of their languages are the quantum particles'."[5] Introspection strongly suggests that consciousness is not just real but that it is a kind of knowledge. Recall the case of Mary from Chapter 1: living her whole life in a black and white room until one day she comes out and experiences colors for the first time. These experiences are not knowledge in the "justified true belief" sense of realist epistemology (which assumes a subject–object dichotomy), but on the other hand it seems odd to say that Mary merely has the "belief" of seeing red. Rather, she now *knows* red in a way that is impossible from a purely third-person perspective. This knowing by experience feels so epistemically secure that most of us would not doubt it even if consciousness ultimately defied scientific explanation altogether. When it comes to knowing our own minds, subjectivity usually trumps objectivity – which is why unlike ghosts, there is a mind–body problem at all. So if a third-person approach to knowing consciousness cannot close the explanatory gap, then we should exploit our unique first-person perspective to see whether we can cross from the other side.

In embracing Schopenhauer's strategy I am swimming against a strong tide of anti-subjectivism in twentieth-century social theory. Positivists, interpretivists, critical theorists and post-structuralists alike have seen subjectivity – and by this I mean its experiential aspect – as a problem to be assumed away,

---

[3] Hall (1995: 85); for good introductions to Schopenhauer's thought see Jacquette (2005) and Hannan (2009). Whitehead's approach to the problem of consciousness (experience in his terms, since he equates 'consciousness' with '*self*-consciousness') might be invoked here as well; for a good overview see Weekes (2009), and more generally Griffin (1998).

[4] The quote is in Jacquette (2005: 84).

[5] Aerts (2010: 2967). I don't know if Aerts was influenced by Schopenhauer, but according to Marcin (2006) and Marin (2009), at least two of the founders of quantum theory who took consciousness seriously, Wolfgang Pauli and Erwin Schrödinger, were.

deconstructed, or worked around. From their perspective an argument that not just acknowledges but privileges consciousness as a basis for knowledge acquisition would be problematic, on the grounds that it's not scientific, that subjectivity is just an effect of language, that the subject is dead, and so on. Since these critiques have been extensively developed in the literature, it is reasonable to expect a defense of my strategy before we move on. If I were able to give one here it would be that subjectivity is only a "problem" in social theory because of the mind–body problem in philosophy: if philosophers knew how to integrate experience into a naturalistic worldview then social theorists would know how to deal with subjectivity as well. But to actually show that would require much exegesis with little return for my argument. So rather than defend my epistemic base, I am going to leave this flank open and press my attack, hoping to make the objection moot.

The question, then, is this: what kind of ontology do we get if, complementing and informed by the third-person knowledge of quantum brain theory, we give full epistemic standing to our first-person knowledge of consciousness? The answer I develop in this chapter joins two doctrines, panpsychism and neutral monism. Closely related yet sometimes seen as rivals,[6] in my view neutral monism presupposes panpsychism but not the reverse. More specifically, panpsychists argue that experience is inherent in the deep structure of matter, so that at the elementary level mind and matter form a duality, but then leave the issue at that – as simply a brute fact of nature. Not satisfied with this solution neutral monists go one step further, by trying to explain the duality by reference to an underlying reality which is neither material nor mental, from which the distinction itself emerges. I shall argue that recent work in quantum theory on "temporal symmetry-breaking" lends credence to the latter view. In so doing neutral monism also offers a provocative view about the origin of time, another "hard problem" for the modern scientific worldview that is of particular interest in the social sciences. Thus, if pushing on to neutral monism helps us make progress on that front then that provides some independent evidence for the utility of a panpsychist approach.

Panpsychism

The philosophical debate about the mind–body problem has featured almost every imaginable position, but in the West it has been dominated by two views, dualism and materialism. Despite deep differences on how the problem should be solved, dualism and materialism share three, almost completely taken-for-granted assumptions about the nature of matter: that it is wholly material; that

---

[6] On the differences between neutral monism and panpsychism see Holman (2008) and Silberstein (2009), though neither category is without internal differentiation as well.

it is a substance, hard and tangible; and that it is passive and reactive. Given these assumptions, which go back to ancient Greece and were supported by classical physics, both sides have conceived of "the problem" as the mind rather than the body. For materialists the challenge is to show how mind can be explained, whether in reductionist or emergentist terms, by an essentially passive substance that contains no trace of mind within. Dualists think this can't be done and therefore that mind is ontologically *sui generis*, but they agree that matter per se is a passive and purely material substance. As we have seen, quantum theory calls these assumptions into question, and in so doing makes the nature of matter as much a problem as that of mind. While there are materialist readings of quantum theory, the only thing they have in common with old-fashioned materialism is that they assume there is no trace of mind within matter; otherwise the behavior of matter at the quantum level is utterly unlike its classical counterpart. If quantum matter be "matter" at all, in short, then it is a thoroughly de-materialized matter, far from what we ordinarily imagine matter to be.[7]

Against this classical view, panpsychists have urged for centuries that mind is intrinsic to matter at the elementary level, for which the bumper sticker might be "no matter without mind, no mind without matter."[8] In other words, mind is neither reducible to matter nor emergent from it, but *in* matter all the way down – which in turn obviates the need to posit dualism's two substances in response to materialism's failure to explain consciousness. In short, if mind and matter are continuous, then the traditional framing of the mind–body problem is "spurious."[9] In a moment I'll explain what the idea of "minded matter" might mean, but since panpsychism is probably less familiar to social scientists than materialism or dualism let me provide some brief background first.

Background

One reason panpsychism is not well known is that for much of the twentieth century most philosophers dismissed it as an absurd doctrine. This is a recent prejudice, however, since as David Skrbina shows in *Panpsychism in the West*, it has a distinguished pedigree in philosophy.[10] Keeping in mind that all metaphysical systems take various forms, Skrbina sees panpsychism

---

[7] This is the main reason why classical materialism morphed into the more ambiguous category of 'physicalism.' For recent reconsiderations of matter in a spirit similar to my own but less far out on the limb, see Fox Keller (2011) on the one hand, and the work of New Materialists on the other, which I discuss in the next chapter.

[8] See Skrbina (2005: 114), paraphrasing Goethe.

[9] Sheets-Johnstone (1998: 260); also see Atmanspacher (2003).

[10] Or even before if the near universal animism of "primitive" cultures is considered a form of panpsychism. On animism see Abram (1996) and Harvey (2006); cf. Sheets-Johnstone (2009).

in the work of, among other ancients, the pre-Socratics, Plato, and Plotinus;[11] in medieval philosophers like Giordano Bruno (who was burned at the stake for it); in Spinoza, Leibniz and Goethe in the early modern period; and in the nineteenth-century philosophies of Schopenhauer, Gustav Fechner, William James, Charles Peirce and Gabriel Tarde.[12] In the early twentieth century, however, panpsychism's importance began to decline – ironically just as the quantum revolution made it more plausible – although Bertrand Russell's neutral monism has a panpsychist aspect, and Alfred North Whitehead's *Process and Reality* is arguably the greatest panpsychist system ever built. But with very few exceptions[13] after 1940 panpsychism disappeared from the Western philosophical landscape, so much so that in 1997 John Searle probably spoke for many in dismissing it as an "absurd" doctrine that "there is not the slightest reason to adopt."[14]

Given this hostile environment just a few years ago, it is therefore striking how much panpsychism has revived lately, particularly after an important 1995 article by William Seager and then the publication a year later of David Chalmers' landmark, *The Conscious Mind*, which explicitly flirts with the idea.[15] Discussion has since grown rapidly, and although far from supplanting the materialist orthodoxy, panpsychism is philosophically respectable today in a way that it has not been for a long time.[16]

The discussion has emerged in three separate disciplines,[17] each with its own starting point and concerns, so what we have today are really three panpsychist literatures that typically make little reference to one another. One is in philosophy of physics, where long-standing interest in the role of consciousness in the measurement process has been elaborated into full-blown panpsychist ontologies.[18] Here the pattern of argument is generally to move straight from the physics to panpsychism, skipping what is going on inside the brain. This

---

[11] See also Malin (2001).
[12] Tarde's (1895/2012) ontology is more "pan-social" than panpsychist, but as the only sociologist in this set and his debt to Leibniz's monadology it seems appropriate to include him here.
[13] The only significant panpsychist contributions of which I am aware for the period from 1940 to 1990 are Teilhard de Chardin (1959), Globus (1976), Nagel (1979), Berman (1981) (who speaks of animism rather than panpsychism), and Sprigge (1983), but these were not cumulative and generated no more than local philosophical interest.
[14] The quote is in Skrbina (2005: 236). [15] See Seager (1995) and Chalmers (1996).
[16] So much so that in a report on the most recent bi-annual "Toward a Science of Consciousness" conference in 2014, Keith Turausky (2014: 234) came to the conclusion that "we are all panpsychists now."
[17] Or four if one counts environmental philosophy, where the question of panpsychism has been on the table somewhat longer; see for example McDaniel (1983), Zimmerman (1988), and today especially Mathews (2003).
[18] For early statements by physicists see Walker (1970) and Cochran (1971), and more recently Bohm (1990), Miller (1990), Stapp (1993; 1999; 2001), Penrose (1994), Hiley and Pylkkänen (2001), Malin (2001), Dugič et al. (2002), Atmanspacher (2003), Nakagomi (2003a; 2003b), Primas (2003), Clarke (2007), Pylkkänen (2007), Gao (2008; 2013), and Tegmark (2014).

pattern is inverted in a second stream coming out of quantum brain theory, which some of its advocates have argued implies panpsychism.[19] Finally, the largest cluster of recent work is in philosophy of mind itself.[20] There it seems to be a growing feeling that materialism will never solve the mind–body problem which is generating interest, epitomized in the title of a recent article on panpsychism – "It Must Be True – But How Can it Be?"[21] Although one hopes this disciplinary fragmentation will eventually be overcome, consumers benefit from having several distinct rationales for panpsychism, which give the idea added force. Since my interest is not in panpsychism per se but in what it might imply for social life, I will treat this diverse body of work as if it were one, drawing freely from its streams for my argument.

In doing so, I can take advantage of something earlier panpsychists could not: the findings of quantum physics. Historically panpsychists had to argue on purely philosophical grounds, since classical physics provided no basis for viewing matter as minded – it was the *inability* of physics to accommodate consciousness that motivated panpsychism, not its ability.[22] Today the situation is different: far from opposing physics, from a quantum perspective there is much to recommend panpsychism.

Defining 'psyche,' aka subjectivity

Panpsychism is a claim about the intrinsic nature of matter. As Bertrand Russell and Kant too observed, physics describes matter only in terms of its properties and behavior, not in terms of what it is inside.[23] Yet we do know what it is like to be inside at least one bit of matter, our own brains, which we know from experience. To project this knowing downward we first need to distinguish those aspects of it that might be essential to the phenomenon in general from those that are merely contingent, features of the human psyche specifically. A prominent example of the latter is self-consciousness, an awareness *that* we are aware, which is unlikely to go very far down the evolutionary ladder.

---

[19] See Miller (1990), Globus (1998), Miranker (2000; 2002), and Romijn (2002); also see Tononi (2008), whose "information integration theory of consciousness" develops a kind of panpsychism from a classical standpoint.

[20] See Seager (1995; 2009; 2010; 2012), Hut and Shepard (1996), Griffin (1998), Bolender (2001), Montero (2001), de Quincey (2002), Gabora (2002), Rosenberg (2004), Skrbina (2005), Schäffer (2006), Strawson (2006), Clarke (2007), Franck (2008), Basile (2010), Coleman (2012; 2014), Robinson (2012), Kawade (2013), and Lewtas (2013a).

[21] See Basile (2010). Then again, 'panpsychism' does not even appear in the index of Koons and Bealer's recent *The Waning of Materialism* (ed., 2010), and is the focus of only one chapter in Göcke's *After Physicalism* (ed., 2012). Both books instead mostly offer new formulations of dualism.

[22] See Seager (2009) for a good overview of the classical arguments for panpsychism.

[23] See Chalmers (2010: 133) and Bolender (2001). As Nakagomi (2003b) puts it, in physics matter has no "inside."

In humans it is easy to conflate this capacity with consciousness per se,[24] since our own experience *is* reflective, but to maintain such a conflation would in effect be to deny consciousness to almost all other organisms, which seems counter-intuitive. (Moreover, the hard problem of consciousness is not about reflexivity, but the more primitive capacity simply to feel.) But then what kind of consciousness could encompass other organisms, all the way down to the nature of matter itself? One finds various answers within the panpsychist tradition, but a review would take me far afield, since who said what, why, and how does not really matter here. Instead, informed by my reading of the literature, empirical observation, and introspection I will offer my own sense of what the panpsychist consensus on 'psyche' might be.

I will also make a terminological move that some contemporary panpsychists might not endorse, but which I believe is warranted both on substantive grounds and given my audience: that what panpsychists mean by 'psyche' is equivalent to what social theorists – in the phenomenological tradition at least – mean by 'subjectivity.' So in advancing panpsychism my claim is that subjectivity is intrinsic to matter, or more precisely "proto"-subjectivity, which I argue below, is the antecedent to subjectivity before the latter is organized into living matter.

The essential features of psyche or subjectivity I take it are Cognition, Experience, and Will.[25] While phenomenologically bound together, they are analytically distinguishable and, as I argue later, map onto different aspects of the quantum formalism. Since I will say more about them with reference to human beings in Part III, here let me just give an intuitive sense of what I mean by each, before making the case for projecting them downward.

'Cognition' refers to all the functions associated with "thinking," including information processing, memory storage and retrieval, and learning. Human beings often associate thinking with self-awareness, but that is anthropocentric: as will become clear below bats and mice think without (presumably) being self-aware, and even in humans most thinking is done sub-consciously. Although cognition is not yet fully understood, cognitive science has shown it to be computational in character, which means that computers can do it. Thus, unlike cognition in the brain, which is difficult to observe without killing its owner, cognition in a computer – or at least a classical one – could at least in principle be directly observed if the machine were big enough.[26] This is partly what makes explaining cognition an "easy" problem of mind, even if in practice it is far from that.

---

[24] See Jaynes (1976).   [25] Cf. Kawade (2009).
[26] Cf. Lodge and Bobro (1998) on Leibniz's "mill" argument. This would not be true of quantum computers if they are ever built, since they depend on maintaining an interiority that can't be disturbed without destroying it.

If Cognition is thinking, then 'Experience' (or 'Consciousness') is feeling, a different phenomenon and the core of the mind–body problem. Feeling here refers not to what social scientists would know as "emotions," which have a large discursive and thus specifically human element, but to "what it is like" simply to feel, at the most basic level to feel pain. As such, unlike Cognition, which might be observable, Experience is intrinsically private, something that can be truly known only from the inside. We get an *ersatz* knowing of others' experience by seeing pain in their face, but that is not knowing what it is like to be in that particular pain right now, yourself. This interiority of experience constitutes subjects as Leibnizian "monads" with a unique perspective on the world, and by virtue of which that world has unique meaning for them.

As aspects of subjectivity Cognition and Experience get most of the attention from philosophers these days, but historically there has also long been interest in Will, of which Schopenhauer, Nietzsche, and Bergson are prominent modern exemplars.[27] In recent years this continental tradition has been complemented on the analytical side by scholarship on "mental causation," which addresses the question why/how consciousness seems to have causal powers like the ability to move our bodies.[28] Cognition and Experience are passive and reactive in the sense that they reflect rather than create reality; their direction of fit to the world is one of world to mind. Will, in contrast, is active and purposeful, a drive that imposes itself upon, and thus changes, the world.[29] This power to re-make the world I take it is a crucial aspect of our sense of agency, though as we will see below I think agency presupposes Cognition and Experience too.

Projecting subjectivity through the tree of life

Stripped of their specifically human content and imagined in their most primitive possible form, how far can Cognition, Experience, and Will be projected downward into forms of matter simpler than our own? Given the private character at least of Experience, from a third-person standpoint the question cannot be answered, and indeed in theory the question might be asked even of our fellow human beings. The Problem of Other Minds is in part that, given the interiority of experience, none of us can be absolutely sure that other people are actually conscious, as opposed to just machines or zombies mimicking conscious behavior.[30] I cannot prove that dandelions and dogs are conscious any more than I can prove *you* are conscious, but yet beyond a reasonable doubt

---

[27] Though their accounts differed in important ways; see, for example, Janaway (2004), François (2007), and Khandker (2013). Note that Schopenhauer's notion of Will includes what I have separated out as Experience; see Hamlyn (1983) and Hall (1995).
[28] See Robb and Heil (2014) for a good introduction to the mental causation literature.
[29] Also see Goethe's concept of "Steigerung" or the "inner drive" of nature (Tantillo, 2002).
[30] See Hollis and Smith (1990) for an accessible introduction to the Problem of Other Minds, with special reference to IR, and also Chapter 12 below.

I know it at least about you. This knowing is not scientific, but the intuitive and practical kind one has from the first-person perspective, which I assume we all share. Attributing subjectivity to non-human organisms is much more difficult, but I will suggest through a combination of empirical evidence and logical reasoning that we can arrive at a rational answer.

Panpsychists say that in some sense subjectivity goes *all* the way down, to the intrinsic structure of matter itself, whether it takes an organic or inorganic form. Given that it is obviously harder to make the case for the inorganic form, let me start with living matter.

The easiest step in projecting subjectivity is to organisms close to us in complexity and genetic makeup. It seems difficult to deny that higher mammals like apes, dogs, and even Nagel's famous bats[31] think, feel, and will; even the scientific community seems to be reaching that conclusion, through a variety of indirect means.[32] What about insects and molluscs? At this level the science of animal subjectivity is thinner, though arguments have been made.[33] Contemporary intuitions might be gauged by where there are animal cruelty laws, since if animals cannot feel pain then the idea of "cruelty" seems incoherent. I do not know where the legal line is drawn (or – an interesting question – on what grounds), but I doubt there is a law against torturing earthworms, or many people who would offer moral objections either; so the easy cases end somewhere in that range.

What then about the hard cases for subjectivity in organic matter? Consider two of the intuitively hardest, single-celled organisms and plants. Can paramecia and bacteria think, defined in the computational sense? It turns out that this one is relatively easy. A recent scientific review article entitled "Bacteria Are Small but not Stupid" reports that in contrast to the old mechanistic view of cells, today we have "abundant results showing that what a cell does is a function of the information it has about itself and its surroundings."[34] Information has replaced matter as the key to explaining cell behavior, such that "bacterial cognition" is taken literally rather than as just a metaphor. What of Will? Again, there seems to be evidence that goal-directedness or "nano-intentionality" is a fact of life at the cellular level.[35] The truly hard question is, do bacteria *feel*. Well, why not? If there were a threshold of organismic

---

[31] See Nagel (1979).
[32] See for example Baars (2004) and Seth et al. (2005). Note that the intelligibility of this inference presupposes our own experience of consciousness – i.e., if we were not conscious it would never occur to us to ask whether other organisms were.
[33] See respectively Carruthers (2007) and Mather (2008); for a comprehensive overview of the literature on animal consciousness see Allen and Trestman (2014).
[34] See Shapiro (2007: 808); also see Ben-Jacob et al. (2005), Hellingwerf (2005), Waters and Bassler (2005), and Tauber (2013) – and Weber (2005) for skepticism that biological information implies any kind of intentionality.
[35] See Fitch (2008), and also Miller (1992), Jonker et al. (2002), Kawade (2009), and Campbell (2010).

complexity above which we could clearly say there is feeling and below which there is not, then one might conclude that the simplest organisms, at least, do not feel. Such a threshold might exist, but it is neither empirically nor theoretically apparent. There would have to be an evolutionary reason for the emergence of experience at a certain level, but evolutionary theorists have no more idea how to explain consciousness than anyone else, since it is not clear what survival function something that (for materialists) is either epiphenomenal or illusory could serve. Given the lack of such a principled threshold, therefore, it seems to follow that, as biologist Lynn Margulis argues in "The Conscious Cell," single-celled organisms are conscious too.[36]

A similar slippery slope applies to plants. The idea that plants are conscious will probably strike most social scientists as silly (it certainly made me laugh at first), more at home in New Age gardening than modern science. But in a recent review in the *Annals of Botany* Anthony Trewavas pulls together an impressive array of evidence to argue that plants are at least "intelligent."[37] Defining 'intelligence' in terms of what I am calling Cognition and Will – capacity for memory and its retrieval, information processing and "adaptively variable behavior during the lifetime of the individual" – Trewavas relays some fascinating ways in which plants are intelligent. They process signals from the environment, sense each other's presence, compete for light and resources, and whereas we respond to pressures from the environment by moving our bodies, plants do so by literally changing theirs (growing new limbs). All this happens so slowly that it is imperceptible to the naked eye, but it is there nonetheless. Does this mean plants are conscious too? Trewavas perhaps wisely does not ask, and I could find little work by other scientists that does either.[38] But once again there is the slippery slope of explaining why plants are *not* conscious, in their own way, if they have the faculties of Cognition and Will? If they are indeed intelligent, then it seems to me the burden of proof is on the skeptic to show that they do not also have Experience.

None of this, of course, is to deny that the content and quality of subjectivity varies hugely across the spectrum of life. But in their capacity to think, feel, and impose upon the world, bacteria and plants I claim *are* just like us, simply by virtue of being alive. In that light one might say that nature is not an "It" but a "They," or with Novalis, even a "You."[39]

---

[36] See Margulis (2001).
[37] Trewavas (2003; 2008); also see Kull (2000), Barlow (2008), Cvrčková et al. (2009), and Affifi (2013). Cvrčková et al. (2009) offers a more qualified view, albeit one published in the journal called *Plant Signaling and Behavior*. See Narby (2005) for a highly readable introduction to Trewavas' and others' work in this vein.
[38] Though see Nagel (1997).
[39] A German Romantic poet of the late eighteenth century; on Novalis' conception of nature as a *You* see Becker and Manstetten (2004).

... And then all the way down

So far I have argued that all organisms are subjects and therefore conscious. But this is not the specifically panpsychist claim, which is that consciousness is intrinsic to the structure of all matter, not just its living form. It is this claim that does the real work on the mind–body problem, since otherwise we still have to explain how conscious life emerges from dead matter, which is the very problem with which we began. So how then to argue that mind and matter are continuous, all the way down?[40] The answer is to exploit the causal gaps in quantum theory by identifying subjectivity (or more precisely, proto-subjectivity, since we are not talking here about living matter) with the formalism of quantum mechanics,[41] only viewed now from the inside, as intrinsic to matter, rather than from the outside. As Chalmers puts it, "[t]his way, we locate experience *inside* the causal network that physics describes, rather than outside it as a dangler... And importantly, we do this without violating the causal closure of the physical."[42]

Consider a single sub-atomic particle in a cloud chamber interacting with an experimenter. Normally physicists think of the formalism as a tool for describing what the outside observer knows about the particle and its likely behavior. Now assume that quantum brain theory is true. Although what goes on inside the brain involves zillions of particles, by virtue of quantum coherence we experience all that complexity in a unitary way, as "*I*." That means that just as for the particle in the cloud chamber, an outside observer could in principle write a single equation to describe our behavior. With this homology in mind, now consider what it means to be on the inside of a human wave function.[43] It does not mean that I will necessarily have more knowledge than an outside observer about what is going on inside my body, except insofar as I can bring it to awareness by observing my wave function from within[44] – but very little in our bodies is accessible in this way. Rather, what being on the inside of our wave functions means is to be a subject, a living being who thinks, experiences and wills. These are all processes that an outside observer cannot know in the first-person way that we know them. Thus, even though Jones and Smith might write the same equation describing Jones' wave function in a given context – and to that extent have similar third-person knowledge about him – Jones has a privileged form of access to this equation from the inside.

---

[40] Also see Kawade (2013). [41] See for example Clarke (2007) and Jansen (2008).
[42] See Chalmers (1997: 29; emphasis in the original), here describing a suggestion of Bertrand Russell's that has since been taken up especially by Lockwood (1989).
[43] See Mould (1995; 2003).
[44] Note, however, that in observing yourself in this way, you would change your wave function for the next moment in the stream of consciousness. These ideas bear a strong resemblance to the idea of "internal measurement," of which Howard Pattee and Koichiro Matsuno are the principal representatives; see Balazs (2004) for an overview.

Now consider how Cognition, Experience, and Will might be mapped onto a single particle.[45] Cognition is perhaps the most difficult case, since the term calls to mind an image of classical computations being performed with many different parts, which hardly seems possible in a single particle. Yet, cognition would not look like that if the mind is a quantum computer rather than a classical one. Although there are plenty of classical things happening inside the brain too, these would be supports for what underneath is a quantum process of thinking inside a wave function – a structure of potentiality which has no separable parts and, by definition, cannot be observed. It is thus noteworthy that most human thinking occurs *un*consciously, which makes sense if cognition takes place inside our wave functions.[46] If the brain is a quantum computer capable of exploring many possibilities at once, it will be able to do that only as long as none of those possibilities is actualized in consciousness. Similarly then one might argue that in the case of a single particle: whatever thought "it" engages in cannot be observed because to do so would collapse its wave function. So who is to say that when a wave function collapses into a particle, there is nothing like thinking involved in how that outcome is reached? It seems to me that the real difference from human beings is that particles are not alive and thus have no continuity of identity or memory over time; if there is thinking going on there, therefore, it is immediately lost to the ether when its wave function collapses.

A particle's experience, in turn, may be identified with the collapse of its wave function. Externally, what is observed in collapse is a reduction of many possible states to one actual state; internally, what happens is a differentiation of phenomenal content, with one conscious state realized from among many possible ones.[47] Although some physicists have argued that the role of consciousness in collapse is causal (see Chapter 4), following others, I propose that experience does not cause but just *is* the collapse, as observed from the inside. Importantly, this does not make experience epiphenomenal and thus ontologically redundant, for two reasons. One has to do with the role of consciousness in bridging the past, present, and future, to which I will return below.[48] The other is that for X to be epiphenomenal, there must be some Y that fully accounts for its properties and causal powers. Yet in wave function collapse we have an X that, when seen from the outside, defies a materialist account altogether. So proposing that when observed from the inside, the collapse of the wave function

---

[45] While there is overlap there is no agreement in the literature on what precisely this mapping should look like; compare for example Malin (2001), Pylkkänen (2007), Vimal (2009), Baer (2010), Martin et al. (2010), and Lewtas (2013a). Some of this variation may be down to semantics, some more substantive. To keep things moving I will only present the view to which I personally have come, which is closest to Chris Clarke's (2007).

[46] See, for example, Dijksterhuis and Aarts (2010).   [47] Mensky (2005: 405).

[48] For a non-quantum argument to this effect see Baumeister et al. (2011).

is experienc*ed* is not redundant with what we already know, since absent this assumption we do not actually know anything at all.

With the wave function as Cognition and its collapse as Experience, Will would then be the force that brings collapse about. Let me unpack this suggestion in a bit more detail.

By the "force" of will I mean that will is a kind of cause, but not the kind that dominates the modern scientific worldview, which Aristotle called efficient causation and social scientists today might call "mechanical." Given the classical prohibition on action at a distance, efficient causation requires an external X to establish physical contact with a separable Y and induce a change in its state. Yet, while this is a good description of some causes, quantum theory shows that efficient causation is missing in wave function collapse – which is precisely why it is so mysterious. It is therefore interesting again that, for human beings, efficient causation, if not missing, is at least an *awkward* way to describe mental causation, the experience of which hardly seems mechanical – a fact which has sustained a long debate in the philosophy of social science about whether "reasons are causes" at all.[49] Yet such awkwardness is precisely what we would expect if the kind of force involved in human Will is the same as in the collapse of the wave function, which is not reducible to a classical conception of cause.

It might be thought that this does not take us very far, precisely because it is unclear whether reasons are causes, but from here we can tack back to the debate in the philosophy of physics, where we find several interpretations of quantum mechanics that support a "will"-discourse. The philosopher most often invoked in this literature is Aristotle, so how might a broadly "Aristotelian" view move us forward in thinking about the causation involved in wave function collapse?

In two ways, depending on whether we approach collapse from the outside or inside. From an external perspective, a number of philosophers of physics including Heisenberg have conceived of wave functions as "dispositions," "propensities" or "tendencies," which in their directed quality are not reducible to probabilities alone.[50] However, while it is helpful that human behavior may also be described in dispositional terms (e.g. the "causal powers" routinely invoked by scientific realists), dispositional readings of quantum theory do not speculate about what is going on inside the wave function to cause its collapse, and as such do not capture the phenomenology of mental causation or willing. For that we need to take an internal perspective, for which

---

[49] Among social scientists Davidson (1963) is often thought to have won this argument for the (efficient) causal side, but the philosophical debate actually has continued. In Chapter 9 I review this debate and use a quantum framework to argue against Davidson's view.

[50] See, for example, Suárez (2007), Dorato and Esfeld (2010), and Bigaj (2012); on how propensities differ from probabilities see Humphreys (1985).

the early quantum theorist Hermann Weyl provides everything one could ask for. Taking his cue from Leibniz and *Naturphilosophie* (and thus derivatively Aristotle), Weyl thought of matter at the quantum level as an "agent."[51] In retrospect his word choice is not ideal, since in social science today the word 'agent' connotes a substance, which is precisely what Weyl was not talking about; what he "should" have said, is that matter is an agen*cy* or *process*.[52] But with that qualification in mind I shall follow Weyl in arguing that the experience of human agency provides a plausible understanding of the kind of causation involved in the collapse of a particle's wave function.

First, whereas in the classical view matter is inert and passive, only moving when pushed by other forces, from an agency view at the quantum level matter is active and spontaneous, moving by virtue of its own internal force. Causation here comes from within matter rather than from without. As Schopenhauer famously put it at the human level, "motives are causes seen from within."[53] Second, whereas efficient causation is retrospective, with causes preceding effects, the experience of agency is prospective and purposeful, directed toward ends in the future. Agency is *teleological*, in short, like Aristotle's final causation.[54] Finally, whereas classical causation is ontologically deterministic, our experience of agency is one of freedom. Given the "hard problem" that free will has long posed for the classical worldview, it is no surprise that from the dawn of the quantum era efforts have been made to explain it by reference to indeterminism at the sub-atomic level – efforts in turn rejected on the grounds that free will must be more than indeterminism to constitute "will." Yet, what these dismissals ignore is the possibility of an internal perspective, in which what appears on the outside to be random could be willed on the inside. As we saw in Part I, some physicists take seriously the idea that wave function collapse is a "choice" by particles in response to measurement.[55] And in "The Free Will Theorem," John Conway and Simon Kochen prove that if "there exist any experimenters with a modicum of free will then elementary particles must have their own share of this valuable commodity."[56]

That said, individual particles do not have internal structure, are not necessarily organized with other particles, and do not persist once their wave

---

[51] See Sieroka (2007; 2010) for an introduction to Weyl's philosophy of physics. Also see Bohm's early "pilot-wave" model of quantum mechanics, and his subsequent development with Basil Hiley of the concept of "active information" (Bohm and Hiley, 1993), both of which have similarly intentional connotations.
[52] Compare Miller (1992).   [53] The quote is from Hamlyn (1983: 457); cf. Miller (1992: 362).
[54] On Aristotle's view of final causation see Gotthelf (1987).
[55] See also Miller (1990; 1992), Mensky (2005), and La Mura (2009: 409), as well as Klemm and Klink's (2008) reading of quantum mechanics as a "theory of alternatives."
[56] Conway and Kochen (2006: 1441). In 1927 Bertrand Russell was among the first to suggest that in light of quantum theory atoms must have free will; see Basile (2006: 220). See Chapter 9 for further discussion of the free will problem.

functions collapse. In all of these respects particles are quite unlike organisms; they are processes *only*, rather than processes that are also substances. This creates something of a problem, since normally we think of subjectivity as being experienced by a subject; consciousness is "what it is like" *for* someone.[57] And Weyl's loose language notwithstanding, it is difficult to argue that particles are actual subjects of experience.[58] This has led some panpsychists to distinguish between mentality (subjectivity in my terms) at the organism level and "proto"-mentality at the elementary level. However, critics see this distinction as "empty," because it leaves us right where we were with materialism, with having to explain the emergence of consciousness out of matter that is not really conscious.[59] This brings us to the "combination problem," which advocates and critics alike have long seen as the central challenge facing panpsychism.[60]

The combination problem and quantum coherence

The combination problem is really two related problems. One is analogous to the "binding problem" in materialist models of the mind: how do the zillions of ephemeral proto-subjectivities of sub-atomic particles combine into the stable and unitary consciousness we experience in everyday life? The other concerns the specificity of life: if consciousness is inherent in the deep structure of matter, what about the difference between macroscopic matter that does not seem to be conscious and matter that does? Does panpsychism imply that rocks and glaciers are conscious too?

With respect to the latter, some panpsychists are willing to bite the bullet and argue that if consciousness is found at the elementary level, then even inanimate objects must in some sense be conscious too.[61] However, in my view such a "strong" panpsychist position should, if possible, be avoided, not just because it is counter-intuitive (hardly a strong argument in the present context!), but because it fails to save the phenomenon, namely the apparent distinction between conscious and non-conscious matter. So ideally a basis for a real, as opposed to just an apparent, distinction should be found. To do that, we need to explain how the experiences immanent in a rock's elementary parts get lost in the rock itself – and *not* in people. Although I will have much more to say about the nature of life in the next chapter, it turns out that the

---

[57] See Coleman (2014).
[58] Though see Lewtas' (2013a) discussion of "what it is like to be a quark."
[59] McGinn (1999: 99).
[60] I believe that Seager (1995) coined the phrase, but the problem goes back to James (1890); for a recent critique of panpsychism along these lines see Goff (2006; 2009).
[61] See for example Chalmers' (1996: 293–297) discussion of thermostats, and Tononi (2008: 237) on other artefacts; for a more skeptical view see Velmans (2000: Chapter 5).

answer to the first combination problem provides an answer to the second as well.

Panpsychists have devoted much effort recently to dealing with the (first) combination problem, though mostly without reference to quantum theory.[62] This is strange, because in light of quantum brain theory the solution seems clear: what distinguishes conscious matter from its non-conscious counterpart is the presence of quantum coherence. Recall that individual sub-atomic particles normally *de*cohere in interaction with other particles, which is why quantum brain theorists have staked their claim on finding structures that can prevent decoherence in the brain. Assuming that experience maps onto wave function collapse, this means that at the elementary level experiences – whether disembodied/subjectless or not – are happening all the time throughout matter, including inside rocks and thermostats, whenever particles interact. In the case of non-coherent matter, however, these experiences are disorganized and fleeting, random sub-atomic events with no memory of the past or continuity of purpose for the future. So while rocks and glaciers have relatively stable structures at the macro-level, because those structures are classical such objects do not have experiences, even as their elementary constituents momentarily do.

In contrast, according to quantum brain theory, the brain has an internal structure that continuously produces quantum coherence, even in the face of its constant decoherence in its interaction with the environment. In this way the otherwise fleeting experiences of its parts can be unified and amplified into the experience of a whole and retained as memory. In short, proto-consciousness is just like ordinary consciousness except that it does not cohere with other elementary experiences across space or time, instead losing its identity immediately to the vacuum. This would seem to solve the combination problem by explaining what happens physically when ordinary consciousness emerges, while also satisfying the intuitive constraint that, even if sub-atomic particles exhibit traces of consciousness, there is still an essential difference between rocks and us.

Neutral monism and the origin of time

Projecting psyche or subjectivity all the way down to sub-atomic particles solves the fundamental problem facing any materialist approach to the mind–body problem, which is to explain how consciousness emerges from non-conscious

---

[62] See for example Griffin (1998), Basile (2010), Shani (2010), Hunt (2011), Coleman (2012), and Jaskolla and Buck (2012). Seager (1995) and Coleman (2014: 34–38) are exceptions to the tendency to neglect quantum theory in this context.

matter. Namely, it doesn't emerge at all, but has been there all along. The key to this solution is that, at the quantum level, matter is no longer old-fashioned "matter" – i.e. clearly devoid of mentality – but an ungrounded potentiality to which, when actualizing, attributions of mentality make sense. Importantly, given that they are described by quantum physics, these potentialities are still *physical* and thus part of the natural order. But they are physical in a broader sense of the term that is not co-extensive with *material*. With this distinction justified by quantum theory, we can see that materialism has failed to make progress on the hard problem of consciousness because it framed the problem incorrectly from the start.

Still, it might be asked, how does projecting mind all the way down really solve the mind–body problem? Does it not merely compound it, since in leaving us with an unexplained distinction between the objective and subjective aspects of reality it not only fails to close the explanatory gap in epistemology, but seemingly enshrines it now as ontology as well?

To be sure, the force of this objection comes partly from an implicit materialist assumption about what constitutes "solving" the mind–body problem, which is to explain how mind arises from mindless matter. From that standpoint my argument is no solution at all, but a convenient moving of goal posts that defines the problem away. However, that assumes that materialism is true: if it is not, then the solution to the problem will necessarily look quite different.

Yet the objection also has a separate, aesthetic force: that by itself panpsychism solves the problem in a sense on the cheap, by the expedient of adding subjectivity to the list of irreducible aspects that are assumed to exist at the fundamental level, and as such it might be said to be neither elegant nor parsimonious. Wouldn't it be better to have just one aspect down there rather than two, and if not then two aspects of *what*?

When the objection is sharpened this way, it seems a response could go two ways. One, starting with the last issue, would be to argue that the "what" in question, the fundamental unit of reality, is information rather than matter/energy (Wheeler's "it from bit").[63] From there, the idea would be that it is a brute fact of nature that information has both objective and subjective aspects, which must simply be accepted. A brute fact is one for which no further explanation is possible, like the existence of matter for classical materialists, which is simply given.[64] In this case the brute fact is not a dualism of substances but of aspects of information. And while in a perfect world there might be just material aspects down there and not mental ones too, if there really is no

---

[63] See for example Wheeler (1990), Bohm and Hiley (1993), Zeilinger (1999), and Vedral (2010); cf. Tononi (2008) and Tegmark (2014).
[64] Fahrbach (2005); cf. Searle (1995).

explanation for the distinction then what is the alternative? Since the physics allows it, positing mentality as a fundamental aspect of reality at least tells us something we did not know before, even if it does not "explain" the distinction in the usual sense.[65]

Properly developed, such a dual-aspect, information-theoretic ontology might be sufficient for social scientific purposes, although I have my doubts about that. However, it comes at the price, I take it, of not fully satisfying our aesthetic sense. Thus, I am intrigued by a second response to the objection, which pushes beyond panpsychism to neutral monism.

Rather than accept the duality of aspects as a brute fact, neutral monists seek to explain the *emergence of the distinction* between the two aspects out of an underlying sub-stratum that is neither mind nor matter. Historically many of the thinkers who today are claimed for neutral monism – the term itself was coined by Bertrand Russell – are also claimed for panpsychism, most notably Spinoza and Leibniz, attesting to the difficulty of disentangling the two doctrines.[66] However, with quantum theory providing potential empirical support for the idea, today we are seeing specifically neutral monist arguments put forth at an increasing rate. Some are by philosophers of mind,[67] but most are by quantum theorists themselves, including Wolfgang Pauli, David Bohm, Teruaki Nakagomi, David Lockwood, Paavo Pylkkänen, Giuseppe Vitiello, and others.[68] Here I address just one of these proposals, by Harald Atmanspacher and Hans Primas, which explains the emergence of the mind–matter distinction as one of "temporal symmetry-breaking."[69] By providing leverage on the "hard problem" of time, their approach offers an independent warrant for a panpsychist perspective, and it may be of interest in its own right to social scientists, who have also long wrestled with how to think about time.

Time is a hard problem because, as McTaggart made clear in a classic 1908 paper, it seems to have two incompatible natures, one mental and one physical, which he named the "A-Series" and "B-Series."[70] The A-Series is the tensed "arrow" of time of subjective experience, which flows from the past to the future

---

[65] See Fahrbach (2005).   [66] See especially Skrbina (2005).

[67] For a comprehensive review of neutral monism see Stubenberg (2014), and for other recent work see Holman (2008), Velmans (2008), Silberstein (2009), Alter and Nagasawa (2012), Robinson (2012), Nunn (2013), and Seager (2013). For a skeptical view see Banks (2010).

[68] Bohm (1990), Nakagomi (2003a; 2003b), Pylkkänen (2007), Lockwood (1989), Vitiello (2001); and for an extension of these ideas into the para-psychological realm see Jahn and Dunne (2005).

[69] See especially Atmanspacher (2003) and Primas (2003; 2007; 2009), who link their ideas to Pauli's reflections on mind and matter (see their joint 2006). I take Franck (2008) and Uzan (2012) to be following in this spirit as well. Nunn (2013) provides a particularly clear introduction to Primas' approach.

[70] McTaggart (1908); for an excellent introduction to McTaggart's distinction see Gell (1992: 149–174).

through a succession of Nows. This is time as Becoming, in that "what time it is" continually changes, depending on where you are in its flow; what was once future will someday be past. From this constant change the Now derives a privileged ontological status as the only time that actually exists. While it may leave material or memory traces, the past has already happened and cannot be changed, while the future has not arrived and so can only be imagined. Neither the past nor the future is real, in short, even though they were once real or will become real in time. The B-Series in contrast is the tenseless time of physics. The only temporal relation there is the symmetrical relation of before/after, which does not favor one direction of time over the other, nor does it provide any basis for a privileged Now. This is time as Being, which never changes; what was before X will always be before X, and similarly with what is after. The problem in all this is how to reconcile the A-Series, the experience of which seems impossible to deny, with the B-Series and the causal closure of physics.[71] It is a problem that apparently worried Einstein, not least because it invites McTaggart's own conclusion that time is "unreal" – or, as materialists might say today, yet another illusion of experience.

Atmanspacher and Primas approach the problem of time by exploiting an important but rarely thematized feature of fundamental physical laws: when those laws are used to describe closed systems they are "time-reversal invariant."[72] That means the equations governing their evolution have two equivalent solutions: a forward-moving or "retarded" solution (so named I think because one solves for effects from causes), and a backward-moving or "advanced" solution (solving for causes from effects). The first corresponds to the familiar modern notion of efficient causation, the second to the teleological concept of final causation. Although equally valid in closed-systems, physicists normally use the first and discard the second solution, on the assumption that final causation does not have any physical meaning. Most approaches to quantum theory are accordingly time-*a*symmetric, in that the state of a quantum system is assumed to depend only on its past. However, as we saw in Chapter 2's discussion of Delayed-Choice Experiments, quantum theory permits time-symmetric approaches as well, in which measurements made in the immediate future are incorporated into descriptions of the system's current state.[73] In an important book on the physics of time, Huw Price therefore argues that our bias toward forward-moving solutions is nothing more than a deep-seated

---

[71] See Primas (2003: 85).
[72] On time-reversal invariance, see for example Savitt (1996), and – in case you're wondering – see Henderson (2014) on the debate about how to square this with the time-reversal *non*-invariance of thermodynamics.
[73] See for example Aharonov et al. (1964); Cramer's (1986; 1988) Transactional Interpretation of quantum mechanics also belongs in this category. Kastner (1999) is a relatively accessible guide to these issues.

convention, reflecting pre-conceived classical ideas about causation.[74] In the quantum realm, where causation of any kind is problematic, there is no good reason to reject advanced causation a priori.[75]

This point about time-symmetry at the fundamental level reinforces the claim that in physics time has no "arrow" – i.e. the sense in which time for us always seems to flow forward, such that events in the past can cause events in the future but not vice-versa.[76] So where then does this arrow of temporal experience come from (or, the other one go)? Atmanspacher and Primas argue that a materialist approach alone will not give us the answer. The arrow of temporal experience is after all an *experience*, so if we cannot explain any experience with a materialist ontology, we will not be able to explain temporal experience either. Thus, they argue the only way forward is to give up on a materialist interpretation of physics and bring consciousness into the equation.[77]

Atmanspacher and Primas make two moves. The first is to shift the frame of reference in physics from standard, time-asymmetric quantum theory, which describes only the material world, to a generalized quantum theory that describes an underlying reality in which there is neither time nor a distinction between mind and matter.[78] That such an underlying reality exists is the main idea of neutral monism. But whereas Spinoza and Leibniz could only speculate, quantum physics has shown that there really is such a reality, the "zero-point field" or "vacuum," which, far from being empty, is a foaming sea or plenum of background energy from which new particles continuously and spontaneously emerge.[79] This holistic reality figures centrally in several contemporary quantum philosophical schemes, as Wheeler's "pre-space," Bohm's "implicate order," d'Espagnat's "Ultimate Reality," and Jahn and Dunne's "Source." Atmanspacher and Primas themselves relate it to Jung's concept of the "unus mundus," which was influenced by Pauli, but they also just call it "X." However it is described, the key point is that at this level neither the wave/particle duality (and with it the mind/matter duality on my reading above) nor the arrow of time exists – everything is symmetric.

From here, Atmanspacher and Primas make a second move, which starts by seeing the emergence of time as a process of "temporal symmetry-breaking" from the timeless unity of X. This symmetry-breaking happens as soon as any

---

[74] See Price (1996).
[75] On the problem of causation in quantum theory, see Price and Corry, eds. (2007).
[76] See Savitt (1996) for a good overview of the problem of the arrow of time (which is actually several related arrows).
[77] Also see Bierman (2006).
[78] Cf. Jones (2014) on Whitehead's "flat ontology," and also see Price (1996), chapter 5, for what to my untrained eye looks like a similar argument.
[79] See Laszlo (1995), Vitiello (2001), and McTaggart (2002) on the vacuum/ZPF, and Bradley (2000) for a thought-provoking discussion of how this links up to social life.

system interacts with its environment (i.e. basically all the time everywhere), and results in two "semi-groups" evolving in opposite temporal directions.[80] Unlike in the idealized closed system case, these semi-groups are not equivalent mathematically, only correlated (entangled), and thus complementary in the quantum sense.[81] One semi-group moves forward in time satisfying the rules of retarded/efficient causality, which is associated with material states. The other semi-group moves backward in time satisfying the rules of advanced/final causality, which – and here is the panpsychism part – Atmanspacher and Primas associate with mental states.[82] As such, the emergence of the arrow of time is the "interface" through which the distinction between mind and matter itself emerges.[83]

In light of Schopenhauer, this makes sense too. For if there is one place in nature where time and causality seem to work backwards it is in human action, with its strongly teleological quality. Thus, while from an external, material perspective our behavior seems "pushed" by the interactions of matter in the past, from an internal, phenomenological perspective it feels more like we are "pulled" by reasons advanced into – indeed, in a sense *from* – the future.[84] As Scott Jordan puts it, "the continuously generated template of the [human] body in space-time is feedforward (anticipatory) in nature." The feeling of "mental time travel" is ubiquitous in humans,[85] and, although the temporal "distances" we can travel are vastly greater than other organisms, the capacity to "act from the future" seems intrinsic to purposiveness and thus on my account to life and ultimately matter itself.

So where does the other arrow of time go? It goes into mind, which along with matter/energy is emerging continuously in pervasive processes of temporal symmetry-breaking from an underlying monistic reality.[86] As organisms, we tap into (and are sustained by) this fundamental process and so we experience both arrows of time – as material bodies, subject to the mechanical effects of retarded causality, and as subjects, who will the teleological effects of our own advanced causality. In short, the other arrow of time has been there in the physics all along, on the implicit mental side, just not seen for what it is.

---

[80] Atmanspacher (2003: 24), Primas (2003: 94).
[81] Atmanspacher and Primas see this correlation as a naturalistic basis for Leibniz's idea of the pre-established harmony of mind and matter; also see Nakagomi (2003a; 2003b) on "quantum monadology."
[82] Primas (2003: 94).   [83] See Uzan (2012).
[84] See Jordan (1998: 173). For further discussion of the physical nature of anticipation and its role in consciousness see King (1997) and Wolf (1998).
[85] See Suddendorf and Corballis (2007); I discuss this idea in Chapter 10 below.
[86] This seems similar to Whitehead's view that "mind, at its most rudimentary, is simply the intrinsic temporality of a physical event" (Weekes, 2012: 40).

Much more would need to be said about Atmanspacher and Primas' proposal, and related ideas from others, for this to be a proper argument for neutral monism as a solution to the mind–body problem. I have brought it up just to show that there are attractive and intriguing ways, coming out of quantum theory itself, to remove the taint of dualism in panpsychism, and thus to bolster my argument. Whether neutral monism is consequential for the social sciences is unclear, but I will return to its ideas about time in Chapter 10.

7 A quantum vitalism

The thrust of the last two chapters was ever more micro, beginning with quantum brains and then projecting the essence of their subjectivity all the way down to the sub-atomic level. While that might seem to have taken us very far from the world of social science, I am working toward just the opposite conclusion: that social life is not essentially different from that of sub-atomic particles. To complete this argument, however, I need to bring matters back to macroscopic reality, which I shall do in this chapter by considering some implications of this ontology for the nature of life. That is a big topic in its own right, and ultimately I am interested here only in one very unusual form of life, our own. Still, there are two reasons to bring up the subject.

One is that from a quantum perspective human life is essentially continuous with, rather than qualitatively different from, other organisms. Apart from its potential ethical implications,[1] this means that in speaking of life in general I will be laying a foundation for the quantum model of man discussed in Part III. The other reason to bring up the nature of life is that the view of it which follows from the ontology above is a kind of vitalism, according to which life is constituted by an unobservable, non-material life force or *élan vital*. Since vitalism today is almost universally rejected as unscientific, and the threat of which I used to motivate my proposal to give intentional explanations a quantum basis, it seems important to follow quantum consciousness theory through to its logical conclusion, and then from there, try to turn the threat of vitalism into a willing embrace.

The key to this strategy is the phenomenon of quantum coherence. Addressing the combination problem for panpsychism, in Chapter 6 I argued that coherence in the brain explains the unity of consciousness at the human level. In this chapter I extend that argument to a broader claim, that quantum coherence is ultimately what distinguishes life from non-life. As Chris Clarke puts it, following Mae-Wan Ho, "coherent states are the essence of organisms."[2] I develop this thesis in relation both to more mainstream notions of life and to

---

[1] See R. Jones (2013) for a recent introduction.
[2] Clarke (2007: 58); also see Ho (1998: 213–214) and today especially Igamberdiev (2012).

the New Materialism, and show that quantum coherence has everything one could ask for as a physical (vs. material!) basis for the *élan vital*.

I will begin with a brief overview of the classical materialist–vitalist controversy, which resulted in the decisive defeat of the vitalists, and then summarize today's standard materialist view of life and its problems. Then, isms aside, I discuss what all organisms might have in common from a quantum perspective, and will end by reconsidering vitalism in this light.

The materialist–vitalist controversy

The question "what is life?" has probably occupied human beings for as long as they have asked "what is mind?" My argument is that quantum coherence is the physical basis of both, but connecting the two problems is not the norm within philosophy, despite what appear to be similar debates. Thus, in philosophy of mind we find materialists who argue that mind can be reduced to mindless matter ranged against dualists, panpsychists, and idealists who argue for more mind-centric ontologies instead. Similarly, in philosophy of biology we again find materialists (often called "mechanists" in this context) who argue that life can be explained by reference to lifeless matter pitted against animists, vitalists, and others who argue that life will never be explained that way and invoke non-material vital forces instead.

Such is the hegemony of materialism today that the idea of "non-material vital forces" might strike most readers as occult and completely unscientific. By the end I hope to have changed your mind about that, but first it is useful to recall that just a century ago vitalism was taken quite seriously.[3] Vitalists like Hans Driesch and Henri Bergson drew on a rich philosophical tradition dating to the ancient Greeks, which re-emerged in the seventeenth century in opposition to Descartes' view of animals as machines.[4] Although today often associated with the counter-Enlightenment – the *Naturphilosophie* of Goethe and Schelling, and later *Lebensphilosophie* of Schopenhauer and Nietzsche – vitalism was an important influence in the Enlightenment as well, on thinkers like David Hume and Adam Smith.[5]

Today vitalism is essentially dead. The scientific literature amounts to a lone Hameroff article on "quantum vitalism," a piece on "molecular vitalism" from 2000, the authors of which invoke the term in a "millennial" spirit, and

---

[3] See Normandin and Wolfe, eds. (2013) for a comprehensive overview of the history of vitalism in relation to changing images of science.

[4] Different aspects of this controversy are covered in Lenoir (1982), Burwick and Douglass, eds. (1992), Harrington (1996), Reill (2005), and Normandin and Wolfe, eds. (2013). On the continuing dominance of the machine model of organisms see Nicholson (2013).

[5] See especially Reill (2005); and for subsequent developments also Huneman (2006). On the vitalism of Hume and Smith see respectively Cunningham (2007) and Packham (2002).

research by psychologists on the "causal placeholder function" that vitalist reasoning plays in the development of young children.[6] In social science and the humanities vitalism arguably maintains more of a hold, implicitly, insofar as social theory cannot dispense with intentional explanations, and explicitly in the work of "neo-vitalists" (aka New Materialists) like Jane Bennett and Bruno Latour.[7] Even so, the disrepute of vitalism runs so deep that, with the partial exception of Bennett, even the neo-vitalists are at pains to disavow its scientific claims, which Monica Greco for one calls "inadequate and philosophically naïve."[8] For them the theory's value is polemical rather than positive – a way, with Bergson, to remind ourselves of our ignorance and thus, a tool for critique.

What changed everything for vitalism was the genetic revolution and subsequent advances in biology. Before 1900 biological knowledge was so rudimentary that neither side had any real warrant for its claims about the nature and explanation of life. However, subsequent scientific progress began to fill in the materialist picture, and that in turn challenged the key contention of the vitalists, that materialism will never explain life. Given that vitalism is ultimately an inference to the best explanation – it is only if materialism fails that we would be justified in positing a non-material vital force – scientific progress seemed to eliminate the need for vitalism, and simultaneously highlighted two of its own weaknesses as a theory. First, the vitalists offered no alternative research program for furthering our knowledge of life, or at least one recognized as such by most biologists.[9] Second, whereas materialist claims about the nature of life can at least in principle be tested, vitalism seemed un-testable and thus non-falsifiable.[10] For both external and internal reasons, then, vitalism essentially vanished by mid-century, such that an idea that just decades earlier was the talk of the Parisian salons is almost taboo today.

Importantly, however, the collapse of classical vitalism did not mean that materialists had solved the puzzle of life, or even necessarily that they are getting closer to solving it today – it may be that the vitalists were right in their critique, even if wrong in other ways. For despite all the progress in biology, a materialist explanation of life has been notoriously elusive.[11] The difficulty goes down to the very definition of life, on which there is no consensus. NASA's

---

[6] Hameroff (1997); on molecular vitalism see Kirschner et al. (2000: 79); on child/naïve vitalism see Morris et al. (2000).
[7] See Bennett (2010) and Latour (2005).   [8] Greco (2005: 18).
[9] See Garrett (2006). The recognition point here is crucial, since as I suggest below in contemporary "biosemiotics" we do see what such a research program might look like.
[10] For a re-statement of this critique see Mayr (1982). Papineau (2001) also reiterates long-standing questions about whether vitalism is consistent with the principle of the conservation of energy.
[11] For a classic statement of the problem see Polanyi (1968). As Bedau (1998: 125) put it thirty years later, "The fact today is that we know of no set of individually necessary and jointly sufficient conditions for life," and if Cleland (2013) and Denton et al. (2013) are right the situation is no different today.

widely used definition – "a sustained chemical system capable of undergoing Darwinian evolution" – is subject to various counter-examples, some simple (mules), others technical.[12] Other efforts have fared no better. The problem is that because we lack an explanation of life we are forced to define it in terms of its observable features, like earlier scientists who tried to define water before they knew about H_2O – with equally confusing results.[13] The issue is so muddled that, despite its new-found practical importance in astrobiology, most biologists have given up on trying to define the subject of their discipline.[14]

They can do that because biologists do have a strong practical grasp on life, from which a rough but sophisticated consensus has emerged on its "observable features" at least.[15] Summarizing brutally, organisms are widely understood to be: (1) *individuals*, in the sense of "spatio-temporally bounded and unique" systems with "a particular history of interactions"; (2) *organized*, structured totalities in which parts and whole are dynamically interdependent and mutually constitutive; (3) *autopoietic*, feeding on energy to sustain their self-production in the face of thermodynamic decay; (4) *autonomous*, in that their behavior is determined not just by the external environment, but also by their internal constitution; and (5) capable of *genetic reproduction* or something equivalent. Exactly how these properties are instantiated is a matter of debate within biology between reductionists and organicists, but judging from the absence of vitalists in the debate all sides agree that whatever the details they must be consistent with a materialist ontology.

So, with this practical description of life in place, even if it is not a formal scientific definition, is there any reason to bring vitalism back? That depends on whether the gap between description and definition is likely to be closed by future scientific progress, without an intervention from new metaphysics. If we think it will be closed, then vitalism has no role to play and materialism carries the day. Conversely, to argue the gap will remain is to say there is a "hard problem" of life for materialism, analogous to the hard problem of consciousness, in which case there is an opening for vitalism analogous to panpsychism in the mind–body problem. Yet, in contrast to philosophers of mind, of whom many accept the existence of a hard problem (if not the need for a non-materialist solution), philosophers of biology generally view life as an "easy" problem, one that will eventually be solved without abandoning materialism – much less reviving vitalism.[16]

---

[12] Luisi (1998: 617). [13] Cleland and Chyba (2002: 391); also see Cleland (2012).

[14] Machery (2012) goes even further, arguing that while it might eventually be possible to define life, it is "pointless."

[15] The following discussion draws especially on Ruiz-Mirazo et al. (2000); also see Robinson and Southgate (2010).

[16] For good measure, Chalmers himself (1997: 5) does not think there is a hard problem of life (see Garrett, 2006).

Such a preponderance of expert opinion obviously needs to be taken seriously; however, it is also important to acknowledge that, despite all our scientific advances, when it comes to life itself we still do not know what we do not know. As Donald Rumsfeld might say, the nature of life is an unknown unknown, not a known unknown. Now, that does not mean it will not have a materialist answer in the end. But given the uncertainty, the case for accepting a materialist view *today* has to be more than just an extrapolation of the past: it also requires a judgment that scientific progress will not itself undermine a materialist view of life. And in this light, two contemporary movements are of particular interest, one scientific and the other more philosophical.

The first, as recently reported in *Nature*, is the "dawn of quantum biology."[17] That there is a connection between quantum mechanics and life is something physicists have speculated about since the 1930s. Niels Bohr thought the principle of complementarity extended into the biological (and psychological) realm; in a quest for free will, Pascual Jordan developed a theory of organisms as "amplifiers of micro-physical indeterminacy;" and most importantly there was Erwin Schrödinger's 1944 classic, *What Is Life?*, which had a lasting impact on the development of biology itself.[18] Still, it is only in the past decade that technology has permitted us to move from theoretical speculation to empirical probes below the cellular level. And what has been found is quite striking: among other things, that birds exploit non-local connections with the earth's magnetic field to help them navigate, that plants exploit quantum effects in photosynthesis, that fruit flies' sense of smell relies on the ability to detect quantum vibrations in smelly molecules, and that quantum processes might even facilitate social learning by protozoa.[19] Importantly, all of these effects involve quantum coherence – which skeptics of quantum brain theory tell us cannot exist in the warm and wet environment of living organisms. So this research raises the question that if birds, plants, fruit flies and even protozoa can do it (the quantum thing), then why couldn't people?

---

[17] See Ball (2011); for overviews see Mesquita et al. (2005), Abbott et al. (2008), Igamberdiev (2012), Bordonaro and Ogryzko (2013), Kitto and Kortschak (2013), and Al-Khalili and McFadden (2015). That an editorially conservative journal like *Nature* would publish Ball's article represents quite a change, given that just a few years ago the distinguished neuroscientist John Hopfield (1994: 53) could claim that "there is absolutely no indication that quantum mechanics plays any significant role in biology."

[18] See Bohr (1933); Čapek (1992) and Beyler (1996) on Jordan; and Schrödinger (1944). Somewhat later Max Delbruck is another important crossover figure (see Domondon [2006] on differences among these physicists on life), as well as Andrew Cochran (1971) and Walter Elsasser (1987). Working from the other direction, Hans Jonas (1984) and Dale Miller (1992) have also explored the potential for a quantum mechanical solution to the problem of subjectivity.

[19] See respectively see Dellis and Kominis (2012), Hildner et al. (2013), Lloyd (2011), and Clark (2010). For an early discussion of the hypothesized use of non-locality by organisms see Josephson and Pallikari-Viras (1991).

The growth of quantum biology does not in itself mean that the nature of life is a hard problem for materialism. As we have seen there are materialist interpretations of quantum theory, so materialists about life could look forward to quantum biology actually vindicating their view. What the penetration of quantum mechanics into biology does mean, however, is that all those weird quantum effects – entanglement, non-locality, tunneling, and the rest – are being introduced into what had previously been purely classical models of life. And as we have also seen, the meaning of quantum theory is itself a hard problem for materialists and as such idealist ontologies have long flourished in that context. As biology goes quantum, will a similar challenge emerge there too? Future biologists may conclude that the quantum effects within organisms do not cross the threshold of "non-trivial" with respect to life. But that remains to be seen, and in the meantime quantum biology is certainly not making the problem of life any easier, and it may actually be making it harder.

The other development is the small but growing number of philosophers of biology who argue that mind and life are co-constitutive, such that where there is one there is the other. To put this in context, note that mind is *absent* from the consensus definition of life that I summarized above; while evidently supportive of human minds, none of those properties is necessarily mental. This is not surprising given the still widely held Cartesian view in biology that mind is either exclusive to humans or at least does not go very far down the evolutionary ladder. Yet, as we saw in Chapter 6, empirical evidence is accumulating which suggests that mind might go all the way down to bacteria, and this work is now being followed by philosophical arguments, that life and mind are actually continuous. Of particular importance here is Evan Thompson's *Mind in Life* (2007), which makes a sophisticated case for such continuity, seeking thereby to recast, and hopefully eventually eliminate, the explanatory gap between biology and phenomenology.[20]

But what is this mind with which life is continuous, and will seeing the two this way vindicate a materialist ontology of life? If we consider the faculties of Cognition, Experience, and Will, then Thompson projects only Cognition and Will down to bacteria, not Experience.[21] As we have seen, these are the "easy" problems of mind, so positing their continuity with life seems relatively unproblematic for materialism.[22] Moreover, Thompson is no old-fashioned materialist, since he rejects mechanical conceptions of life that see only its externally observable functioning in favor of the theory of autopoiesis, which highlights the self-organization of interiority within organisms – but which in

---

[20] Other work in this vein would include Hoffmeyer (1996), Stewart (1996); Bitbol and Luisi (2004), and Kawade (2009; 2013); and, farther back, especially Jonas (1966).

[21] Thompson (2007: 159–162); also see Hoffmeyer (1996) and Swenson (1999); cf. Margulis and Sagan (1995) and Sheets-Johnstone (1998).

[22] Though see Robinson and Southgate (2010).

his view is only a precursor to consciousness, not conscious itself.[23] Yet, even with the self-organization focus, and even if we take his further "enactivist" view of mind, which shifts attention from what is happening inside the brain to how mind is enacted in embodied encounters with the world, at the end of the day I suspect Thompson would insist that – lest he be accused of vitalism! – life can be explained by the organization of otherwise lifeless and thus mindless material elements.[24] And in that case the old question is still there: if matter is intrinsically mindless then what accounts for the emergence of experience, whether with life itself or only at higher levels of complexity? Given the materialist view of matter, in short, how could *any* life be conscious, rather than none?

Not that I am unsympathetic to Thompson's argument; while I think even the simplest organisms are conscious, that only underscores his view that the problem of life and the problem of mind are essentially the same. The point is that, given the failure of materialists to explain consciousness, making mind constitutive of life has the perverse effect of multiplying hard problems: if there is a hard problem of consciousness then there is also one of life. Given this subversive ramification, one could understand Thompson's less radical colleagues wanting to keep the problem of life simple, by keeping mind out of the picture altogether.

So is there a case for reviving vitalism? The quantization of biology and the life–mind continuity thesis at most constitute negative arguments, as signs that despite (or because of) the progress of science, the problem of life might actually get harder for materialists in the future, not easier. Yet there is also the question of vitalism's own problems, which contributed to its demise a century ago. If life cannot be reduced to material processes, then what *is* this mysterious life force, and how can it be reconciled with modern science? Since the argument for vitalism will have to answer these questions first, let me take up what I see as the implications of quantum panpsychism for the nature of life, and then return to vitalism per se.

Life in quantum perspective

My claim is that life is a macroscopic instantiation of quantum coherence. As Mae-Wan Ho puts it,

[W]hat is it that constitutes a whole or an individual? It is a domain of coherent, autonomous activity. The coherence of organisms entails a quantum superposition of coherent activities over all space-time domains, each correlated with one another and

---

[23] See Thompson (2007: 222–225).
[24] It is perhaps instructive that Thompson does not discuss quantum theory, although a brief mention on p. 439 suggests openness to a move in this direction.

with the whole, and yet independent of the whole. In other words, the quantum coherent state, being [non-]factorizable, maximizes both global cohesion and local freedom. It is that which underlies the sensitivity of living systems to weak signals, and their ability to intercommunicate and respond with great rapidity.[25]

Such a view does not make the classical consensus about life "wrong" any more than quantum physics made classical physics wrong. Indeed, from the above list of life's properties – individualized, organized, autopoietic, autonomous, and capable of reproduction – it seems that all but the last could play central roles in maintaining an "internal quantum state" at life's core.[26] Far from being opposed, then, when it comes to the physics of life the bumper sticker might be "classical outside, quantum in."

This approach has clear implications for how we conceive of the boundaries of life. Internally, it means there are no essential differences among life forms, of the kind that would allow us to say that some are just machines and others are more than that. Life is life, and so the variation it exhibits is quantitative rather than qualitative. Externally, and conversely, the idea that life equals quantum coherence means there is a clear difference between life and non-life. Given the otherwise inevitable fact of decoherence, quantum coherence can only be maintained in very special, highly protected physical conditions – the conditions of life. To be sure, the panpsychist aspect of my argument sees wisps of subjectivity in sub-atomic particles, which means that matter is "pregnant" with life.[27] But as long as matter is not coherent, life is not actually "born." I will return to some implications of this sharp life/non-life boundary below.

All quantum systems have two, complementary descriptions, and the same goes for life, where I have argued that the "wave" description and its collapse are the physical correlates of subjectivity. This correlation gives social scientists an epistemic advantage over physicists, since whereas the latter have no access to what is going on inside sub-atomic particles, we do know what it is like to be on the inside of human wave functions. So while we wait for more detailed quantum physiologies to arrive, we can use the rich vocabulary human beings have developed to talk about our wavy insides – folk psychology – as a complementary but equally valid road to knowledge about life. In the rest of this section I expand on my earlier discussion about Cognition, Experience and Will, with special reference this time to their potential quantum bases.

---

[25] See Ho (1998: 213–214). The actual text says "factorizable," but my assumption is that this is a misprint.
[26] The phrase is Igamberdiev's (2012), whose book is a must (if difficult) read in this context.
[27] A view known as "hylozoism;" also see Bennett (2010).

Cognition

Three implications stand out if quantum cognition is a defining feature of life. First, and most obviously, whatever computation (thinking) is going on inside organisms will be quantum mechanical. That would give to even the simplest organisms vastly more on-board computing power than would classical models of cognition, and allow for a correspondingly greater ability to navigate their environment.

Second, an interesting duality emerges in the organism's interaction with the environment. On the one hand, quantum coherence requires shielding from the environment to prevent its permanent collapse. To that extent a boundary between subject and object follows necessarily from the theory. On the other hand, to survive in the face of thermodynamic decay organisms also need energy from the outside world, which requires an open system able to perceive and interact with its environment. The interface where this duality is sustained is consciousness, which is the subjective manifestation of wave function collapse in the moment, but which is also reconstituted as a stream of such moments by the protective shielding of the organism's body. As such, the organism's coherent state may be seen as an "attractor, or end state towards which the system tends to return on being perturbed."[28]

Third, and most concretely, if organisms are quantum mechanical then the senses connecting them to the external world should have a non-local aspect. That is the upshot of the quantum biological findings about birds, plants, and fruit flies noted above, and in Chapter 12 I take up the case of vision in humans at greater length.

Will

Earlier I equated Will with the force that collapses wave functions into particles. In living matter this force is given identity through time by the structures that protect quantum coherence – and which is, in turn, directed toward the end of reproducing those structures itself. This is to say: all organisms share a will to survive.[29] Of course, we don't need quantum theory to tell us that, but a quantum perspective puts two questions about the will to live in a new light.

First, historically there has been much debate about whether life's apparent purposefulness is irreducibly purposive – i.e. whether it can be reduced to materialist ideas about causation, which leads to the machine view of organisms; or whether the Aristotelian idea of final causation or teleology is also necessary to make sense of Will, which leads to a view of organisms as "natural

---

[28] See Ho (1998: 214). [29] See Kawade (2009: 211).

purposes."[30] Modern scientists generally regard teleological explanations as being unscientific. At the same time, most biologists would probably also agree that organisms are purposeful in some interesting sense, and perhaps even that teleological language is a practical necessity in their discipline.[31] But following Ernst Mayr's influential argument, they would argue that what looks like teleology in organisms is actually "teleonomy," a process fully in accord with materialism requiring no mysterious causes to explain.[32]

I will not address whether the concept of teleonomy can make sense of organic purposefulness; instead, I want to highlight Mayr's premise: that for biology to be a science, the appearance of teleology in organisms must be rendered *just* an appearance, with no ontological significance. This is "politically" necessary only under a classical physics constraint, since quantum theory is open to teleological reasoning. Although most of its interpretations do not invoke ideas of final causation, as we will see in Chapter 9, some do.[33] So while a quantum perspective does not entail a teleological view of organisms' will, it allows for it, within what is still a scientific approach to life.

Second, there is a related question about how Will is constituted internally, whether through the bottom-up aggregation of the organism's parts ("reductionism"), or top-down through "downward causation" exerted by the whole on the parts ("holism" or "organicism").[34] For most of the twentieth century the reductionists had much the better of the debate, partly because technological advances allowed for ever deeper probing down to the micro-level, culminating in the recent triumph of sequencing the human genome. But the debate has also been influenced by the (classical) philosophical assumption that there is something intrinsically dubious metaphysically about holism, which puts the burden of proof on its advocates to explain why reductionism should not be the default option.

In recent years the debate has evolved in ways that seem to give both sides more ammunition – or might bring them together. Thus, on the one hand, scientists are now learning that the expression of genes depends not only on genes themselves but also on their larger context, which favors the top-down approach; on the other hand, as a result of the development of self-organization and complexity theories, bottom-up approaches have become much more sophisticated

---

[30] On the classical debate, see especially Lenoir (1982). For recent discussion of organisms as natural purposes see Weber and Varela (2002), di Paolo (2005), Walsh (2006; 2012), and Zammito (2006), and for a critique see Teufel (2011).

[31] See Barham (2008) and Toepfer (2012) on the constitutive role of teleological reasoning for biology, and Birch (2012) on its pervasiveness throughout the sciences.

[32] See Mayr (1982; 1992).

[33] On the question of teleology in early quantum approaches to biology see Sloan (2012).

[34] For recent organicist critiques of reductionism see Gilbert and Sarkar (2000) and Denton et al. (2013).

in handling the emergence of macro-structure and its effects, including possibly even downward causation.[35] Yet, the metaphysical suspicion of holism is still there, suggesting a continued acceptance that the problem should be conceived in classical terms, with parts given some kind of ontological priority over wholes.

A quantum approach to Will would change this burden of proof. In systems characterized by quantum coherence parts lose their individuality as fully independent "parts," which is the basis for their privileged ontological status in the classical worldview. Quantum coherence is irreducibly holistic, in short, yet unlike classical holism does not raise metaphysical worries about wholes existing independent of their parts. This is because in quantum coherence the whole exists merely as a potentiality (a wave function), and as such is not "real" in the usual sense. It only becomes real in its expression (collapse), which actualizes it into something classical. That would give a quantum basis for Schopenhauer's claim that Will "objectifies" itself in the world, and in so doing put the reductionism/organicism debate about purpose in an entirely new light.[36]

Experience

In Chapter 6 I argued that experiences are the inside manifestation of what on the outside appears as the collapse of wave functions into particles. From here I would like to highlight three implications for life.

First, experience is the origin of meaning, understood as "information for someone."[37] In quantum terms, if experience is one side of a process of temporal symmetry-breaking, then its role in collapse is to encompass non-locally the future actuality that Will creates, and project it backward to conform mathematically to the past potentialities of the system.[38] Or, in plainer terms: experience confers meaning by anticipating future information in relation to an organism's evolving purposes through time. The effect of this appropriation of the future is to transform objective information into subjective meaning – and it is on the basis of the latter that people act. This transformation is the essence of semiosis, and as such perception and action are fundamentally semiotic processes.

Probably few social scientists would disagree that human beings act on the meaning information has for them, rather than on information per se; what is of more interest is the claim that this is true of *all* organisms. While perhaps not widely shared by biologists, this is the foundation of the small but

---

[35] On downward causation see P. Andersen et al., eds. (2000), and Bitbol (2012) for a quantum approach; I take up downward causation at more length in Chapter 13.
[36] Also see Schrödinger (1959) on "objectivation."
[37] See Mingers (1995) and Markoš and Cvrčková (2013: 62).
[38] See especially Cramer (1986; 1988), King (1997), and Wolf (1998).

long-standing discipline of "biosemiotics." Tracing their roots in biology to Jakob von Uexküll (a contemporary of Driesch, also associated with vitalism), and to Thomas Sebeok in semiotics, Claus Emmeche, Jesper Hoffmeyer, Kalevi Kull and others have pursued research based on the idea that processes of signification – of creating and communicating meanings or "signs" – are pervasive in organisms, including plants.[39] Interestingly, while emphasizing the subjective quality of signification, biosemioticians have not much engaged the mind–body problem (or quantum theory for that matter). Mostly just taking it as given that organisms are conscious, their focus has been on what this tells us about organisms' behavior and functioning. So whether biosemioticians would want to be associated with my argument is unclear, but for better or worse I think it points toward their view of life as essentially semiotic. And that matters in the present context because if meaning is on the table in all the life sciences, then the human sciences are really just a subset of a much larger domain.

A second implication is that every organism's experience, "what it is like" to be them, is intrinsically private – not necessarily without a public sign (e.g. pain), but inaccessible *as such* to an observer. This privacy of experience follows directly from the claim that quantum coherence is the physical basis of life.[40] Coherence can only be sustained if it is shielded from the environment by a wall; breach that wall in an effort to get inside an organism's experience and you will kill the organism. So while we can know *that* an organism is in pain, and, if it is sufficiently like ourselves, even know vicariously what that might feel like, to be *in* a particular organism's pain is something that only it itself can truly understand. We on the outside can have only an ersatz, objective knowing of its experience.

Finally, the world of private experience in and through which each organism lives – what von Uexküll called its *"umwelt"* – is unique.[41] Putting this together with the previous discussion, this suggests a picture of organisms reminiscent of Leibniz's monads, each living in, and making choices based on, its own, inaccessible bubble of experience – though unlike in his model, organisms here have "windows" that allow them to interact with the world.[42] Nevertheless, the uniqueness of *umwelten* gives this ontology an irreducibly subjectivist aspect.

That might raise a question about how life's bubbles of experience are coordinated and made stable enough for organisms to make choices, since without some predictability in the environment it will be impossible for them to relate will (ends) to behavior (means). Leibniz solved an analogous problem for his philosophy with his doctrine of "pre-established harmony" provided

---

[39] The classic contemporary text is Hoffmeyer (1996), and for a recent overview of the field see Hoffmeyer (2010), as well as the journal *Biosemiotics*.
[40] See Georgiev (2013).   [41] See von Uexküll (1982).
[42] For a "quantum monadological" approach to cognition and life see Nakagomi (2003a; b); cf. Tarde (1895/2012).

by God. From a naturalistic perspective one would look instead to the material constitution of organisms, which creates cognitive stability in two ways. First, their perceptual apparatus provides organisms with access to only a tiny cross-section of the information available in their environment – presumably the kind most relevant to their survival. These cross-sections vary across species (dogs hear things we can't, we see colors they can't), but within species they are quite uniform, and as such confer some cognitive order on the world. Second, organisms also have the ability to exchange signs with other organisms, most obviously with con-specifics – a big help in reproduction one would suppose! – but even to some extent with other species, like dogs and people. While semiotic exchanges always contain the potential for misunderstanding, as we know from human experience they can be made predictable enough for communication to occur. If biosemioticians are right that all life involves semiosis, then the exchange of signs serves to structure the information environment such that all organisms can generally realize their purposes.[43]

To sum up, to mainstream biology's list of life's five observable properties I have added a sixth, partly *un*observable, property: subjectivity, defined as Cognition, Will, and Experience.[44] The quantum contribution to this proposal is twofold: quantum brain theory (and now biology) provides a naturalistic reason for thinking that quantum coherence is the physical basis of life; and the fact that quantum theory is open to a panpsychist interpretation makes it possible to link that coherence to subjectivity. I have also briefly suggested some implications of this view for thinking about organisms in general, some of which I will develop at greater length below with reference to the special case of human beings.

Why call it vitalism?

Among biologists today the term 'vitalism' is anathema, used solely as an epithet to dismiss work that does not sit comfortably in the mainstream as unscientific.[45] For social scientists, in contrast, the problem is probably not that we think vitalism is pseudo-science (which makes sense if it is latent in any theory that appeals to intentionality as an explanatory force). The problem is that we tend to associate vitalism with fascism and other forms of irrational politics. I will address this political question in Chapter 14, but the short answer is that while fascists used vitalist ideas, so did almost everyone else, and so the unique contribution, if there is one, of vitalism to politics lies elsewhere.

---

[43] Even at the cellular level; see for example Fels (2012).
[44] Though since my definition of subjectivity encompasses some of the other five the ultimate list would be shorter.
[45] For a good illustration of this tactic in play see Oyama (2010).

However, since these worries might affect the reception of my argument, and since I might avoid the trouble simply by not using the term 'vitalism,' let me explain why I think it is necessary.

Principally, because my argument *is* vitalist, though the issue is complicated by the variety of forms vitalism has taken historically, some of which overlap with other doctrines.[46] Of particular relevance here is organicism, a form of biological holism which is the main contender to reductionism in the biological sciences today.[47] Organicism has taken both materialist and vitalist forms over the years, which has forced its materialist advocates to work hard to distance themselves from vitalism to avoid its fate. As it happens, thinking that life is a function of quantum coherence is as holistic as holistic can be, so there is a close connection in my argument to organicism here as well. In social science that term has its own fascist connotations, but for now the question is whether the ontology here is vitalist in the thinnest possible sense, which I take to involve two claims: negatively, that life cannot be explained by reference to lifeless matter; and positively, that life can be explained by an unobservable and non-material life force.

Quantum coherence is just such a force. It may sound strange to call coherence "non-material," since it is a physical phenomenon and as such a far cry from metaphysical speculations about entelechies and an *élan vital*. But that is precisely the point: coherence is *physical but not material*. Moreover, it cannot be observed because doing so would by definition collapse its wave function and thus render only its particle manifestations visible. As I see it this is the ultimate contribution of quantum theory to solving the problems of consciousness and life, since it provides an opening within the causal closure of physics (CCP) for the naturalistic but non-materialist doctrines of panpsychism and vitalism.

Having bitten the vitalist bullet, then, what kind of vitalism is this (beyond being quantum), and can it avoid the pitfalls of classical vitalism? Since there is no space here to review the varieties of vitalism, let me just highlight two features of my proposal that I think distinguish it from classical and contemporary neo-vitalism alike.

The first is that, unlike classical vitalism, which was primarily a negative doctrine and non-falsifiable, and also unlike neo-vitalism, which sees itself as a critical rather than positive theory, quantum vitalism is based on a physical hypothesis that can in principle be tested. Not easily, to be sure, since quantum coherence cannot be observed directly. But we can imagine at least two indirect tests. One is the current debate about quantum brain theory, which although

---

[46] See Benton (1974) on the historical varieties of vitalism.
[47] See Elsasser (1987), Allen (2005), and Denton et al. (2013), and also Garrett (2013) on "emergent materialism."

focused on humans is centrally concerned with whether quantum coherence is possible in living tissue. If for theoretical and/or empirical reasons it were shown that quantum coherence cannot be sustained in the brain then that would falsify not only quantum consciousness theory, but quantum vitalism as well. Conversely, if such a possibility can be demonstrated, then given the additional findings of quantum biology and quantum decision theory, we would have grounds to argue that the factor explaining all these findings is quantum coherence. The other indirect test is more prosaic. An interesting implication of the idea that life is constituted by quantum coherence is that if one could build a machine with such coherence – i.e. a quantum computer – then it should be alive.[48] How then could we tell if it is *actually* alive and not just alive by definition? Since adults might be biased by their preconceptions, my slightly tongue-in-cheek suggestion would be to give the computers to four-year olds, the age at which kids are known to be able to reliably spot the difference between life and non-life. If the kids say they are alive, then that should count for the theory. Neither of these tests will be available soon, but they do point to a scientific rather than purely metaphysical proposal.

The second distinctive feature of this quantum vitalism is that it takes the essence of life to be subjectivity, understood not only as Cognition and Will but as Experience. With some exceptions like Bergson and von Uexküll, classical vitalists emphasized Will to the neglect particularly of consciousness. The kind of vitalism I am proposing also emphasizes Will, but sees it as intrinsically connected to Experience. Recall that I have conceived of the collapse of the wave function as a process of temporal symmetry-breaking in which Will corresponds to the "advanced" semi-group, in which time moves in a teleological fashion backward from the future, and Experience corresponds to the "retarded" semi-group in which time moves in the traditional way forward from the past. As such, it is only through the symmetry-restoring process of Experience that the potentiality of Will can be actualized.

Although for this aspect of my argument I find less affirmation in the vitalist tradition than I do in the phenomenological one, it is precisely due to its focus on Experience that I think the term 'vitalism' is apt. For what distinguishes the view of life on offer here most clearly from a materialist one is its emphasis on consciousness. The mistake of the classical vitalists was to focus mostly on the "easy" problems of life, which, as in the case of mind, became increasingly tractable as materialist science progressed.[49] But consciousness has been impervious to that progress, which suggests that if there is anything uniquely

---

[48] Or at least have subjectivity, since the requirement of reproductive capacity would presumably not be met. Insofar as we have already made extremely simple quantum computers, therefore, we have already created life.

[49] See Garrett (2006).

"vital" in life, then it is experience. From the standpoint of conceptual clarity it therefore seems important to re-cast vitalism in this way, as a doctrine clearly distinct from materialism, both physically and metaphysically.

This ontology has both affinities with and differences from an important new movement in critical social theory, associated with Jane Bennett, Gilles Deleuze, Graham Harman, Bruno Latour and others, known variously as New Materialism and/or Neo-Vitalism.[50] A common starting point for this otherwise heterogeneous body of scholarship is a re-thinking of the nature of matter, from the inert and passive substance of classical physics to a productive and active force in nature. A provocative effect of this move is to reveal the essential continuity, not of living matter with dead (as in Old Materialism), but of dead matter with living, such that, in varying degrees, we can attribute to inanimate objects (sic) many of the intentional qualities we normally associate just with human beings. Thus, Latour conceives of material things as agents or "actants," with varying capacities to resist and mess with human projects.[51] Similarly, Bennett argues that "thing-power" is an "impersonal agency" on a continuum of agency that cuts across the usual life/non-life divide. While not explicitly panpsychist, this is at least interestingly reminiscent of the strong panpsychism of David Chalmers or Giulio Tononi discussed in Chapter 6, according to which artefacts like thermostats and computers are to some degree conscious.

My proposal for a quantum vitalism has important elements in common with New Materialism. It too aims to re-think matter into a less "material" and more active force. In its panpsychist basis it also sees an essential continuity between living and dead matter. And in its claim that all organisms are subjects it shares a non-anthropocentric, post-humanist view of reality, which would deny to human beings a privileged ontological position from which to justify abusing nature.[52]

However, there are also at least three important differences.[53] First, my proposal foregrounds consciousness and sees it as a defining feature of life. In contrast, New Materialists hardly mention consciousness, like older critical theorists apparently seeing in a concern with it a Cartesian anxiety that ontology can do without. In her "vital materialism," for example, Jane Bennett explicitly brackets subjectivity (consciousness in my terms), the search for the physical

---

[50] For a good introduction see Coole and Frost (2010b).
[51] See Latour (2005); for a good discussion of what precisely this claim amounts to see Sayes (2014).
[52] Also see Wendt and Duvall (2008).
[53] A proper engagement with New Materialism would require a much more elaborate discussion (and homework on my part) than I can offer here, so what follows is only a set of signposts for how I might respond to this very interesting line of argument.

basis of which she says is normatively problematic and "quixotic."[54] Second, my proposal draws much more explicitly and extensively on quantum theory. Although New Materialists have invoked "post-classical" physics to justify their re-thinking of matter,[55] by this they also mean complexity theory and non-linear dynamics (which ultimately embody a classical ontology), and to my knowledge their engagement with quantum theory itself has only been fleeting.[56] Finally, my proposal implies a sharp distinction between life and non-life, both substantively and normatively. While in my panpsychist view matter at the quantum level is latent with life, it only *becomes* life when organized into quantum coherent wholes. When there is no quantum coherence, as in thermostats or computers, matter is dead, and as such does not have agency or other intentional properties – causal powers yes, but not agency.[57] Normatively, in turn, I worry that an ontology which elides the distinction between life and non-life, while usefully denying humans a ground for abusing nature, could also have the opposite effect: if rocks are on a continuum of agency with people, then why not treat people like rocks, as if they were merely objects rather than also subjects?

The ultimate problem here is that by failing to come to grips with the hard problem of consciousness, the New Materialists/Neo-Vitalists remain caught up in the limits of the classical worldview – in short, by the Old Materialism. Making a quantum panpsychist turn enables us to abandon materialism once and for all in favor of a broader, vitalist physicalism that can accommodate that which is most distinctive about life, namely its subjective aspect.

---

[54] Bennett (2010: ix). [55] See for example Coole and Frost (2010a: 10–14).
[56] The big exception here is Barad (2007), whose quantum worldview shares much with New Materialism, but to date has not figured in most of this literature.
[57] See Vandenberghe (2002) and Cole (2013) for non-quantum critiques of New Materialism that point in a similar direction.

Part III

A quantum model of man

Introduction

The central claim of this book is that all intentional phenomena are quantum mechanical. That goes both for the private thoughts inside our heads and for public or collective intentions like norms, culture, and language, which we might generically call institutions. I suggested in Chapter 1 that by virtue of their dependence on consciousness, a classical, materialist ontology cannot explain these phenomena, and as such from that perspective they must be epiphenomenal or illusions. In the rest of this book I show that a quantum, panpsychist ontology can provide a physical basis for what we all know to be true, which is that both private and collective intentions are part of the natural order.

The elementary parts of social ontology are human beings in their biological individuality,[1] which in this Part I consider in abstraction from their lived social contexts. By taking humans out of their natural habitat I aim to focus attention on what we bring to the social table simply by virtue of being organisms of a certain kind. The abstract individual is a common enough starting point in classical social theory (think Hobbes and the state of nature). However, it might seem an odd place to begin a quantum social theory, given the holism of quantum phenomena, both "all the way down" in having no elementary parts, and "all the way across" in universal non-locality. What look like separate organisms are just local decoherence effects of quantum fields; everything really is related to everything else. In that light the abstract individual appears not just as an abstraction, but as positively occluding what "the individual" really is.

Today one might prefer 'Person' to 'Man,' who in models of man discourse feminist theorists have argued really was a man in the past and thus not representative of human beings. However, 'Person' is clumsier, and I also don't think the model that emerges below is vulnerable to the feminist critique, since in his essentially relational character Quantum Man is if anything a Woman. Either way, for the sake of balance, I will refer to Classical Man as 'he' and his Quantum partner as 'she.'

[1] Though see the discussion of collective consciousness in Chapter 14.

However, there is a difference between what is true of a universal ontology and what is practically relevant to specifically social ontologies, which are constrained by the properties of their constituent organisms.[2] Unlike some organisms, humans live in highly interdependent societies, and most of what is interesting about us is socially constituted as well. While that might amplify worries about abstract individualism, it does not change a basic material fact. By virtue of processes at the biological level and below, every organism is given to the world as a free standing subject. As we will see in Part IV, the quantum nature of language creates non-local connections that far transcend these subjects. Nevertheless, to be a human or any other subject is to be individuated as an organism with quantum coherence, the physical integrity of which is constitutionally independent not just of society, but in a sense of the universe as a whole. Unlike the physicist's particles, which literally *do* come from nowhere, our elementary units are given at birth by nature, and as such impose a "rump individualist" limit on a holist argument.

With the human body as our starting point, the first question to ask is of its nature: what essential properties and dispositions do we have that might enable and constrain social possibilities? Some may doubt whether a human nature exists, but in my view, there is no less reason to think that humans have a nature than horses or honeybees do. We all come from the same evolutionary process and in each case our behavior is differentially empowered by its material and mental gifts. Horses and honeybees can't talk because it is not in their nature, and we can because it is in ours. In addressing this essential nature, what follows may be seen as an exercise in "philosophical anthropology" in the broad sense, understood as an inquiry into "the unchanging preconditions of human changeableness."[3]

Still, it is one thing to talk about the content of human nature and another to discuss what I will call its form, and I hope to avoid at least some controversy by focusing on the latter. By the content of human nature I mean what most of the debate is about, namely fundamental behavioral dispositions: the extent to which, compared to animals or each other (men vs. women), people are naturally selfish, altruistic, aggressive, sociable, and so on. These are mostly empirical issues and much has been learned about them by cognitive scientists, evolutionary psychologists, and others. Since this work has been carried out within a classical framework, it would be interesting to speculate how a quantum

---

[2] Cf. Fodor (1974) on the "special sciences."

[3] Honneth and Joas (1988: 7). I say "in the broad sense" because philosophical anthropology also refers to a particular school of continental social theory associated with Arnold Gehlen, Helmuth Plessner and Max Scheler (for introductions to this tradition see Honneth and Joas [1988], Fischer [2009], and Rehberg [2009]). Although I will not take up these thinkers here, my sense is that there is considerable resonance below to their ideas, for which Schopenhauer was also an important influence (Honneth and Joas, 1988: 42).

approach might advance it further, but I will not do that here.[4] Instead, my interest is in how a physicist might think about human nature, which, I take it, would concern its form – how the expression of our dispositions is constrained and enabled by the physics of our bodies. In the past there was no reason to ask this question,[5] since it was assumed the body was classical. And regarding the body specifically I partly agree. The physics of the mind, however, is quantum, which I show could resolve some long-standing controversies that surround human subjectivity.

In the following three chapters I take up some implications of quantum consciousness theory for human Cognition, Will, and Experience respectively. In each chapter I start with current debates and their implicit classical assumptions; then review theory and evidence favoring a quantum approach; and conclude by showing how this would move the debates forward. However, I first want to situate my argument in relation to an over-arching classical alternative, different elements of which will be in question at various points below.

If the goal here is to think the form of human nature through the quantum, then what would it mean to think it through classical physics instead? Given the origins of the social sciences in the classical worldview one might think a clear answer would exist, but it does not. Having lost interest in physics long ago, contemporary social theory is little help, since it has integrated intentional phenomena into its ontologies that have no place in the classical worldview. A pure classical foil might be found in the work of the nineteenth-century social scientists who first "thought through physics."[6] However, their discourse is obscure today and lacks our hindsight advantage to put it in perspective. So I will start from scratch instead, though with a low level of ambition. Rather than propose a full model of Classical Man, I identify five constraints that such a model should satisfy if it is to be consistent with a strict, classical ontology. I'll be brief, since we will see these constraints at work in various contexts below.

First, Classical Man is completely material, in the sense that, ontologically speaking, mental states are nothing but brain states. This does not necessarily mean that people cannot have consciousness, but if they do then it must be epiphenomenal, with no mysterious powers that cannot be explained by movements of classical matter and energy. That may seem a high bar, but it is the bar that materialists themselves have set for solving the mind–body problem, and as such seems right in this context.

---

[4] Quantum models of evolutionary processes would provide a starting point; see for example McFadden (2001) and Gabora et al. (2013). Fry (2012) offers a thoughtful reflection on whether evolutionary theory is based on an implicit materialist ontology.
[5] Ecological psychology is the big exception here (see Gibson [1979]), and interestingly is also the site of some early reflections on quantum cognition; see Shaw et al. (1994).
[6] For good overviews see Mirowski (1988), Cohen (1994), and Redman (1997).

Second, Classical Man is completely separable, in the sense that nothing about his physical identity as an organism – whether as a body or a mind – depends constitutively on other people. That is not to deny that we depend hugely on each other in a causal sense for survival, but in the classical view we do so as independent beings, with properties wholly contained within our skins. Intuitively that makes sense, since where else would our properties *be* if not inside our skins?[7] But whereas in the classical case biological separability implies mental separability, in the quantum case mental states can depend non-locally on other minds.

Third, the properties of Classical Man are real and well-defined at the micro-level. At the macro-level people often experience ambivalence, doubt, and other states of mind which feel indefinite, as if we were not fully "in" one state rather than another. Classical Man might have such feelings, but deep down inside he is always in a fully specified state. Classical logic tells us that a material state cannot be both A and not-A, so if ambivalence is a material state, at any moment in time we must be actually in one state *or* the other, even if it does not feel that way.[8]

Fourth, Classical Man responds to local causal forces only, whether external forces from the environment entering his body or perceptual field, or internal forces set in motion by these stimuli that eventually culminate in reasons and behavior. Since there is no non-local causation in classical physics, if reasons matter in explanations of human behavior then it must be through local causes, just like everything else.

Finally, Classical Man's behavior is determined by the conjunction of internal and external causes operating on his body, which I understand to mean that he does not have free will.[9] Note that this does not necessarily mean we could ever know all those causes; the point here is ontological rather than epistemological. In the classical worldview free will would be like thermodynamics, unknowable in its details but ultimately deterministic when it comes to ontology.

Pulling these five requirements together, the first thing to notice is that they omit experience, meaning, and purpose in a teleological sense. That is because they are not consistent with a strictly classical ontology – experience because materialism cannot explain consciousness; meaning because it cannot exist without experience; and purpose because teleology is incompatible with a mechanistic worldview. So given these constraints, what is Classical Man like overall? Two answers suggest themselves: a machine or a zombie.

---

[7] On the skin as a boundary relevant to social science see Farr (1997).

[8] This is a realist view of psychological attributes, which is arguably presupposed by efforts to measure them; see Michell (2005) and Maul (2013).

[9] This requirement is more debatable, since some philosophers think that free will is compatible with classical determinism. I address this literature in Chapter 9.

The idea that people are just very complicated machines has a long pedigree, and became dominant in cognitive science and beyond with the advent of the computational theory of mind in the mid-twentieth century.[10] In this picture we are walking computers, constantly crunching data from the environment to realize pre-programmed objectives. It is not hard to see this image in rational choice theory, which is praised and condemned alike for its "mechanical" view of man. The zombie is less familiar in social theory, but it seems an apt characterization of *homo sociologicus*, in his purest form an unreflexive dope habitually re-enacting social norms.[11] Compared to the machine the zombie seems less inspired by classical physics, since he is at least animated (if not "alive"!), but the two share a lack of subjectivity and free will.

Many social scientists might be uncomfortable fully embracing man as a machine, much less his zombie friend, and as such see Classical Man as a straw man. But recall my purpose here: it is not to describe models of man in contemporary social theory, which routinely incorporate intentional phenomena that have no place in the classical worldview. Rather, it is to think the human being from scratch through classical physics. The result I submit is deeply counter-intuitive, for what we get is a very complex but essentially lifeless object. Indeed, if the feminist objection to models of man discourse is that it is about *men*, then the objection to Classical Man is that he is *dead*. And why shouldn't he be, given that classical physics was invented to study lifeless objects?[12] In a materialist worldview, whatever constitutes the specificity of subjectivity will necessarily be absent. Quantum Man, in contrast, is very much alive.

---

[10] On the machine metaphor in the social sciences see Menard (1988), Mirowski (1988), and Maas (1999), and for its continuing importance in biology see Nicholson (2013).

[11] The classic critique of this "over-socialized" model of man is Wrong (1961); for a good discussion within IR see Sending (2002). Zombies, however, do feature prominently in philosophy of mind; e.g. Chalmers (1996).

[12] See Wigner (1970).

8   Quantum cognition and rational choice

Of the three faculties of subjectivity, the empirical case for quantum cognition is the strongest. In the past decade a rapidly growing body of scholarship has emerged at the intersection of mathematical psychology and physics that not only models cognition in explicitly quantum terms, but has rigorously tested these models as well. While it remains to be seen how mainstream psychologists will respond, "quantum decision theory" has clearly come of age.[1] Moreover, at the same time an almost completely independent literature has emerged on quantum game theory as well. This work is more in a purely formal vein and thus less well tested against real data, although its predictions about how people should behave in quantum games appear to conform to what experimental game theorists have found. I briefly take up quantum game theory at the end of this chapter, but my main focus will be on its decision-theoretic cousin.

Quantum consciousness theory suggests that human beings are literally walking wave functions. Most quantum decision theorists would not go that far, and indeed – perhaps wary of controversy – they generally barely mention quantum consciousness, and then only to emphasize that they are making no claims about what is going on deep inside the brain (much less about consciousness), but are only interested in behavior.[2] Instead, they have motivated their work in two other ways.[3] First, they have highlighted the intuitive fit between quantum theory's indeterminism and the probabilistic character of human behavior.[4] If we were founding the social sciences today and started – without metaphysical prejudice – by looking for useful models in physics, then quantum models would

---

[1] The oldest precursors to quantum decision theory of which I am aware are Bohm (1951: 168–172) and Dobbs (1951), though the latter to my knowledge has never been cited in the contemporary literature. More recent but still early contributions would include Orlov (1982) (written while he was a prisoner in the Gulag), Shaw et al. (1994), Aerts and Aerts (1995/6), Bordley and Kadane (1999), and Deutsch (1999).

[2] This disclaimer is almost de rigueur in the literature; for examples see Aerts et al. (2011: 137), Yukalov and Sornette (2009b: 1075), Busemeyer and Bruza (2012: 24; and 349–357 for further discussion), and Wang et al. (2013: 673).

[3] Busemeyer and Bruza (2012: 1–8) actually give six reasons, but they seem to be of two basic varieties.

[4] See Glimcher (2005) on indeterminism at various scales of human functioning.

be a much more obvious import than classical ones. Second, quantum decision theorists have argued that their model can explain long-standing anomalies in human behavior under uncertainty. Beyond that they have remained agnostic about quantum consciousness, which makes quantum decision theory a form of what is called "weak" or "generalized" quantum theory, in which the formalism is detached from its physical basis and used simply as a tool to describe phenomena.[5] It should be emphasized, therefore, that by pressing quantum decision theory into the service of my ontological thesis I am taking it much further than most of its advocates are currently willing to go.

Still, while separable in principle, it is clear that quantum brain theory (at least) and quantum decision theory are related. If the former is true at the micro-level, then at the macro-level we should expect to see the behavior predicted by the latter. So if experiments do not support quantum decision theory, then that is a serious problem for quantum brain theory too. The connection is not as tight in the other direction but still there. If experiments do support quantum decision theory, then, at this stage of our understanding, it is still possible that classical mechanisms in the brain might explain this result.[6] However, as noted at the end of Chapter 5, this would open up an entirely new front in the quantum brain theory debate, by putting a burden of proof on skeptics to explain why, despite its claimed impossibility at the micro-level, we observe precisely what the theory would predict at the macro-level? It *could* be that reality is quantum at the sub-atomic level, classical at the brain level, and then quantum again at the behavioral level – but how likely is *that*? Nobody knows, but it is certainly less elegant than a uniform quantum picture from top to bottom.

Quantum decision theorists frame their work in relation to expected utility theory (EUT), the axiomatic version of what is more broadly known as rational choice theory.[7] EUT is the most widely accepted formal model of man in the social sciences today, the foundation not only of neo-classical economics but increasingly political science and sociology as well. It is actually an amalgam of two theories: utility theory, used to model actors' preferences, and (classical) probability theory, used to model expectations under uncertainty. (In this basic structure EUT replicates the more general "Desire × Belief = Action" model of behavior, the dominant meta-psychological framework in philosophy and discussed below.) Given an actor's preferences and probabilities, the rational choice is defined as that which maximizes expected utility. Note that as a

---

[5] See Atmanspacher et al. (2002).
[6] See for example Glimcher (2005: 35), beim Graben and Atmanspacher (2006), de Barros and Suppes (2009), Khrennikov (2011), and de Barros (2012); on the other hand, Hameroff (2013) argues that quantum consciousness theory will ultimately be necessary to explain these results.
[7] Savage (1954) is the locus classicus of EUT; for an excellent discussion of rational choice theory in philosophical context see Hollis and Sugden (1993).

theory of choice EUT encompasses not just Cognition but also what I am calling Will, in the form of a (mostly implicit) assumption that the combination of preferences and expectations ("reasons") are causes. I argue in the next chapter that this assumption is problematic, but since the literature treats the probability and utility elements together, I will do so here as well.

Although often used in a descriptive and/or explanatory capacity, EUT is in the first instance a normative theory of what it means to behave rationally in a given situation. To that end, it defines a set of axioms that impose logical constraints on how actors may calculate probabilities and organize their preferences (the familiar requirements of commutativity, transitivity, independence, and so on). Crucially, in EUT it is assumed – to my knowledge without ever explicit justification – that 'logical' means *classical* logical, even though (coincidentally) classical and quantum logic were formalized in the same year (1933).[8] Classical logic is the familiar "either-or" logic of our schooling and the material world, in which things have definite properties that cannot be incompatible. Not just the axioms of EUT, but the whole interpretive edifice of rational choice theory takes this logic for granted, which quantum theory puts into question.

But before quantum decision theory came along, there were decades of experimental research by psychologists, showing that human decision-making was subject to systematic "biases" relative to EUT predictions. The pioneering work was done by Daniel Kahneman, Amos Tversky, and others in the 1970s on the non-rational "heuristics" that people appear to use in judging probabilities, for which Kahneman received the Nobel Prize. But the empirical problems for EUT appeared at the start with the Allais and Ellsberg Paradoxes and today encompass a number of anomalies.[9] These findings have spurred many creative attempts to account for them, whether by dropping or modifying an axiom from EUT or working outside the theory altogether, and are the basis for many contemporary research programs. However, these efforts all have three things in common: they continue to operate within a classical frame of reference; they are partial, addressing certain anomalies only; and they are ad hoc, in not being derived from an alternative axiomatic foundation. Hence, the interest in quantum decision theory, which not only offers such a foundation, but looks likely to explain all the anomalies of classical decision theory in one fell swoop.

---

[8] The latter by no less than John von Neumann, who later co-authored the foundational text of (classical) game theory (1944) – one wonders if he ever considered applying the one to the other! See Primas (2007: 9–15) for a good comparison of classical (Boolean) and quantum logic in the context of the mind–body problem.

[9] See Shoemaker (1982) for an early but still useful overview, and more recently Rieskamp et al. (2006).

Quantum decision theory

The quantum decision theory literature is highly technical, already too large to summarize here, and complicated further by the fact that authors do not always agree on mathematical details.[10] So, to convey just a sense of the approach, I will briefly characterize three prominent lines of research – on order effects, probability judgment, and preference reversals – and then turn to what all this means for the nature of rationality.

Order effects in quantum perspective

One of the key differences between classical and quantum physics is that in the former interactions between objects and measurement devices are "weak" whereas in the latter they are "strong."[11] The distinction turns on whether the act of measurement is linked to changes in the state of the system in question, which can be discerned by measuring it again. If there is no change then the two measurements are said to commute, and if there is a change they are non-commutative. Commutativity is a basic principle of logic and mathematics that holds when the order in which operations are done does not affect the outcome, so addition is commutative ($8+5 = 5+8$) while subtraction is not ($8-5 \neq 5-8$). Applied to a sequence of physical measurements, if commutativity holds then the state of the world is independent of the observer, who is therefore just registering a pre-existing fact (the classical case); if non-commutativity holds then the subject–object distinction has broken down, and so the observer is somehow participating in establishing what the facts are (the quantum case). The exemplar of the latter is what happens when you measure the momentum or position of a particle; change the order of measurements and you get different results.

"Order effects," where the order in which information is presented leads to inconsistent results, are well known and widespread in the social sciences.[12] They are pervasive in public opinion surveys, and often affect the updating of beliefs over time and as such can play a significant role in medical and jury decision-making. From a practical standpoint they are therefore a problem that needs to be managed, and because they violate the commutative axiom of classical probability theory they are also irrational and thus an anomaly that needs to be explained. Within the mainstream literature this has been

---

[10] For short introductions to quantum decision theory see Pothos and Busemeyer (2013) and Wang et al. (2013); Busemeyer and Bruza (2012) is currently the most comprehensive text.
[11] This paragraph paraphrases Atmanspacher and Römer (2012: 274).
[12] The literature is enormous; see Hogarth and Einhorn (1992) and Moore (2002) for illustrative discussions.

attempted entirely from a classical perspective.[13] Bayesians have dealt with order effects by making presentation order just another piece of information that subjects need to assimilate, but that is ad hoc and merely re-describes the phenomenon with no explanatory insight.[14] "Adding," "belief-adjustment," and Markov models are more substantive but still ad hoc, since they never question the classical assumption that observations *should* commute. But why take this for granted? On intuitive grounds alone, Atmanspacher and Römer suggest that in psychology, "where virtually every interaction of a 'measuring' device with a 'measured' mental state changes the state uncontrollably and where mental states are often literally established by measurements, it is highly plausible to argue that *non-commutativity should be the ubiquitous rule*."[15] In this light it is curious that the possibility that "the problem" with order effects may lie in the commutativity assumption itself has to my knowledge never come up in the mainstream literature, which may attest to how deeply ingrained is the classical model of man within psychology.

Unlike classical physics, quantum mechanics makes no *ex ante* stipulation about whether commutativity holds, treating it instead as contingent on whether, in a given empirical context, "basis vectors" are thought to be "compatible" (see below).[16] If they are compatible, then the quantum model reduces to the classical model and no order effects will be observed, whereas if basis vectors are incompatible then non-commutative results – a signature of quantum phenomena – are expected. Since order effects are defined by their non-commutativity, it is therefore not surprising that in accounting for them quantum decision theory performs very well. It predicts both extant and new experimental data at least as well and usually better than classical models.[17] Zheng Wang and Jerome Busemeyer have also tested an a priori prediction, derived from quantum probability theory, and found that it too was born out in existing data.[18] Atmanspacher and Römer have predicted further order effects that are common in physics but have not yet been observed empirically in psychology, which if found would be the strongest evidence to date that the approach is sound.[19] And to cap it all off, in contrast to the ad hoc and partial nature of the classical models that try to explain order effects, quantum decision theory offers a unified, axiomatically well-founded account. It strongly suggests, in short, that in cognition "non-commutativity is no exotic special case but the rule, from which commuting operations ... are the exception."[20]

---

[13] Atmanspacher and Römer (2012: 275).   [14] Trueblood and Busemeyer (2011: 1522).
[15] Atmanspacher and Römer (2012: 275); emphasis added.
[16] See Trueblood and Busemeyer (2011).
[17] See Trueblood and Busemeyer (2011) and Wang and Busemeyer (2013).
[18] See Wang and Busemeyer (2013); though see Yukalov and Sornette (2014: 89) for doubts that their results *necessitate* a quantum interpretation.
[19] Atmanspacher and Römer (2012: 277).   [20] Atmanspacher and Filk (2014: 34).

Paradoxes of probability judgment

A different stream of evidence for the importance of non-commutativity in cognition comes from research on how people assign probabilities to uncertain events. Classical probability theory requires that people do so in particular ways. Among these rules is that in considering the conjunction of two events, one of which is a subset of the other, the probability of the less inclusive event cannot be greater than that of the more inclusive one. Intuitively that makes sense; indeed, one might wonder how it could be otherwise? Yet psychologists have shown that people routinely commit the "conjunction fallacy" of violating this rule.

The paradigmatic case is "Linda," a fictional subject about whom test subjects (judges) are asked some questions.[21] Linda is first described as having been a philosophy major in college, bright, and concerned with discrimination and social justice. Then, the judges are asked whether it is more likely that Linda is (a) a feminist bank teller or (b) a bank teller. Classical probability theory tells us that B is more probable, since it encompasses A, but includes other possibilities as well. Yet, judges tend to pick A, that it is more likely Linda is a feminist bank teller than just a bank teller. Further research showed that the effect only occurred when the two options were an unfamiliar pairing (as opposed to woman and teacher for example), which suggested that uncertainty about how to relate such pairs plays a role. This led Kahneman and Tversky to the "representativeness" heuristic, which assumes judges reason on the assumption that Linda is more representative of feminists than of bank tellers. As a description it was brilliant and helped generate an entire research program on cognitive heuristics,[22] but it was also ad hoc, and since there is no consensus on the origin of the fallacy, as an explanation it only goes so far.

In contrast, the conjunction fallacy is readily explained by quantum decision theory, which uses quantum probabilities instead of classical. The details are formidable, but there is an excellent narrative discussion in an article by Jerome Busemeyer and his co-authors upon which I shall rely.[23]

Begin by representing a subject's beliefs and knowledge in an n-dimensional vector (Hilbert) space. Different "basis vectors" in this space correspond to combinations associated with different concepts, events, and situations in social life, which all co-exist in superposition as potentialities within the mind. Concepts or events that are inconsistent or not typically experienced together will have basis vectors that are "incompatible." In quantum mechanics 'incompatible' refers to observables that cannot be measured at the same time, such as the

---

[21] For a classic statement see Tversky and Kahneman (1983).
[22] See Gigerenzer and Gaissmaier (2011) for a recent overview.
[23] Busemeyer et al. (2011: 196–197); there are also useful narrative fragments in Franco (2009) and Yukalov and Sornette (2009b).

position and momentum of a particle.[24] However, since the first measurement will affect the outcome of the second this means that their joint probability cannot be defined – in violation of the commutative rule of classical probability theory. Now consider the judges, who have been asked to decide which of the two descriptions of Linda is more likely to be true. When answering the bank teller question alone, the details of the story make it very difficult to think of Linda as a bank teller. When answering the conjunctive question, the judge has to consider both feminist and bank teller features, which are represented by incompatible basis vectors. Since they therefore have to be addressed sequentially, the mind projects itself first into the feminist sub-space (say), assessing the likelihoods, then "rotates" to consider the alternative – bank teller – sub-space (these rotations correspond to adopting different basis vectors or points of view on the question). From the feminist projection, the mind turns to the bank teller question, the consideration of which will include the judgment that Linda is a feminist, which eliminates some of the story details and thus makes it easier than before to imagine that she happens to have a career as a bank teller than that she was a bank teller alone.

Phenomenologically this is plausible, and quantum decision theory also supplies a mechanism that explains the observed bias: interference arising from the mind's internal measurements on incompatible states.[25] Quantum interference always has a characteristic effect, in clear violation of classical probability theory, which is that "the probability of the union of two possible paths can be smaller than each individual path alone"[26] – which is precisely what we observe in the Linda studies. Indeed, based on this mechanism quantum decision theorists can even predict the size of the bias using parameters supplied by existing studies of the effect. This has allowed them to rigorously test their model of probabilistic fallacies such as this, and the result is a very close fit with the data.

Other anomalies in probability judgment follow a similar pattern. Consider the "disjunction effect," in which subjects violate Savage's "Sure Thing Principle."[27] Since an excellent accessible summary of this effect and its quantum interpretation is provided by Jerome Busemeyer and Peter Bruza, let me quote their discussion almost in full.[28]

---

[24] 'Compatible,' therefore, means that "the answer to any set of questions can be known concurrently;" in classical probability theory all questions are assumed to be compatible. See Pothos and Busemeyer (2014: 2).
[25] See Franco (2009), Yukalov and Sornette (2009b: 1088–1093), and Busemeyer et al. (2011); for a comprehensive overview of interference effects in psychology see Haven and Khrennikov (2013: 124–154).
[26] Busemeyer et al. (2006: 220). [27] See Savage (1954: 21–23).
[28] See Busemeyer and Bruza (2012: 8–10). For other quantum treatments of the disjunction effect, see Khrennikov and Haven (2009), Yukalov and Sornette (2009a; 2009b; 2011), and Asano et al. (2012).

[The disjunction effect] was discovered in the process of testing a rational axiom of decision theory called the sure thing principle (Savage, 1954). According to the sure thing principle, if under state of the world X you prefer action A over B, and if under the complementary state of the world not-X you also prefer A over B, then you should prefer A over B even when you do not know the state of the world... Tversky and Shafir (1992) experimentally tested this principle by presenting students with a two-stage gamble; that is, a gamble which can be played twice. At each stage the decision was whether or not to play a gamble that has an equal chance of winning $200 or losing $100 (the real amount won or lost was actually $2.00 and $1.00 respectively). The key result is based on the decision for the second play, after finishing the first play. The experiment included three conditions: one in which the students were informed that they already won the first gamble, a second condition in which they were informed that they lost the first gamble, and a third in which they did not know the outcome of the first gamble. If they knew they won the first gamble, the majority (69%) chose to play again; if they knew they lost the first gamble, then again the majority (59%) chose to play again; but if they did not know whether they won or lost, then the majority chose not to play (only 36% wanted to play again).

...

Researchers working with quantum models see this finding as an example of an interference effect similar to that found in the double-slit type of experiments conducted in particle physics... Both cases involve two possible paths: in the disjunction experiment, the two paths are inferring the outcome of either a win or a loss with the first gamble; for the double-slit experiment, the two paths are splitting the photon off into the upper and lower channel by a beam splitter. In both experiments, the path taken can be known (observed) or unknown (unobserved). Finally, in both cases, under the unknown (unobserved) condition, the probability (of gambling for the disjunction experiment, of detection at D1 for the double-slit experiment) falls far below each of the probabilities for the known (observed) cases... One cannot help but wonder whether the mathematical model that succeeds so well to explain interference statistics in physics could also explain interference statistics in psychology.

Quantizing preference reversals

In decision theory, people can be irrational in at least two distinct ways: in how they judge probabilities and in how they form preferences. On the preference side, rationality is defined by the assumptions of utility theory, three of which figure most prominently in the present context.[29] First, rational actors must have complete, fully determined preferences over all outcomes prior to choice. (Note that this does not mean that others know an actor's preferences, just the actor herself.) Second, preferences must follow the rules of classical logic and thus be transitive, such that if actors prefer A to B and B to C then they must also prefer A to C. Third, because they can't be observed, if preferences are determinate and thus really *there*, then they must be "invariant" over different

---

[29] See especially Slovic (1995), Lambert-Mogiliansky et al. (2009), and Alfano (2012).

procedures that might be used to elicit them. As with the probability side of EUT, utility theory is clearly consistent with the constraints we identified above on classical models of man.

The phenomenon of "preference reversals" challenges all three assumptions, suggesting that preferences are not "revealed" at all, but "constructed" in the process of choice and as such highly sensitive to framing and context effects.[30] The key issue at stake in this literature is procedure invariance – preferences should not depend on how they are elicited, since that could violate transitivity.[31] The classic experiments testing this assumption compared two procedures: (1) asking subjects to choose between two gambles, one with a good chance of winning a small amount and the other with a smaller chance of winning more; and then (2) asking them how much they would pay to play each bet. The experiments are set up so that these procedures – choices vs. expected cash values – are formally equivalent, so rational actors should pay more for the bet they prefer. Yet, in practice, it was found that "choices between pairs of gambles were influenced primarily by the probabilities of winning and losing, whereas buying and selling prices were primarily determined by the dollar amounts that could be won or lost,"[32] which leads to violations of transitivity. These experiments have been repeated many times, and subjected to intense scrutiny especially from neo-classical economists, who, not surprisingly, saw them as a serious threat to their theories.[33] If Jones' preferences over X depend on how we measure them, then not only is his rationality in question, but who's to say Jones even *has* preferences over X – and then where would utility theory be?

While the mainstream preference reversal literature has not considered a quantum interpretation of the phenomenon, it has generated useful (if ad hoc) classical theories to account for it.[34] However, in quantum decision theory we now have an alternative axiomatic framework that not only explains preference reversals but can predict them as well.

An excellent overview is provided by Ariane Lambert-Mogiliansky and her co-authors in an article suggestively entitled "Type Indeterminacy: A Model of the KT (Kahneman-Tversky) Man."[35] Playing off "Harsanyi Man," whose "types" (preference orderings) are fully determined and mutually exclusive (even if unknown to others), Lambert-Mogiliansky et al. start with a representation of a person's "state" as a superposition of all the possible types relevant to

---

[30] See Tversky et al. (1990) and Slovic (1995: 365), and Smith (2012) for a recent review.
[31] Also see Michel (2005) and Maul (2013). [32] Slovic (1995: 365).
[33] See Guala (2000) on economists' resistance to and eventual acceptance of the preference reversal phenomenon.
[34] See for example Alfano (2012).
[35] Lambert-Mogiliansky et al. (2009); see Pothos and Busemeyer (2009), Yukalov and Sornette (2009a: 543–545), Khrennikov (2010), and Khrennikova et al. (2014) for other quantum treatments of preference reversals.

a situation. Whether these types are compatible is then the key question. If they are, then EUT takes over; if they are not, then, like the position and momentum of a particle, types cannot have well-defined values simultaneously.[36] The superposition state does not reduce to an actual or "eigen" type until a measurement is made. Then, depending on the context, one basis vector becomes preferred, and a single type emerges in wave function collapse. Note that this is the exact opposite of the standard view. Rather than being an expression of underlying pre-existing preferences, "[d]ecision-making is modeled as the *measurement* of the preferences."[37] Whereas Harsanyi Man's problem is merely uncertainty about others' types, KT Man does not even know her *own* type until she makes a choice.

Interestingly, this conclusion is strongly reminiscent of the "performative" model of agency due to Judith Butler, a post-structural feminist theorist.[38] In developing this model she draws on language philosopher J. L. Austin's concept of performative utterances, which in contrast to descriptive speech acts, like "this tie is red," are statements that themselves do something, like "I pronounce you husband and wife" at a wedding. Butler's interest is less in language than in gender, which she argues people have only insofar as they perform it. However, although her foil is identity theory rather than rational choice, her critique of the former applies to the latter, which is that gendered performances are not enacted by a preexisting subject/rational actor with a set of gendered preferences. Rather, gendered performances actualize gendered preferences and in that moment make someone a gendered subject in the first place.

In suggesting that performativity is a "quantum" theory of agency, I do not want to implicate Butler in other aspects of my approach, which she might reject on epistemological grounds because it is naturalistic, and on ontological grounds because my concern with consciousness reflects a humanist conception of subjectivity of which Butler is highly critical.[39] However, on the specific issue of how she conceives the relationship between agents and agency, there is a strong parallel to the quantum reading of the preference reversal phenomenon. Moreover, one of the issues Butler struggles with in response to critics is the age-old tension between voluntarism and determinism, both of which she wants to avoid.[40] As we will see in the next chapter on free will, a quantum model of agency provides a way to thread this needle, and as such could contribute to the further development of her approach.

---

[36] Lambert-Mogilianksy et al. (2009: 353).
[37] See Lambert-Mogilianksy and Busemeyer (2012: 103), emphasis added.
[38] See Butler (1990; 1993). A distinct but related tradition of performativity theorizing has been developed in the sociology of economics by Donald MacKenzie (2006) and others.
[39] By not thematizing consciousness, Karen Barad (2007: 59–65 and passim) provides an alternative quantum reading of performativity more in keeping with Butler's theory.
[40] See Allen (1998) for a particularly useful discussion.

Lambert-Mogiliansky et al. contrast their model to classical Markov and random utility models, but they do not test any predictions, nor am I aware of such tests by others.[41] But one interesting factoid is that in "back to back" measurements people are known to reveal the same preferences again, which is contrary to the prediction of random utility models, where choices are assumed to be discrete and probabilistic. In contrast, due to the "Quantum Zeno Effect," in which repeated rapid measurements can slow or even halt the evolution of a wave function, this repetitive pattern is actually predicted by quantum decision theory.[42] That's just one case, but given what quantum decision theorists have found in probability judgment, there seems every reason to expect further success here.

In sum, the situation in cognitive science today seems similar to physics in the early 1900s. In both domains rigorous testing of classical theories had produced a string of anomalies; efforts to explain them with new classical models were ad hoc and partial; and then a quantum theory emerged that predicted them all with great precision. From a Lakatosian perspective one could hardly want more evidence of a progressive problem shift, which in physics was consolidated at Solvay in 1927. Whether that will happen in psychology remains to be seen, but perhaps a Bloomington Conference will one day do the same for quantum decision theory.[43]

Rationality unbound?

The Kahneman-Tversky results led to a large wave of pessimism about human rationality. If people did not behave rationally even in simple laboratory tasks, what hope was there for us being rational in the far more complex problems we face in everyday life? However, this skepticism soon met with substantial pushback, though less from defenders of rational choice than from scholars arguing that the latter's definition of rationality – which the psychologists had taken as their standard – was too narrow and therefore itself the problem. In what have become known as the "rationality wars" two competing views of what it means to be rational have emerged.

The "coherence" view[44] is the EUT orthodoxy, and despite its empirical problems, it remains dominant not only in its economics birthplace but probably most of the social sciences. On this definition, being rational is about having properly organized beliefs and preferences, and then following the rules

---

[41] On the non-Markovian nature of quantum decision theory also see Asano et al. (2012).
[42] See Atmanspacher et al. (2004) and Franck (2008: 135–137).
[43] Bloomington, Indiana, that is, which has been an epicenter of this research; Diederik Aerts' program on quantum cognition at the Free University of Brussels is another.
[44] Note that coherence here has nothing to do with coherence in the quantum mechanical sense (see Chapter 5).

of expected utility maximization in making choices.[45] If people's minds are incoherent or they don't follow the rules, then so much the worse for them: it just means they are irrational, and vulnerable to exploitation and/or otherwise likely to incur unnecessary costs. In contrast, the "correspondence" view, which is due especially to the work of Gerd Gigerenzer and his colleagues, is success-oriented.[46] On this definition, being rational is about whether one succeeds in achieving goals, which depends on the correspondence or fit between one's mind and the environment. Rationality on this view should be defined in adaptive and ecological terms, not a priori and intrinsic ones.[47]

Both sides have a point, as Andrea Polonioli makes clear in an excellent review.[48] On the one hand, there are good theoretical reasons to think that following coherence criteria will not always be adaptive: in both evolutionary and social contexts being inconsistent may sometimes be better. And empirically, heuristics are better predictors of actual behavior. These considerations argue for not making coherence the sole standard of rationality. On the other hand, it is not clear that correspondence criteria can simply replace coherence ones either, since they face substantial difficulties of implementation. Identifying the goals against which success should be judged is not always easy; sometimes criteria of correspondence to the world may conflict; and even if goals are identifiable, it may be hard to assess performance if information is unavailable or hard to interpret. In short, giving up on a universal standard of rationality seems, in practice, to mean giving up on any normative standard at all. That argues for retaining the orthodox criterion even if people do not always abide by it.[49]

To date this debate has played out entirely within a classical frame of reference. This privileges the coherence view, because if correspondence is unworkable in practice then defenders of the orthodoxy can maintain that it should remain at least a normative aspiration. If people are classical decision-makers dealing with a classical world, then having a well-ordered mind is simply the optimal way "to harness the uncertainty in our environment and make accurate predictions regarding future events and relevant hypotheses."[50]

Quantum decision theorists have not yet waded directly into the rationality wars. While recognizing that their work bears on the meaning of rationality, they have not engaged the philosophical literature and mostly reserve their reflections to a concluding paragraph or two, and then, like quantum theorists

---

[45] On rationality as utility maximization see Cudd (1993: 103–110).
[46] See for example Gigerenzer and Gaissmaier (2011).
[47] This seems resonant with the Aristotelian notion of "practical" rationality, though to my knowledge this connection is not made by advocates of the correspondence view.
[48] See Polonioli (2014), whose discussion I rely on in this paragraph; also see Wallin (2013).
[49] See Wallin (2013: 474).   [50] Pothos and Busemeyer (2013: 270).

everywhere, in moving from theory to interpretation they do not always agree.[51] Still, it seems clear that their work strengthens the case for the correspondence view, not only by providing a physical basis for the Kahneman-Tversky effects as I showed above, but by undermining the privileged dialectical position of the coherence view.

First, quantum decision theory calls into question the whole idea of utility-maximization as a criterion of rationality, which presupposes that a normal human mind has well-defined beliefs and preferences which can then be maximized in choice behavior. If we are quantum systems, then a normal human mind will be in a superposition rather than well-defined state, and as such *"there exists nothing to be maximized."*[52] Rationality cannot mean relating means to ends if ends do not even exist prior to the choice of means. To be sure, classical beliefs and preferences may emerge from time to time, but only when there is no uncertainty or no incompatible basis vectors affecting an uncertain decision – which in real life is likely to be the exception rather than the norm.

Second, quantum decision theory challenges the coherence view's assumption that having a well-ordered mind is the optimal basis for harnessing environmental uncertainty and making accurate predictions. That might be true if subjects are fully separable from the objects of their perception, as in the classical worldview, but in the quantum world this is not the case. Recall that when physicists measure particles they become entangled with them, such that while we cannot say measurement "causes" the result, it creates a non-local correlation between subject and object that "influences" it. This non-separability is the basis for the holism of quantum processes, and if human beings are quantum systems too then this holism would carry over into our dealings with the macroscopic world.

So consider again that KT Man is highly sensitive to the context of her environment. Viewed from a quantum perspective, this means not just that context affects her behavior (that is true in EUT as well), or even that over time it might change her beliefs or desires (as in Bayesian updating or endogenous preference formation), since as long as separability is assumed these processes would be causal. Rather, KT Man's context sensitivity means that there is an entanglement between the two, such that her mind is in effect extended beyond her brain and communing with her context. As I will argue later in my discussion of visual perception and language, this gives her insight into her environment not by the causal transmission of information into a previously well-ordered

---

[51] See for example Lambert-Mogiliansky et al. (2009: 356), Yukalov and Sornette (2009a: 537). The longest discussions of rationality I have found in the literature are by Pothos and Busemeyer (2013: 270–271; 2014), though even these are quite abbreviated.

[52] Slovic (1995: 369), quoting D. Krantz on the preference reversal phenomenon; emphasis added. Also see Whitford's (2002) critique of means–ends rationality from a Deweyian perspective. An explicitly quantum version of this point is made by Yukalov and Sornette (2009a: 537).

mind, but non-locally by collapsing her superposed potential states of mind into an actual state of mind. Not only, I submit, does non-local or "direct" perception provide a better correspondence between mind and uncertain environments than the "optimal" approach of the classical view of rationality, it positively depends on the mind *not* being coherent – except in the quantum sense.[53]

Third, quantum decision-making is holistic in another way, which is that it encompasses the whole brain including the emotions and the sub-conscious.[54] Emotions do not follow classical logic and so in EUT not only are they differentiated from reason, but treated as essentially "un-model-able." That might be fine if each kept to its own domain, but when faced with KT Man's behavior we are forced in EUT to conclude that emotion is colonizing a domain of reason from which properly it should be excluded. Neuroscientists have confirmed that reason and emotion are indeed deeply intertwined in decision making under uncertainty,[55] making the classical definition of rationality increasingly moot – but they offer no physical basis as a substitute. Quantum decision theory provides that alternative, enabling us to model the emotional aspect of choice in an integrated way. While from a classical standpoint this may seem to weaken the concept of rationality,[56] to my mind it makes rationality stronger, by incorporating the emotions as a decision-making resource.

The coherence view of rationality only makes sense as a normative standard if our brains are separable from their environment and limited to classical computations. Quantum brains are non-separable and orders of magnitude more powerful, and capable of feats that for classical brains are inconceivable. Thus, whereas classical rationality is mechanical and forces us to purge our every inconsistency, the quantum model allows us to think incompatible thoughts simultaneously and to exploit non-local connections to our environment. This seems a vastly more flexible and supple kind of rationality that would be of particular benefit in an uncertain world. So why base the definition of rationality – our highest standard for behavior – on what our brains do only in the classical limit? From this perspective, "bounded" rationality is a more apt characterization of *classical* decision theory, while the ostensibly irrational processes explained by quantum decision theory constitute a kind of super- or "*unbounded*" rationality.[57]

If the coherence view of rationality is denied its privileged normative position by virtue of its classical underpinnings, does this then mean that we should embrace the correspondence view? With no expertise in this area I hesitate to

---

[53] Also see Pothos and Busemeyer (2013: 271).
[54] See especially Yukalov and Sornette (2009a; 2009b). Other quantum decision theorists have not thematized this connection, so I do not know if they would agree.
[55] In IR scholarship see especially Mercer (2010) and the citations therein.
[56] See La Mura (2009).
[57] Cf. Rieskamp et al. (2006) and Pothos and Busemeyer (2014: 2).

venture a conclusion, but my gut feeling is a quantum one so to speak, or both yes and no. Yes, we should as a better description of how people actually behave, but not as a new normative standard. In part this is because of the questions raised by Polonioli about its workability in practice, which when given a quantum reading point to a deeper problem. Quantum physics teaches us that the behavior of sub-atomic particles is extremely sensitive to the experimental context; measure them in one way and you get one result, in another way you get another. The implication is that we cannot speak of a stable, objective reality at that level. Quantum decision theory suggests that the same is true of human beings: who we are at a given moment cannot be separated from our context. And given that our contexts are vastly more complex, subtle and varied than those in physics, that means that compared to sub-atomic particles our behavior will be vastly more complex, subtle and varied as well.

Moreover, keep in mind that even if heuristics and quantum decision theory make more accurate predictions overall than EUT, these predictions are about how a sampled population will behave, not particular individuals. As such they are necessarily statistical predictions, with variation around the mean. So does that mean that individuals who are not at the mean are behaving irrationally?[58] That doesn't seem right; it seems more appropriate to say that each actor is doing the best they can under their own unique personal and contextual constraints. In other words, perhaps the reason correspondence is problematic as a normative standard is that it cannot be a normative standard at all, because in most cases it has no objective basis.

That in turn suggests a broader conclusion given the quantum critique of the coherence view: when questions are incompatible and thus classical rationality is impossible there simply *is* no normative standard of rationality. All rationality in such situations is contextual and particular.[59] The parties to the rationality wars are fighting over something that, most of the time at least, does not exist.

Thus, in real-world decision-making, rather than defining rationality objectively perhaps it should be done *subjectively*, by reference to how people themselves define success as they try to solve problems in their lives. This is not to say that cognitive scientists should not continue to do research on how successful people are, in some sense "objectively," in solving problems, from which generalizations might emerge that could be useful in public policy. However, we should resist the urge to turn such generalizations into normative standards. For what is their point, if not to discipline people into conformity with *society's* definition of success, which as an intersubjective phenomenon is just as contextual and particular as an individual's? It may be that from

---

[58] Also see Martinez-Martinez (2014: 43).
[59] Cf. Pothos and Busemeyer (2014), who seem to hold out hope for a new normative standard, but in the end, fall back on the descriptive superiority of the quantum approach.

society's point of view some behaviors – illicit drug use say, or joining a gang – are counter-productive, but does that make them irrational?[60] Even if someone later regrets his decisions, to the charge of irrationality it is always open to him to reply that "it seemed like a good idea at the time." And who are you or I to say, if we had their brains and were situated in their context, that we would not have had the same good idea?

Quantum game theory: the next frontier

Given the preceding discussion it is natural to ask what would happen if quantum decision-makers are placed in a strategic setting, where they are playing against each other rather than nature. This is the focus of quantum game theory, which originated in 1999 in papers by David Meyer and Jens Eisert, Martin Wilkens, and Maciej Lewenstein.[61] The literature has since exploded, with a review article in 2008 listing 177 references, the growth of which shows no signs of tapering off.[62] Interestingly, the theory has evolved completely independently of quantum decision theory, due to its distinct roots in the fields of quantum computation and cryptography, where the intellectual concerns are quite different than those of cognitive science.[63] As a result, very little of this work has been experimentally tested or even makes reference to behavioral game theory; almost all of it is formal, exploring the mathematical properties of different games and assumptions.[64] And also unlike quantum decision theory, its practitioners have made little effort to give its ideas substantive meaning relevant to social science, which makes it much less accessible to outsiders.[65]

However, quantum game theorists do share at least one sensibility with their decision-theoretic cousins, which is that they want nothing to do with quantum consciousness theory. Indeed, they are reluctant to claim even that quantum brains are involved in quantum games, on the grounds that quantum decoherence "forbids" such a possibility.[66] Instead, in the literature it is generally assumed that quantum games are populated by classical decision-makers who possess a special device that enables them to play quantum strategies. While that might make sense in cryptography, it obviously doesn't in social science.

---

[60] Cf. Wallin (2013: 472).
[61] See Meyer (1999) and Eisert et al. (1999); Edward Piotrowski and Jan Sladkowski were also early contributors and provide a useful introduction in their (2003).
[62] See Guo et al. (2008).
[63] Martinez-Martinez (2014) is the first explicit attempt I know of to bridge the gap, though see Pothos and Busemeyer (2009), who apply quantum decision theory to strategic interaction.
[64] For a rare exception see Chen and Hogg (2006), and on behavioral game theory see Camerer (2003).
[65] For exceptions see Arfi (2005) and Hanauske et al. (2010).
[66] See Hanauske et al. (2010: 5092) and Chapter 5 above.

Thus, I shall follow F. M. C. Witte in his unorthodox view of quantum game theory, which assumes quantum rather than classical players.[67] He does this in order to respond to some critiques of quantum game theory, whereas I do so in light of my larger ontological argument and also to help make the theory more relevant to social science. Either way, it should be emphasized that, as in my treatment of quantum decision theory, I am going beyond the standard reading.

The concept from quantum theory that plays the most important role in quantum game theory is entanglement, specifically of strategies. Take a classical game with two players – Emma and Otto – each of whom has two strategies, Cooperate and Defect. These strategies are independent or fully separable in two senses. First, they are binary choices, in that Emma and Otto can only play one in a given round; even if they decide to play a mixed strategy of randomizing their choice, only one strategy will actually be played each time. In information theoretic terms each strategy set may therefore be seen as one "bit," which can take one or the other value but not both simultaneously. Second, Emma and Otto have no control over each other's bits. Although if they are rational they will try to anticipate the other's move, each retains complete sovereignty over their own choice.

To quantize this game we now replace Emma and Otto's bits with qubits, which are linear superpositions of their strategies. This means, first, that for each of them considered individually, their two strategies are now entangled, such that each can – in a sense – play both at once.[68] Note that this is not equivalent to playing a mixed strategy: "in a quantum superposition the decision maker is not randomizing in the sense of mixed strategies. Rather, all pure strategies not only equally contribute to shape the decision-making process, but also either sub-additively or super-additively interfere with each other's contribution to weaken/enhance each other's contribution."[69] The "in a sense" qualifier refers to the fact that the result of playing both strategies at once is still either Cooperate or Defect, just as in physics the result of wave function collapse is one actual particle. But because Emma and Otto's minds are in superposition states, that result can only be known – even by them – through the result, not *ex ante*.[70] (Recall the performative model of agency: it is only in the performance itself that preference, in this case for a strategy, is actualized.) Until then, because of entanglement with the actual choice, the choice(s) *not* made also play a role in the process.

Second, thinking about strategy sets as qubits means that entanglement is not limited to what is going on inside Emma and Otto's minds, but may extend

---

[67] See Witte (2005). Khrennikov (2011) and Asano et al. (2011) take an intermediate position in suggesting that a quantum-"like" process in the brain is involved, but one that in their view has a classical basis.
[68] See Goff (2006) on "quantum tic-tac-toe" for a very clear illustration.   [69] Arfi (2007: 795).
[70] Also see the discussion of the Transactional Interpretation of quantum theory in Chapter 6.

to their minds considered jointly,[71] constituting them as a single quantum system. This doesn't mean they lose their individuality altogether, any more than entangled sub-atomic particles do: in neither case do the units of the system become literally the same. But nor are they fully separable either, as in the classical case, and that has an important implication: each now has access to a shared entangled state with which both of their strategy sets are correlated, which gives them some control over *each other's* decisions.[72] Moreover, this control exists without benefit of communication, whether in the form of a pre-game agreement, costly signaling, third-party mediation, or cheap talk – all of which involve the transmission of information by local causal means.[73] The joint control in quantum game theory works through the manipulation of non-local correlations, in much the same way that when the spin of one particle in an entangled pair is measured it induces a state change in the other. This gives strategy in quantum games an irreducibly collective aspect, such that players are at least partly in "We-mode" rather than just "I-mode."[74]

It is one thing to imagine strategic entanglement in quantum cryptography, but what does it look like, concretely, in social life? The literature is frustratingly vague on this, but Matthias Hanauske and his co-authors provide some useful hints.[75] In an evolutionary quantum game-theoretic analysis of the 2008 financial crisis they are concerned with how traders might be induced in future to play more "Dove" strategies rather than the aggressive "Hawk" strategies which helped cause the crisis. They interpret the difference between the two as one of degrees of entanglement: Hawks exhibit low entanglement with others and thus behave classically, whereas Doves exhibit higher, leading to more quantum behavior. Hanauske et al.'s solution is therefore to look for ways to increase traders' entanglement with each other and with the general public. They argue that this could come, on the one hand, from improving moral standards, shared experiences in training, legal reforms, and above all education, and on the other, from reducing the material incentives for classical trading behavior that existed before.[76] Importantly, these changes would move behavior in the desired direction without communication between traders. In short, entanglement in quantum games corresponds to what sociologists would call the shared normative order that constitutes us as members of a society rather than as animals in a state of nature.

There are of course also ways to build norms into classical game theory, but with notable differences. First, the "mechanism" by which norms work

---

[71] I say "may" because the degree of entanglement can vary, in principle down to zero.
[72] See Chen and Hogg (2006: 52); cf. Rovane (2004). Note that this is not the same thing as a "correlated equilibrium" in classical game theory; see Arfi (2007: 795) and Brandenburger (2010).
[73] Arfi (2007: 795).   [74] Ibid: 795.
[75] See Hanauske et al. (2010).   [76] Ibid: 5099–5100.

is different: instead of an external constraint that changes the cost-benefit calculus for independent individuals, in quantum game theory norms are not only internal but also connect individuals in a non-local, holistic way, making them no longer fully separable. Second, instead of an ad hoc addition bolted onto classical game theoretic models, norms as entanglement are an integral part of quantum game theory. And third, as a result of both of these differences the formalization will differ as well.

There is also an important consequence for "rational" strategic choice: in general quantum game theory predicts more cooperation in a given type of game than its classical counterpart, and as such players can achieve more efficient or Pareto-optimal outcomes. In the classical one-shot Prisoner's Dilemma, for example, players should always defect. In quantum PD the formal expectation varies with the level of entanglement assumed, but in experiments behavioral game theorists have found that people cooperate in one-shot PD "about half the time," which is consistent with a relatively high degree of entanglement.[77] This suggests that, even though these experiments are ostensibly designed to assess behavior in classical games, what subjects are *actually* playing are quantum games, if, as members of society, they bring a shared normative background with them into the lab. Indeed, from this perspective it would seem difficult to assess human behavior in truly classical games at all, since test subjects are presumably members of society and thus inherently entangled!

Or perhaps we should say that classical games are the result of playing quantum ones, which I take to be one lesson of Barad's intriguing concept of "intra-action."[78] As is well known, classical game theory is about *inter*-action. Although social scientists tend to take the idea of interaction for granted,[79] it only makes sense if agents are fully separable; in that case, social relationships will indeed be properly seen as exchanges between pre-existing independent actors. Drawing on her quantum reading of Butler's theory of performativity, Barad first argues that human beings only become who they are through the collapse of our wave functions into well-defined states, which happens as a result of continuous measurements on and by our environments. She then points out that as quantum systems we are entangled with the social world and thus not fully separable from each other. This vitiates the premise of "inter"-action and by the same token motivates the neologism of intra-action, since who we become through measurements on each other is internal to our shared relationships – our entanglement – rather than something that happens outside

---

[77] See Camerer (2003: 46), and for further discussion of results in quantum PD and other games see Chen and Hogg (2006) and Guo et al. (2008).
[78] See Barad (2003; 2007 passim).
[79] An exception is the "transactional" approach developed by John Dewey and Arthur Bentley in their (1949), which similarly conceives of action as something that spans more than one actor; see Khalil (2003) for a useful overview.

them. However, the effect of those measurements is, if only for a moment, to constitute us as independent agents. In short, "intra-actions enact *agential separability*."[80] Barad does not mention quantum game theory, but her argument suggests that the effect of playing a quantum game is to create the separability requirement of a classical game – even though that game can never actually be played.

I have called quantum game theory the "next frontier" in quantum cognition because much more work needs to be done relating it both to the empirical findings of behavioral game theory and to rival hypotheses put forward by classical game theorists. But by way of conclusion I want to point out that the picture of human sociability that emerges from quantum game theory illustrates in microcosm a larger point about the nature of reality itself. As Frederick Zaman puts it,

[classical] forces are fundamentally non-cooperative because they are blind and mechanistic, and everything that happens ... occurs through the external imposition of forces that are unwilled and without purpose. [Quantum] forces, on the other hand, are potentially and often truly cooperative because everything that happens occurs through the mutual dissemination of information amongst the forces involved.[81]

The "cooperative" nature of quantum processes is due to entanglement, which gives reality a holistic dimension that is completely foreign to the atomistic classical worldview.[82] As long as social scientists are captured by the latter we will expect competition and conflict to be the default human condition – and continue to be surprised when real people confound our pessimism. In contrast, were social scientists to embrace the quantum worldview, then "instead of seeing human beings as separate elements in causal interaction, we ought to see them too as correlated projections of a common ground."[83] From such a standpoint the human proclivity to cooperate is not an irrational mistake or anomaly, but precisely what we should expect.[84]

---

[80] See Barad (2007: 140); emphasis in the original.
[81] Zaman (2002: 368); also see Brandt (1973: 67).
[82] And indeed, over the years this has frequently led physicists to invoke *social* metaphors to interpret their findings; see Kojevnikov (1999).
[83] Pylkkänen (2007: 145).   [84] See Martinez-Martinez (2014: 43–44).

9 Agency and quantum will

Quantum decision theory is a theory of choice behavior, and as such deals not just with Cognition but also with what I have separated out as Will. However, its advocates have concentrated on demonstrating that quantum preferences and beliefs, conceived as a kind of input, can predict observed behavioral outputs. As a result, they have not thematized what is going on in-between, the mechanism (sic) by which the quantum mind makes choices in the first place. In this way the theory, like its classical forerunner, selects a "basis vector" for thinking about people that is more on the terrain of Cognition than of Will. That makes sense as a first step, given that if the theory can't predict behavior then the rest is moot, but to get a more rounded quantum person we need to "rotate" to the basis of Will.

Will is the essence of agency, a power to animate and move the body – and the mind, in the form of attention[1] – from the essentially passive stance of Cognition to active, purposeful engagement with the world. In Chapter 6, I equated this power with an aspect of wave function collapse, viewed as a process of temporal symmetry-breaking, in which advanced action moves through Will and retarded action through Experience. (Note that this means Will is not straightforwardly conscious, to which I return below.) If that is right, then Will and Experience are complementary in the quantum sense – incompatible, yet jointly necessary for a complete description of the collapse process – and so it will only be after reading the next chapter that the meaning of this one will be fully realized.

If Will is an aspect of wave function collapse – one of the most incomprehensible features of quantum mechanics – then that may help explain why, compared to Cognition and even Experience, its standing among philosophers seems more tentative. After nineteenth-century high points in Schopenhauer and Nietzsche, the concept fell into disrepute by the mid-twentieth, with Gilbert Ryle famously denying the existence of Will as another metaphysical "ghost

---

[1] See Vermersch (2004) and Stazicker (2011) on the role of attention in psychology, and Stapp (1999) for a specifically quantum perspective.

in the machine."² Since then philosophical interest in Will – now more often termed 'volition' – has recovered,³ though doubts remain about what it refers to and whether the concept is needed at all. In particular, the recent emergence in the philosophy of action of the concept of 'intention,' understood as a commitment to act, seems to capture some of the sense of 'volition' as well, suggesting that the latter might be reducible to the former. Others disagree,⁴ to which I will add a quantum point. No matter how firm, an intention remains a superposed state of mind, and as such does no causal work until the body is set in motion by *acting* (willing) on the intention. So, while related, the two concepts do not refer to the same thing. And whereas in the classical worldview the force of Will may appear mysterious and/or eliminable, from a quantum perspective it has a natural place in wave function collapse.

The contemporary literature pertaining to volition/Will addresses a number of issues, of which I will discuss just two of particular importance to social ontology. One is whether will is free, which has long captured both the philosophical and popular imagination. However, in focusing on the "free part," much of this work seems to take the "will part" for granted, as if the nature of willing were unproblematic.⁵ Since that is by no means clear and bears on the question of freedom, I will start with the second question, of what kind of force will actually is. In short, what is agency?

Reasons, teleology, and advanced action

How do we get from states of mind to actions in the world? In philosophy this is known as the problem of mental causation, which in a nutshell is how to reconcile our apparent ability to use our minds to move our bodies with the causal closure of physics.⁶ Within the classical frame that he takes as given, Michael Esfeld argues that there are two basic ways to try to solve this problem.⁷ On the one hand, we can assume that mental states are irreducible to brain states (in an epistemic sense anyway, as in supervenience), and then argue that behavior is "systematically over-determined" by mental and physical causes, with each contributing to the result. But this is problematic because according to the CCP every physical event has a sufficient physical cause, so what additional explanatory work is left for mental causes to do? Alternatively, we can assume that mental states are literally identical to brain states, and then argue that mental and physical discourses are simply epistemically different ways of describing the same (physical) process. But why then do we need the

---

² See Ryle (1949). ³ See Burns (1999) and Zhu (2004a) for good overviews.
⁴ See Zhu (2004b) for a sense of the debate.
⁵ An important exception is Robert Kane's (1996) work, which defines will in terms of "trying."
⁶ For a comprehensive introduction to the literature see Robb and Heil (2014).
⁷ See Esfeld (2007: 207–208).

mental discourse at all? In different ways both approaches seem to make our experience of mental causation epiphenomenal.

The mental causation literature has come a long way since 1963, but in the philosophy of the social sciences the dominant solution to the problem is still that of Donald Davidson, who in that year published a classic article arguing that "reasons are causes."[8] Importantly, although Davidson does not distinguish explicitly between the two approaches above, he understands 'cause' in the physical sense as efficient causation, making his a variant of the second approach above. Davidson was challenging the then widely held Wittgensteinian view that reasons do not cause actions but *constitute* them, by giving meaning to what is otherwise mere behavior. The paradigmatic case is the difference between a wink and a twitch, where having an appropriate reason is what constitutes the one out of the other. Davidson did not deny that actions are constituted by reasons, but pointed out that this does not explain why people act in the first place. And when we look in the structure of reason-giving for why John did Y, the answer is *because* he had reason X.[9] Since according to Davidson this relationship satisfies the criteria for a causal explanation he concluded that reasons must indeed be (efficient) causes. His argument was highly influential in the social sciences, where it helped consolidate the emerging positivist view that social science is not essentially different from physical science.[10] Even though social scientists' reliance on unobservable mental states and folk psychology might make our work seem less than scientific, through Davidson we can claim to be making causal explanations too.

Yet despite its long dominance, the idea that reasons are causes remains controversial. Interpretivists still argue that making sense of action *is* different than explaining the behavior of objects.[11] Others more in the philosophy of social science mainstream have challenged Davidson's framing and/or analysis of the issue, with various implications for his conclusion.[12] And then, of particular interest to me, there are some who agree that reasons are causes, but not the efficient causes assumed by Davidson, but *final* causes.[13] By emphasizing the irreducibly purposive nature of action these "New Teleologists" in my view best capture the causal force of Will, and I propose to build on their ideas here. However, whereas they see their approach as justifying anti-naturalism in the social sciences, I will give it a physical, naturalistic basis through quantum theory.

---

[8] Davidson (1963); for an excellent history of the debate see D'Oro and Sandis (2013).
[9] Davidson (1963: 691).
[10] See Gantt and Williams (2014) on the influence of Newtonian thinking on the debate about reasons as causes in psychology.
[11] See for example Schroeder (2001) and Brinkmann (2006).
[12] See Tanney (1995) and Risjord (2004).
[13] See Stout (1996), Schueler (2003), Sehon (2005), and Portmore (2011), and for an earlier defense of teleological explanation in social science see von Wright (1971).

Let me first unpack Davidson's conclusion in relation to an Aristotelian view of causation, which is more pluralistic than what is generally assumed today. Aristotle held that there are four kinds of causes in the world: formal, material, efficient and final.[14] Formal causation refers to the way in which the structure of an object or process gives it form and identity (much as reasons do on the constitutive account), whereas material causation refers to the sense in which an entity or process exists by virtue of having a certain composition. While relevant to my discussion later, formal and material causes do not describe changes of state and as such cannot explain how or why the body moves; they account for Being, not for Becoming.[15]

Aristotle's other two causes are all about change. Efficient causation refers to a local transmission of energy or force from X propagating forward in time to Y, which has the effect of changing Y's properties or behavior. The classical worldview reduces all causation to this type, and it is how most philosophers and scientists today instinctively think about "cause." Although calling it 'mechanical' does not do justice to its possible forms,[16] efficient causation clearly drives the machine model of man, as well as the search for "causal mechanisms" that has gripped wide areas of social science. In contrast, final or teleological causation, the most distinctive element of Aristotle's scheme, is anathema to advocates of the classical worldview and widely dismissed today as unscientific.[17] It refers to the way in which the ends or purposes of a system, which refer to its future, relate to its behavior or development in the present.[18] So while efficient causation explains in a temporally forward manner, final causation "explains backward."[19] Making that suggestion intelligible is the hard part for advocates of final causation, since they agree that the future does not cause the past in anything like the way that the past causes the future. But if not literal backwards causation, then what precisely is the causal role of finality, such that it cannot be reduced to efficient causation?

Before addressing that question, however, it should be noted that there is also a hard question for critics of final causation, which is that purposiveness – in the form of organisms' goal-directed behavior – seems to be all around. Thus, despite the defeat of vitalism, which one might think would have also vanquished teleology, philosophers of biology continue to struggle to explain organismic purposiveness in efficient causal terms. A particularly

---

[14] See Hennig (2009) for an extended discussion of Aristotle's four causes.
[15] See Short (2007: 136).
[16] See ibid: 94–98 on the evolving meaning of 'mechanical' and its relationship to teleology.
[17] Cf. Nagel's (2012) defense of teleology in the natural world, for an open-minded review of which that engages the physics of final causation see Bishop (2013).
[18] See Gotthelf (1987) on Aristotle's approach to final causation. Final causation was an important part of Leibniz's metaphysics as well; see Carlin (2006).
[19] See Jenkins and Nolan (2008).

influential analysis is due to Ernst Mayr, who argued that what appear to be teleological processes in organisms are really "teleonomic," a benign kind of end-directedness that is reducible to efficient causation.[20] However, many are not convinced that such reductions are possible, and thus continue to defend the earlier, teleological view of organisms as "natural purposes," so the biological debate goes on.[21]

Within the philosophy of action and social science the debate about teleological reasoning seems even more intractable, since intuitions there are stronger in its favor (at least at the individual level). While we have no inside experience of animal behavior and thus might plausibly argue that its causes are purely mechanical, we do have experience of the causes of human behavior (reasons) which certainly do not *feel* mechanical. Indeed, probably most social scientists today would agree that human behavior is fundamentally goal-directed, and that "intentional explanations" are not only valid but practically speaking distinct from causal explanations.[22] Despite this apparent acceptance of purposiveness in human behavior, however, most naturalists in social science would probably also disavow any connection to final causation. That is because naturalists take for granted that materialism is true, which means that ultimately, reasons must be efficient causes just like everything else.[23] Here they can follow Davidson, who notes that reason explanations often take a teleological form, but then argues that such accounts are surrogates for causes in the standard physical sense.[24]

The New Teleologists in the philosophy of action argue that such reductions are impossible to make, and that teleological explanations of human action are therefore epistemically necessary. But surprisingly, when it comes to ontology they too assume materialism is true (or at least I have not seen them advocate anything else). They simply draw the opposite conclusion from naturalists, which is that the irreducibility of teleological explanations vindicates the autonomy of the human sciences from the physical ones. In my view, this conclusion is problematic and premature – problematic because it seems inconsistent with the CCP; and premature because the New Teleologists have not framed their approach against a quantum backdrop, only a classical one.

---

[20] See Mayr (1982), and Perlman (2004) for a good overview of contemporary efforts to make sense of teleological reasoning without embracing Aristotelian final causality.

[21] See Weber and Varela (2002), Short (2007), Griffiths (2009), and Toepfer (2012).

[22] On intentional explanations see Elster (1983), Dennett (1987), and Searle (1991).

[23] See Rosenblueth, Wiener, and Bigelow (1943) for an early reductive attempt, and Jenkins and Nolan (2008) for an argument that any such effort is doomed to fail.

[24] See Davidson (1963). Despite the appearance since of sophisticated teleological approaches to reasons, scholarship in a Davidsonian vein still tends to take the standard view of causation for granted, and thus sees no threat to naturalism from goal-directed behavior; see for example Mantzavinos (2012).

Agency and quantum will 179

As this brief review suggests, the debate about reasons as causes has been conducted almost wholly in folk psychological terms, without explicit reference to how reasons are instantiated in the brain.[25] That makes sense, given how little we know about the brain, yet in characterizing this "black box" everyone has assumed that it is classical, which makes it hard to imagine reasons as anything other than efficient causes. But what if quantum theory provides a physical basis for final causation?[26] Assuming that quantum brain theory is true, that would change the discussion entirely, by allowing us to have it both ways – naturalism with teleology.[27]

Although I have already identified will with an aspect of wave function collapse, in developing this further it should be noted that most interpretations of quantum theory have no more truck with teleology than does the classical worldview – so it is not obvious the theory does provide a physical basis for final causation.[28] However, there is one interpretive strain in which final causation does play an at least implicit role, namely the time-symmetric approach that includes Atmanspacher and Primas' neutral monism (see Chapter 6), John Cramer's Transactional Interpretation, and others.[29] Of particular interest in this tradition is the concept of "advanced action."[30] Although time-symmetric quantum theorists do not always link this to final causation – some actually seem more comfortable with the idea of "retro-" or "backwards" causation[31] – I will argue that advanced action has everything one could want from a teleological force and thus must be what Aristotle "had in mind."

Recall the starting point for time-symmetric interpretations of quantum theory: that all fundamental physical principles are symmetric under time-reversal, meaning that in closed systems their equations can be solved forwards or backwards. However, the backwards solutions are normally discarded because they are not thought to have any physical meaning. This leaves us with a temporal asymmetry that, on the one hand, has the virtue of corresponding to our experience of time moving only forward, but which, on the other, cannot explain

---

[25] Though see Gustafson (2007), who emphasizes the irony that whereas causalists about reasons presented themselves as defenders of naturalism, recent neuroscientific findings reviewed below seem to call causalism into question, and suggest that the constitutivists were right after all.

[26] Whether the language of 'causation' is appropriate at all in quantum contexts is an issue in itself (see Price and Corry, eds. 2007), but I will set aside this larger question here and focus just on final causation.

[27] See especially Barham (2008; 2012).

[28] Though see Bohm (1980: 12–15), Primas (1992: 27–29), Costa de Beauregard (2000), Helrich (2007), and Castagnoli (2010: 313), who argues that "[q]uantum algorithms, being partly driven by their future outcome, provide well formalized examples of teleological evolution." Within biology several eminent physicists have wrestled with the possibilities for grounding teleology in a quantum view of the organism; see Sloan (2012).

[29] See Cramer (1986; 1988), Price (1996), Atmanspacher (2003), and Primas (2003; 2007).

[30] For a comprehensive discussion see Price (1996: 231–260).

[31] See for example Price (1996; 2012), Dowe (1997), and Berkovitz (2008).

that experience and is aesthetically unappealing. Yet, while most of quantum physics is time-symmetric, this is not the case in wave function collapse. There temporal symmetry breaks down, such that we cannot work the equations in either direction. In effect, it is in collapse that "Now" emerges, and with it the distinction between past and future. Advocates see this temporal symmetry-breaking as the key to interpreting quantum theory.

Cramer's approach to wave function collapse is particularly helpful here. In his view, quantum events may be described as a "transaction" between two state vectors or waves, a retarded wave proceeding in the normal, forward direction, and an advanced wave proceeding backwards from the measuring apparatus or "destiny state." These waves are not symmetric but they are correlated – and not just in the obvious sense that one might expect the future to be correlated with the past, but also in the sense that what will become the future enforces correlations with the past.[32] Measurement induces a "two-way contract between future and the past," such that the collapse is not complete until the future "confirms" the past. Note that because this confirmation process only involves enforcing correlations it cannot be used to transmit information to the past, and so does not conflict with causation in the efficient sense. Rather, what we are dealing with here is a temporal non-locality like that observed in Wheeler's Delayed-Choice Experiments, which Cramer argues we can see every night:

> When we stand in the dark and look at a star 100 light-years away, not only have the retarded light waves from the star been traveling for 100 years to reach our eyes, but the advanced waves generated by absorption processes within our eyes have reached 100 years into the past, completing the transaction that permitted the star to shine in our direction.[33]

In this temporally non-local sense we may say that "the present is literally created out of influences both from the past *and from the future*."[34]

Time-symmetric arguments like these were originally advanced to explain the behavior of sub-atomic particles, and later contributions have been mostly on the terrain of fundamental physics as well.[35] In my view, this may be one reason why they have remained marginal in the debate over quantum theory. Even if the concept of advanced action restores an aesthetically pleasing temporal symmetry to our picture of the world, in physics it is hard to see concretely what it could refer to, which is why it is usually discarded as meaningless. The difficulty here may stem in part from the lingering materialism that most physicists

---

[32] See Cramer (1988), Price (1996: 242–243), and Castagnoli (2009; 2010); and Kastner (2008) for a good overview of this literature.
[33] Cramer (1988: 229).
[34] See Tollaksen (1996: 559); emphasis in the original. On temporal non-locality see Price (1996) and Filk (2013).
[35] Tollaksen (1996) and Wolf (1998) are exceptions that I draw upon below.

bring to the table, and especially the belief that matter at the fundamental level is dead. Despite the fact that after quantum theory old-fashioned materialism is no longer an option, in the interpretive debate most philosophers seem more willing to accept the utterly bizarre, but at least materialist, conclusions of the Many Worlds Interpretation than to admit that mind or purpose is a brute fact of nature. With that possibility closed off by fiat, whatever advanced action might be final causation can't be it, leaving us with ideas like retro- and backwards causation that are even more counter-intuitive.

Consider then how differently things look if we approach advanced action instead from the standpoint of a quantum biology and specifically psychology, and thus on the assumption that human beings have direct experience of quantum effects like advanced action. Psychology deals with organisms whose behavior feels as much pulled by a future they intend as it is pushed by a past they inherit. In everyday life we usually explain such behavior in terms of actors' reasons, and the causal force of those reasons is typically narrated in teleological terms. Materialists assume that despite this appearance of final causation, what is really going on here just efficient causation. However, they have produced little evidence for this view, either from the brain or philosophically, and if people are actually walking wave functions then it is clearly problematic. For one thing that is not debated in quantum theory is that, when it comes to wave function collapse, nothing like efficient causation is involved – which is precisely why collapse is so hard to explain. This suggests the following syllogism: if, (1) final causes cannot be eliminated from folk psychological accounts of human behavior, (2) human beings are walking wave functions with direct experience of advanced action, and (3) advanced action has a peculiar backward working quality that seems counter-intuitive in physics, then (4) advanced action in physics just *is* final causation in people. In short, reasons are causes, but quantum final causes rather than Davidson's efficient ones.

What then can we conclude about the opening question of this section, of how human beings get from states of mind to behavior? As I see it, quantum will plays two key roles in agency. First, Will makes decisions, understood in quantum terms as the reduction of indeterminate reasons to determinate choices. As such, the actual reason for why an agent does X does not exist before doing X, but emerges with the latter. Note that Will is important in this sense even when all of an agent's reasons are compatible in the quantum sense and thus choice reduces to a classical problem, since an act of will is still necessary to collapse her wave function. In bridging the gap between our superposed insides and the classical world outside,[36] decisions create for a moment order out of chaos, and combined with the experience with which it

---

[36] On "gaps" in practical reasoning to which different aspects of Will respond see Searle (2001: 14–15) and Zhu (2004b: 177–180).

is correlated, therefore meaning for the subject. This may be seen as a form of "downward causation," since decisions are made at the level of the whole, to which the body's parts respond. But it is downward causation "without foundations," since, as a superposition, the state from which decisions emerge is a potentiality rather than actuality.[37]

Second, Will controls the direction of the body's movement over time by harnessing temporal non-locality, potentially over long "distances." As advanced action, Will projects itself into what will become the future and creates a destiny state there that, through the enforcement of correlations with what will become the past, steers us purposefully toward that end.[38] Note that this projection is not necessarily conscious; what we experience in consciousness is retarded action not advanced, with time moving forward in the usual way. However, will can become conscious, in the form of an explicit intention, in organisms that can project themselves self-consciously into alternative futures, which is essential to planning behavior and known as "mental time travel."[39] Although the literature treats this idea only metaphorically, in the next chapter I suggest it be taken literally, but in a non-local sense. By maintaining a conscious intention through repeated acts of will, agents enforce correlations backwards on their behavior over time, which gives it coherence not just in the sense that it is consistent with the past, but in the teleological sense that the future gives meaning to and completes the past. In human beings such correlations can reach very deeply into "the future," and be shared by many individuals as well, which will make temporal non-locality an important theme below.

Free will and quantum theory

But is the power of Will "free"? Free will is another "hard problem" that seems impervious to resolution, and the literature is correspondingly enormous, about the only thing common to which is a classical frame of reference.[40] Even its definition is contested, contaminated by debatable assumptions and/or the conclusions that protagonists want to reach. So let me begin with some widely accepted common sense: your will is free if in doing something you in principle could have done otherwise. Precisely how to unpack this suggestion is what sets philosophers to work, but most at least agree that the freedom here should be understood in an existential rather than practical sense.[41] In the latter, "could have done otherwise" implies an absence of external constraints on one's scope

---

[37] See Bitbol (2012).   [38] See Tollaksen (1996: 562).
[39] On mental time travel see Suddendorf and Corballis (2007).
[40] Kane, ed. (2011) provides an excellent and comprehensive introduction. While quantum arguments for free will sometimes now receive their own chapter (e.g. Hodgson, 2011), they are generally ignored in the literature as a whole.
[41] See O'Connor (2014: 1–2).

for action, which does not speak to free will per se. A prisoner has no freedom of action, but still (perhaps) the existential freedom to face his fate as he chooses.

This distinction matters for social science, where 'agency' is often defined in terms of how much control or power actors have over events or structures, which is clearly about freedom of action.[42] Reams have been written on agency in this sense, but as for free will, while psychologists have shown some interest[43] among social scientists there is surprisingly little[44] – even though many of us talk about agency as if it presupposes free will. The disconnect for some may be that if the goal of social science is to find law-like generalizations, then a phenomenon like free will that does not vary and cannot be measured will inevitably fall in the "error term."[45] So insofar as the job of the social scientist is to reduce her error terms as much as possible, the attitude seems to be – best leave free will to philosophers. Yet the question is of obvious importance in real life, given that modern law and morality assume that people actually have free will in the existential sense.[46] And of obvious importance for social theorists too, since agency without free will hardly seems like agency at all.

The philosophical literature

So what precisely is the problem? Well, not everyone agrees that there *is* a problem, but intuitively it is this: we mostly experience our actions as freely willed, as if we could have done otherwise, but it is not clear how to reconcile this feeling with the ontology of modern science. Of course it would help if science could explain *any* experience, but whereas consciousness overall poses a problem for the orthodoxy's materialism, the specific experience of free will challenges its (ontological) determinism, the assumption that – at the macro-level of the brain at least – every event has an efficient physical cause.

When I entered into this literature I expected the dominant view to be that free will is inconsistent with determinism, and so our experience of free will is an illusion. That is indeed the prevailing discourse of neuroscientists (see below), but to my surprise it is much more contested by philosophers. Many, perhaps the majority, of the latter are "compatibilists," who, while conceding that the brain operates deterministically, argue that the details of brain function are simply not relevant to whatever, practically speaking, we might want a concept of free will for – like justifying the idea of moral or legal responsibility for

---

[42] See Ortner (2001) for a good discussion of agency as power, which she distinguishes from agency as intention.
[43] See, for example, Sappington (1990), Baer et al. (2008), and Bertelsen (2011).
[44] For exceptions see Dray (1960) and de Uriarte (1990).
[45] Though see Sappington (1990) for a more subtle analysis.
[46] See Habermas (2007) (with rejoinder by John Searle), Bertelsen (2011), and Hodgson (2012) for some recent, particularly thoughtful discussions of this issue.

our actions. In other words, for compatibilists there is no real problem of free will, since it is "compatible" with determinism.[47] "Incompatibilists" disagree, claiming that this sets the bar too low for a naturalistic theory of free will, and that the details of brain function are crucial to the issue.[48] Since quantum brain theory challenges the shared premise of this debate – that brain function is deterministic – compatibilism in particular seems moot in the present context, and so I will set it aside in the remaining discussion. But that hardly settles the issue in favor of incompatibilism.

By definition there are two ways to be an incompatibilist: reject free will or reject determinism. Most incompatibilists do the latter,[49] such that 'incompatibilism' is often equated with anti-determinism. Since that can be confusing I will use 'libertarianism,' a synonym for anti-determinist incompatibilism, instead. The key challenge for libertarians is again a question of "compatibilism," but now of free will with *in*determinism. Namely, free will implies the ability to control one's actions, whereas indeterminism implies randomness – which seems more like a basis for madness than willing. In response to this problem, libertarians have focused on making sense of a kind of causation – which is what free will must be if it is to move the body – that is not deterministic, such as "agent-causation" or "event-causation."[50] Some have invoked quantum theory to justify these ideas (though to my knowledge never quantum brain theory), but, with a few exceptions, only in passing.[51] Like most philosophical work on free will, libertarian discourse remains on a folk psychological level.

In contrast, quantum theorists have from the start wondered about the relevance of their work to the free will problem.[52] In recent years their discussion has become more systematic and even produced a formal result, the "Free Will Theorem" – though the name is a bit misleading.[53] Rather than proving that human beings have free will, the theorem proves that *if* human beings have free will in some degree, then sub-atomic particles must have it in the same degree as well. That result scores a point for panpsychism, and also for my larger argument that what is going on at the sub-atomic level and the human level are

---

[47] Frankfurt (1971) is a classic compatibilist treatment of free will; for a quantum critique of Frankfurt see Georgiev (2013). Note that this literature is highly anthropocentric and thus it is hard to see its arguments going far down the evolutionary ladder.
[48] Kane (1996) is perhaps the most prominent incompatibilist today, with O'Connor (2000) another important contributor. See Warfield (2003) for a good overview of the debate.
[49] Though see Honderich (1988) and Balaguer (2009) for a critique of his approach.
[50] See Clarke and Capes (2014) for an excellent survey of these accounts.
[51] The main exceptions of which I am aware are Kane (1996) and Hodgson (2012); cf. O'Connor (2000), Balaguer (2004: 402–403).
[52] For early treatments see Eddington (1928), Compton (1935), and Margenau (1967); and most recently Suarez and Adams, eds. (2013).
[53] Conway and Kochen (2006); for non-technical introductions see Valenza (2008) and Hodgson (2012: 121–128).

connected, but it does not prove the reality of free will. That inconclusiveness captures my sense of the philosophical debate among physicists overall, which is that some think quantum mechanics does not help with the free will problem, and others think it does.[54] While disappointing, this is not surprising given how contested the interpretation of quantum theory is; even associating it with something as mystical-sounding as "Will" at all would be problematic to some. So rather than summarize the physicists' debate, let me turn to neuroscience, where an important experimental finding, due to Benjamin Libet, is widely taken to disprove the existence of free will, period. I then take up a quantum re-interpretation of the Libet research that supports the existence of free will and as a plus, uses the idea of advanced action to make its case.

The Libet experiments

Libet's experiments, dating from the 1970s and '80s, produced among the most well-known neuroscientific results of recent decades. Yet, while the results themselves are broadly accepted, and an interpretive orthodoxy has developed, their meaning has been contested from the start.[55] Libet sought to pin down the timing of the relationship between conscious awareness, on the one hand, and electrical activity in the brain called the "readiness potential" (RP), which had been previously observed and known to correlate with awareness. He looked at the question both for voluntary actions, like lifting a finger, and for involuntary experiences like a pinprick on the skin. The former have garnered most of the attention because they seem to bear more on the issue of free will, but the latter also yielded a striking result that I will come back to below.

For the voluntary case, Libet asked his subjects to lift their finger whenever they chose, and then measured three quantities with very precise clocks: when the RP built up, when the subjects reported first being aware of their intention to lift a finger, and when their finger actually moved. What he found is that the RP begins to build up 350–400 milliseconds *before* awareness of the intention, and fully 550 ms before the act itself. At least among neuroscientists and biologists,[56] this is widely seen as evidence against the reality of free will because consciousness only enters the picture after most of the action is over, making it epiphenomenal. Libet offers a more nuanced interpretation, arguing

---

[54] For quantum skepticism about free will see Loewer (1996) and Esfeld (2000), and on the other side, see Ho (1996), Pestana (2001), and Hodgson (2012).

[55] See Libet (1985) for an early statement of his work, with peer commentary, and Sinnott-Armstrong and Nadel, eds. (2011) for the contemporary state of the art.

[56] Biologist Anthony Cashmore probably speaks for many in thinking that belief in free will is "nothing less than a continuing belief in vitalism" (quoted by Walter, 2014a: 2216). Philosophers are more divided about whether Libet's work speaks to the free will problem at all, given how he operationalizes the concept; see Schlosser (2014) for a diagnosis of these contrasting reactions.

that, while free will per se does not exist, the gap between awareness and act enables consciousness to "veto" action, giving us "free won't."[57] Skeptics have in turn countered that a conscious intention to veto would itself have to be preceded by an RP, starting an infinite regress that fails to save free will. Thus, scholars today seem increasingly inclined toward Daniel Wegner's important argument, using Libet's results and others, that the experience of free will is ultimately an illusion.[58]

Still, the deterministic interpretation of the Libet results has been challenged in at least three ways. One questions the implicit assumption that to count as free, decisions must be initiated by (and thus temporally co-extensive with) consciousness. The idea here is that ultimately free will is not about consciousness but about control, and so we should allow for a "pre-conscious free will," which would render the temporal discrepancy moot.[59] A second response is that the orthodoxy neglects the role of the experimental context in producing the Libet results, in particular the prior conscious choice by each subject to be tested in the first place, and then to follow experimental instructions to pay attention to bodily signals that are normally sub-conscious.[60] Finally, no one really knows what the RP is for, or how it causes behavior. It might, for example, only serve to bring about the occurrence of a choice rather than the choice itself, leaving space for free will.[61]

These responses would gain further traction from a quantum perspective, but the idea has almost never come up in the Libet debate, where it has simply been assumed that the brain is classical and therefore deterministic. Outside the literature, however, I am aware of four critiques of this assumption, three by physicists.[62] Roger Penrose thinks the results point to inadequacies in the classical conception of time, and hints that they might be explainable through the kind of retro-causation or advanced action that is permitted by quantum theory, though he does not develop the suggestion at length.[63] Henry Stapp draws a connection between the Libet results and the EPR paper in quantum theory, which first put the issue of non-locality on the table.[64] However, while he concludes in favor of free will, most of his article deals with other issues,

---

[57] See Libet (2004) for the book-length version of his research and views.
[58] See Wegner (2002). For a broad ranging early discussion see Wegner et al. (2004), and for skeptical treatments of "willusionism" see van Duijn and Bem (2005), McClure (2011), and Walter (2014a). Ironically, Libet (2004: 152–156) himself rejects Wegner's interpretation of his work.
[59] See Velmans (2003), Levy (2005), and Pacherie (2014), and Sheperd (2013) on the contemporary state of play on this issue; on unconscious goal-directed behavior more generally see Dijksterhuis and Aarts (2010).
[60] See Zhu (2003) and Schlosser (2014: 258), and for a further critique focusing on issues of experimental design see Radder and Meynen (2012).
[61] See Balaguer (2009; 2010).
[62] None to my knowledge has ever elicited a response from within the Libet community.
[63] See Penrose (1994: 383–390).   [64] See Stapp (2006).

and the Libet discussion is not extensive. Stuart Hameroff brings his quantum brain theory to bear on the issue, on the basis of which he concludes that "temporal non-locality and backward time referral of quantum information [advanced action] can provide real-time conscious control of voluntary action."[65] Hameroff's article makes a useful connection to Delayed-Choice Experiments (Chapter 2), but most of it is devoted to summarizing his other work and so the discussion of free will per se is again relatively short. So let me rely here on the most systematic quantum response to Libet, by Fred Alan Wolf, which also highlights temporal non-locality.[66] Wolf deals with the other aspect of Libet's research I mentioned above, on sensory awareness rather than free will, but he proposes a general quantum model of mind that I believe could be applied to the voluntary case as well.

Libet found that, upon stimulation with a pinprick it takes about 500 ms for an RP to build up that is of sufficient strength for conscious awareness of the stimulus. Yet, paradoxically – and as we might expect from our own experience – his subjects reported awareness of the prick almost immediately.[67] Libet explains this temporal discrepancy by arguing that people subjectively "antedate" the timing of a sensory event, even though they only actually experience it later. He likens this process of "backward referral" to the "spatial referral" that happens in our perception of external objects, which we experience as being "out there" even though what we are actually seeing (in his view) is light on the back of our retinas. The analogy is suggestive (see below), but Libet does not propose a neural mechanism through which antedating might occur, and the idea has been anathema to critics, who dismiss it as physically impossible.[68] Libet held his ground, however, and at least to some the question is still alive.[69]

Wolf shows that Libet's otherwise anomalous finding of subjective antedating can be explained by Cramer's Transactional Interpretation of quantum theory, and specifically by advanced action and temporal non-locality. Since I just discussed these I will not repeat Wolf's treatment here, but note three specific contributions of it instead. First, Wolf adds an evolutionary consideration to the issue, arguing that being able to feel out what will become future experiences would have obvious survival value over waiting a full half-second before becoming conscious of a dangerous stimulus.[70] Second, he offers a suggestive image of the brain as a giant "delayed choice machine," which I will pick up in the next chapter. Finally, and most importantly here, Wolf shows that a time-symmetric quantum perspective can account for a key anomaly in the Libet results – subjective antedating – and, not just account for but *predict* it, much

---

[65] See Hameroff (2012a: 14).   [66] Wolf (1998); see also Wolf (1989).
[67] See Libet (2004), passim, and Chiereghin (2011).
[68] See for example Churchland (1981) and Pockett (2002).
[69] See Libet (2004) and Chiereghin (2011).   [70] On this point also see Castagnoli (2009; 2010).

like quantum decision theory predicts anomalies of irrational behavior that we observe. This stands in sharp contrast to the orthodox, classical approach, which is so puzzled by subjective antedating that it thinks the phenomenon must be an illusion.[71] Like the supposed illusion of consciousness itself, the appeal of such an argument depends heavily on there being no possible physical explanation that saves the appearances – since otherwise there would be no need to explain them away. Quantum theory offers such an explanation, which therefore needs to be considered before we embrace "willusionism."

All that being said, it seems fair to conclude that a fully developed quantum theory of free will does not yet exist. However, the elements of a theory are clearly there, and could answer the fundamental question for libertarianism of how indeterminacy could be willed. Namely, in wave function collapse, what is indeterminate when viewed from the outside is, on the inside, purposeful determination in the final causal sense.[72]

This suggests a distinct perspective on the conundrum of agency in social science, for which free will seems at once essential yet consigned to the error term. Normally we think of the latter as non-systematic variance in observed behavior, and therefore as not playing a causal role (making free will not just an error term but epiphenomenal as well). The causes of behavior are assumed to lie instead in whatever is *non*-random, both in the external situation and within the agent himself. While such an assumption makes sense in a classical context, it need not be made in a quantum one. Conceived as collapsing wave functions, the choices we make, our actual agen*cy* is not generated by an underlying (efficient) causal order at all, but is precisely a rupture or "perforation" of that order.[73] In this light, the error term in social scientific explanations is less a measure of our outsider's ignorance than a sign of what makes causal explanations of action possible in the first place – the force of free will. Indeed, this is true even when behavior is quite predictable, as in a deeply ingrained habit, since even then agents retain the quantum freedom to frame the choice problem differently.[74] So my point is not that human behavior cannot be made more predictable – more "classical" – through coercion, discipline, or incentives, but rather that no matter how successful such schemes are there is a spontaneous vital force in the human being that fundamentally eludes causal determination.

[71] See especially Wegner (2002).   [72] Cf. Kane (1996: 151) and also Hodgson (2012).
[73] I believe the term 'perforation' is due to Michel Bitbol, but I cannot find the reference.
[74] See Holton (2006) on how even habitual or non-conscious choices can still be free; cf. Balaguer (2009: 2).

10 Non-local experience in time

No model of human beings is complete that does not have room for the experience of *being* human, of what it is like to be you or me. This feeling, consciousness, is such an essential feature of the human condition that a life without it would be hardly worth living at all.[1] Yet as suggested in Chapter 1, for fear of Cartesian dualism the main currents of twentieth-century social theory, both mainstream and critical, have run away from experience, seeking to reduce, displace or otherwise marginalize it in their models of man. Human beings are rendered instead into machines or zombies, both ultimately material systems which are able to think and behave but not to feel – transformed, in short, from subjects into objects.

Recently there has been more pushback against these materialist tendencies, in the form of sustained efforts to "bring the subject back in."[2] Scholars in the continental tradition have drawn on insights from German idealism and phenomenology to critique critical theory, for failing to place the experiencing subject at the center of its ontologies. Feminist theorists have brought experience to the fore by arguing that women's experiences differ from men's. And in the analytic tradition the founding of *Journal of Consciousness Studies* and the work of David Chalmers and many others has given subjective experience a mainstream philosophical standing that it has not had in a long time. If there is still no consensus on what to do positively with experience, the taboo on subjectivity has at least finally been broken. This book seeks to contribute to this subjectivist revival by giving consciousness a quantum basis in wave function collapse, understood as a process of temporal symmetry-breaking driven by will. As such, "part of what it is like to be a thinking human being is to have direct experience of the effects of quantum theory."[3] In Chapter 6 I argued that this is true of not only human beings but all organisms and, for an instant, even sub-atomic particles too; however, here I will consider only the human case.

---

[1] On the value of consciousness see Siewert (1998).
[2] Here in the narrow sense of subjective experience, rather than the broad sense encompassing all three faculties of mind.
[3] See Pylkkänen (2004: 183).

This effort is complicated in two ways. First, experience is inherently particular and as such by definition hard to generalize about, which is what a model of man is supposed to do. However, there are at least two kinds of experience that are universal, the experience of time and the experience of space. By this I do not mean that we all experience time and space in the same way,[4] but that by virtue of sharing the same physics of the body our experience of time and space has a universal aspect. In the case of time, what is universal is that we feel it moving forward, from the past through Now to the future; with space it is that we experience objects as being "out there" in the world rather than on our retinas where the photons land. Since these two kinds of experiences are constitutive of social life at a very deep level, I propose to focus on them as a way of saying something general about Experience. It turns out that both are problematic when viewed classically, and that a quantum approach not only resolves these problems, but helps us see them in a new way, as intrinsically non-local phenomena. In this chapter I take up the experience of time, reserving the experience of space for my discussion of language in Part IV.

A second difficulty in bringing quantum theory to bear on Experience is that the content of human experience is in part constituted by language and in particular by narratives, which give a succession of fleeting Nows meaning and coherence. This centrality of narrative may seem to pose a problem for my effort to thematize experiences per se, since it is precisely because of the constitutive role of language that followers of the linguistic turn think we can dispense with the category of experience altogether.[5] However, acknowledging the role of language does not mean experience is reducible to language. Whether or not it is accepted that all organisms have experiences, it seems clear that at least some non-linguistic animals do; and in the case of human beings, without experience language would have no meaning and thus not really be "language" at all. So there is a gap between temporal narrative and experience, and by focusing on the latter I hope to suggest a new perspective on the former.

More specifically, I argue that – in certain important respects – it is possible to literally change the past. My starting point will be a debate among philosophers about some peculiarities of temporal narrative that, on the surface at least, seem to suggest that the past is indeterminate and can be changed. Most of the parties to this debate favor an epistemological reading of this claim, arguing that it is only our descriptions of the past that can change, not the past itself. I suggest that this tendency in the literature stems from a failure to consider the physics of the past, and especially from an implicit assumption of temporal locality, which is part of the classical CCP. Drawing on the quantum concept of delayed

---

[4] See Gell (1992) on the variability of time consciousness across cultures.
[5] See Grethlein (2010) for a thoughtful discussion of how experience and narrative relate that avoids reducing either one to the other.

choice, I argue that experience is temporally *non*-local, and that this supports a stronger, ontological interpretation of changing the past.

Finally, it should be noted that most of the philosophical literature on changing the past is about collective narratives ("history"), which differ from individual ones in two important respects: (1) we have first-person access to memories of our own pasts that we lack to the impersonal, historical past; and (2) collective narratives of the past are social facts upheld by thousands if not millions of people, which gives them an added source of staying power relative to personal narratives. Nevertheless, the specific issues with which I will be concerned below are present in both cases, and since there is more material to work with on the collective side, I shall draw freely on the latter even as my primary concern here is with personal experience.

The qualitative debate on changing the past

Time is normally thought to have a clear direction, from past to future, and while most social scientists probably consider the future to be open,[6] the past is routinely assumed to be closed, over and done with. Yet at least among philosophers there has been considerable debate in recent years about the fixity of the past – and not just among philosophers of physics, for whom "retro-causality" is just one more weird possibility that quantum mechanics puts on the table, but among philosophers of history. On one level what are at stake in this debate are purely academic questions like the nature of history, causality, and events. However, on a practical level these issues are intimately bound up with questions of identity – of who we are – at both the collective and individual levels, and as such our answers may have far-reaching social and political consequences.

The philosophical debate about changing the past is three-cornered. In one corner stands "realism" about the past, the common-sense view that all statements about the past are either true or false, and so the task of the historian – or for that matter, trial juries – is to determine "what really happened."[7] However, judging from journals like *History and Theory* and *Journal of the Philosophy of History*, realism about the past is not where the philosophical action is these days. The real debate is about how far we can push a "constructivist" view of the past. On this there is a spectrum of opinion clustered into two main camps, which I will call the Epistemological and Ontological views. Since the former is the commonsense and by far majority position, I will focus most of my attention on the intuitions that motivate the latter.

---

[6] It is unclear, however, how this could be true under a classical physics constraint, given its deterministic ontology.
[7] See Roth (2012: 324).

One thing all sides in this debate share is that almost no one has invoked physics. I will remain faithful to that neglect in my review of the literature, but keep in mind that my agenda here is to lay the groundwork for arguing in the next section that the Epistemological view makes an implicit classical assumption of temporal locality (to be defined below). In contrast, the Ontological view I take it is assuming temporal non-locality, for which I intend to provide a quantum rationale.

The Epistemological view

The epistemological view on changing the past is that only our descriptions of the past change, not the past itself. Most of the scholarship in this vein has been in response to the work of two scholars, Arthur Danto and Ian Hacking, although because it is separated by almost twenty years the result has been essentially independent literatures.[8]

In *Analytical Philosophy of History* Danto gives a number of examples of statements that are now true about the past but which could not have been known at the time by even an "ideal chronicler," someone who knew and transcribed everything that happened anywhere at each moment.[9] "The Thirty Years' War began in 1618"; "Aristarchus anticipated in 270 B.C. the theory which Copernicus published in 1543"; "Petrarch opened the Renaissance"; and so on. In each of these instances, which Danto notes are typical of historical inquiry, descriptions of what happened in the past only became true after the fact. Although Danto flirts with an ontological reading of these cases, suggesting that "[t]here is a sense in which we may speak of the Past as changing,"[10] he ultimately takes an epistemological line, arguing that what changes are only our descriptions of the past today, not what really happened at the time.

Ian Hacking's widely discussed examples refer to what he calls the "indeterminacy" of the past.[11] This occurs when new concepts emerge and are then applied retroactively to people who lacked them, like 'child abuse,' 'sexual harassment,' 'PTSD,' and so on. He considers the case of Alexander MacKenzie, a celebrated explorer who in 1802 at the age of forty-eight married a girl of fourteen. Was MacKenzie a child abuser? Hacking's own initial answer was a somewhat convoluted yes, although in response to critics he clarified his

---

[8] Though with Roth's (2012) recent synthesis, this may be beginning to change. There is however yet another island in this archipelago that seems to be completely undiscovered, which is the literature on "temporal externalism"; see for example Jackman (1999; 2005) and Tanesini (2006; 2014). Since I will be discussing externalism in its more generic form in Part V I will set this interesting work mostly aside here.

[9] See Danto (1965).   [10] See ibid: 155.

[11] See Hacking (1995), and also Tanesini (2006: 199) on how the "open texture" of legal, epistemic, and linguistic practices enables the future to "contribute to the determination of what the past has always been."

position, arguing that a clear answer to the question simply does not exist.[12] Or take sexual harassment, the concept of which was invented around 1950. On the one hand, it seems doubtful that the coercive and degrading behavior that we call sexual harassment today was any less prevalent before 1950. Yet, absent the concept both its perpetrators and victims might have thought it was perfectly normal. So what "actually happened" before 1950? Again, it seems indeterminate, in the sense that there is no true or false answer to the question, at least relative to our discourse today.

Still, like Danto, Hacking and those responding to his work take the only plausible interpretation of past indeterminacy to be epistemological. As such, Hacking retains an implicit assumption that the past *itself* is over and done with. Indeed, the idea that the past itself could change is barely mentioned in this literature, except as a reductio that constructivism about the past must clearly avoid.[13] As one critic puts it, the ontological claim is just too "mind-boggling" to be taken seriously.[14]

The Ontological view

Yet there are a few brave souls who do just that. I will draw in particular on an unheralded article by David Weberman to develop the intuition,[15] and then turn briefly to a piece by Jeanne Peijnenburg, who is also the only person in this literature of whom I am aware to bring in physics. In both cases, what is at issue is the ontology of events.

Taking Danto's analysis as his foil, Weberman argues that events in the historical (or human) past can acquire new properties in response to subsequent events – not just new meanings for us today, but properties of the events themselves. His first move is to distinguish two ways in which events can be individuated as "events."[16] One is in physicalist terms: a trigger is pulled, a gun fires, and a man dies. Events constituted in this way are ontologically discrete and as such make no intrinsic reference to other events – so in this case there are three distinct events surrounding the man's death. Although only implicit in Weberman's discussion, his emphasis on discreteness suggests that by 'physical' he means classical (or material) physical. Classical physics is atomistic, so that events in such a world are connected at most in causal

---

[12] See Gustafsson (2010) and Roth (2012) for excellent syntheses of the debate.
[13] However, I read Roth's (2012) pragmatist analysis as coming close to an ontological interpretation.
[14] See Gustafsson (2010: 312).
[15] See Weberman (1997), and also Ni (1992); I say "unheralded" because to my knowledge it has only been cited once, in a footnote by Roth (2012).
[16] See Weberman (1997: 753). Weberman's distinction parallels that made by others between "hard" and "soft" facts; see Hoffman and Rosenkrantz (1984) and Todd (2013). Unlike hard facts, soft facts "depend on or hold in virtue of the future" (Todd, 2013: 830).

terms, not intrinsically. The second way to individuate events is "in terms not restricted to basic physical states and movements."[17] While this subsumes material happenings, what he is really getting at here are events constituted in intentional and/or relational terms: "the Senate ratified the arms-limitation agreement," "striking workers forced the plant shutdown," and so on.

Based on this distinction we get two descriptions of the past, what Weberman calls the "skeletal" past and the "thick" past. From here two important implications follow. The first is that, from the standpoint of the purely physical, skeletal past, most of the events of interest to historians – the Renaissance, World War II, or Cuban Missile Crisis – were not "events" at all. This is not to say the latter have no material basis, since of course part of what constituted WWII as an event is that many people met violent deaths. But by themselves, those deaths were just that, separate deaths, which in no way add up to "WWII." The only way to constitute the latter as a single event is to give those countless physical events a unitary meaning. This leads to the second implication, which is that meanings are internal relations among intentional states. It is how people *think* about physical happenings which constitutes them as events, not their physicality per se.[18] Such thinking is inherently relational, in the sense that, for both individuals and groups, it is how physical events are situated within a web of meaning that matters for what those events *are*.

Weberman next invites us to consider now how this perspective on the constitution of events might inform our thinking about change, first synchronically and then retroactively with respect to the past. A widely discussed example of the former is what happened to Xantippe when her husband, Socrates, was forced to commit suicide: she became a widow.[19] Described in the material terms of the skeletal past the only one of them to whom anything happened at that precise moment was Socrates, since Xantippe herself was not there. Moreover, Socrates' death did not cause Xantippe's widowhood either (at least in the efficient sense of causation), since no force was transmitted that changed her properties. To be sure, after his death other people treated her differently, grieved with her, brought goods to her home, and so on, but that is not what made her a widow. Rather, what made her a widow can only be described in the relational terms of the thick past: if your husband dies then you become a widow *by definition*, in virtue of the shared intentions that constitute the meaning of the physical happenings we call "marriage." On this view, the change in Xantippe's status is a relational property attached to the skeletal fact of Socrates' death, just as being part of WWII is a relational property attached

---

[17] Weberman (1997: 754).
[18] On the constitutive conception of events see T. Jones (2013), and for an interesting discussion of the discursive work that went into constituting 'WWII' see Reynolds (2003).
[19] See Kim (1974).

to the deaths of millions between 1939 and 1945. Philosophers disagree about whether such non-causal changes – known as "Cambridge Changes" – are real changes. Hard-core materialists might think they are not, but many philosophers accept their reality – and I suspect few social scientists would dispute that WWII was a real event.

With this framework in place, we are now in a position to consider the idea that events can change after the fact, not just in an epistemological but ontological sense. Importantly, Weberman limits his claim to the thick, relational past; the skeletal past he agrees (as do I) cannot be changed. That obviously imposes limits on how much the past can be changed: we cannot raise the dead, go back in time to kill Hitler's grandparents, and so on. However, as we have seen the skeletal past tells us so little about what happened in the human past that most historical events would not be events at all, so there is still plenty of room for a provocative argument. So the question becomes: can events in the thick or historical past acquire new, relational properties in virtue of later events?

Consider some of Weberman's examples. Amir shot Rabin at 10 a.m. and as a result Rabin dies at 1 p.m.: the later event changed the properties of the earlier one from a shooting to an assassination.[20] "Smith submitted the winning entry in the poetry competition:" the event of submitting the poem acquires a new property by the judges' subsequent decision. "A man unknowingly becomes a father:" although no physical properties of the man have changed, our legal system recognizes that he is a different man than he was before. (From Danto), "The Thirty Years' War began in 1618:" the localized fighting in that year becomes part of a much larger event after 1648.[21] And so on. Weberman argues that in all of these cases, while the material properties of the original happenings remain unchanged, the latter gained new social or *relational* properties – what he calls "delayed relational properties" – due to later events. Moreover, he notes that these changes need not be only the addition of new properties, but can erase the past as well, as when "a man's life may go from successful to unsuccessful as a result of later actions and events. Or an event may go from forgotten to remembered as well."[22]

Are these changes merely in our description of earlier events, or changes in the events themselves? Weberman spends much of his article rebutting various formulations of the Epistemological view, the bottom line of which I take to be the following. Most of the properties that make people who and what they are, are intentional and relational rather than material and intrinsic. Husbands, wives, masters, slaves, citizens, soldiers, and almost all other social roles are

---

[20] Cf. T. Jones (2013: 79–81).
[21] On the retrospective constitution of the Thirty Years' War see Steinberg (1947) and Mortimer (2001) for an opposing view.
[22] Weberman (1997: note 34).

defined by convention, by individuals' positions in a shared web of meaning. Of course, in one sense these are "descriptions" of people, but they are not *just* that. While describing a flower as beautiful does not affect its actual properties, role descriptions constitute us as kinds of people, as who we *are*. If that does not count as ontological then nothing in social life has ontological status except the purely material properties of individuals and their interactions. The same goes for most social events. When people get married they acquire new properties that change not just how we describe them, but who they are in society. Similarly for soldiers who served in the Gulf War, who are now veterans of the war, entitling them to benefits they otherwise would not have received. And then there are concatenations of events like the Gulf War itself, which would not be "an event" at all but for all the relationally constituted individuals and actions that went into it. In short, given that historical events are constituted relationally by shared intentional states, it hardly seems a stretch to say that they could literally become events after the material facts by which they are also constituted are long gone.

Where this leaves us is with a kind of temporal holism: the idea that when it comes to the thick past, history is not a succession of completely separable events, but a succession of events that are internally or logically related to one another. What happened yesterday depends in part on what happens today, and by implication (though I have not emphasized it here), also on what will happen tomorrow.[23] This dependence is not causal but constitutive, in the same way – only now in a temporal sense – that the master depends constitutively on the slave to be a master.[24] Note that this is perfectly consistent with saying that what happens in the past helps cause events in the present and future, just as the internal relation between master and slave is sustained over time by causal processes of coercion and resistance. Indeed, for events to be related constitutively across time they *must* be connected by a causal chain, which means that not every event in history is internally related to every other later event. But what the concept of temporal holism adds is that because events in the thick past are constituted by the meanings attached to them, and because those meanings are internally related to subsequent meanings, what comes after those events plays a role not just in how we describe them today but in making them what they actually were.

Weberman's examples are mostly at the collective level, and he does not use physics to support his argument; in both respects Peijnenburg's article provides a helpful way to complete this review.[25] Her interest is in how the character

---

[23] For arguments in a similar spirit that emphasize the role of the future see Parsons (1991), Jackman (1999), and McSweeney (2000).

[24] Weberman (1997: 760–762) emphasizes that his claim is not one of backwards causation.

[25] See Peijnenburg (2006).

of an action can be determined by actions that a person takes after it has been performed. Most of her examples involve pairs of dispositions for which behavioral manifestations overlap and thus are hard to distinguish, like bravery and recklessness, miserliness and prudence, and self-consciousness and vanity. In these cases she argues it is only through the repetition of behaviors that past actions become constitutive of one disposition or another. Moreover, she points out that to some extent this is a matter of choice, and not just in the obvious sense that actions today affect who we will be tomorrow. Rather, by choosing to repeat (or not) certain behaviors we also affect our past, not in a causal but a constitutive sense.[26] This is made clearer in another Peijnenburg example, which is not about dispositions.[27] Consider a happily married woman who goes to a party, drinks too much, and wakes up in the morning in bed next to a stranger. Although she cannot change the material fact that she slept with a stranger the night before, by her subsequent actions she can change its *character* – by continuing to see him, "what happened" was the start of a love affair, by not doing so it was a one night stand. Finally, drawing on Huw Price's time-symmetric view of quantum theory, Peijnenburg argues that these cases involve not just re-descriptions of the past, but implicate ontologically the nature of the original actions themselves: "there are actions that are better accounted for by saying that their character is *determined* by later actions than by saying that their character *acquires a new description later on*, or by claiming that their character *is revealed through, or discovered by, the observations of later actions*."[28] This is a bold claim, and in a response to Peijnenburg, Cornelis van Putten challenges it in a way that is instructive for my argument.[29] Like Weberman, van Putten argues that the past can be understood in two different ways: (1) in material terms, as "a constellation of physical events, things that actually happened"; and (2) in psychological terms, as subjective understanding of what happened, which he links to personal narratives. He does not dispute that narratives of the past can change; indeed, he thinks it is commonplace, citing the example of a war veteran who later becomes a peace activist, and now views actions he once deemed heroic as crimes against humanity. However, van Putten claims that this change in narrative has nothing to do with what actually happened, much less with exotic quantum physical notions of retrocausality, and therefore it can take place only in the present.

Van Putten's argument is commonsensical, but it begs a crucial question in its distinction between material and psychological facts. He does not comment on this dualism, but like most social scientists I suspect he would argue it is

---

[26] Cf. Varga (2011: 72)
[27] Her final example is from improvisational music, the succession of notes which give previous notes their key retroactively. Bohm (1980: 198–200) also invokes the example of music to illustrate the holism of quantum theory.
[28] See Peijnenburg (2006: 248), emphasis in the original. [29] See van Putten (2006).

only epistemological. In other words, just because we do not yet know how to explain psychological facts by reference to brain states, they must still be ultimately material in an ontological sense. So let us assume that he would accept a classical physical constraint on what psychological facts are possible. However, in that case, given that the veteran's re-thinking was conscious, and given the hard problem of consciousness for the classical worldview, what is the *physical* basis of his new mental state? If the physics constraint on human beings is classical, then it seems that the veteran's new state could have no physical basis at all, and thus is either epiphenomenal or an outright illusion. In short, van Putten's easy appeal to a common-sense dualism to defeat Peijnenburg's argument does not go through, because it brings in its wake a fundamental problem of how new psychological facts could be "facts" at all.

In this philosophical debate the burden of proof has been heavily on advocates of the Ontological view, so much so it seems that they can barely get a hearing. That's understandable, since the idea of changing the past *is* "mind-boggling." Yet, it is not obvious that Weberman and Peijnenburg's analyses of the ontology of events are wrong either. In turning now to the physics of time and memory, I propose to give their intuitions a quantum basis. In doing so I hope, if not to convince you that the Ontological view is right, then at least to raise the Epistemological view's burden of proof.

The physics of changing the past

Implicit in the epistemic view of changing the past is an assumption of time as a linear succession of points moving through Now. Although I have not seen it characterized this way in the literature, this assumption is one of temporal locality. Recall that locality in space is a foundation of the classical worldview, one that critics of quantum theory like Einstein fought hard to save in the face of the theory's apparent non-locality. However, thanks to experiments confirming Bell's Theorem, we now know that non-locality in space is part of quantum reality. The key assumption at stake in those experiments was "separability": that events located in different places cannot be connected by faster-than-light influences, and as such are intrinsically separate.[30] By the same token, locality in time assumes that events at different times are separable points in a temporal sequence and as such have no intrinsic connection either; yesterday was yesterday, today is today. Temporal *non*-locality refers to a loss of such separability, to an entanglement or "superposition of states at different times."[31]

---

[30] See Healey (1991; 1994).
[31] Filk (2013: 535). Filk's article is an excellent overview of various aspects of temporal non-locality, although unfortunately it does not make a connection to Wheeler's Delayed-Choice Experiments, which I try to do below on my own.

Moreover, just as non-locality in space makes quantum theory holistic spatially, non-locality in time makes it holistic temporally.

Now consider McTaggart's distinction, discussed in Chapter 6, between the "A-series" and the "B-series" views of time: the tensed time of past/present/future and the tenseless time of earlier/later. The A-series is the familiar subjective time we all experience, which moves forward like an arrow toward the future through a succession of Nows. On this view "what time it is" is relative to Now and constantly changing (though it is always "Now"); what was once the future will eventually become the past. This is the conception of time implicit in the epistemological view of the past. The B-series is the tenseless objective time of physics, which does not recognize a privileged moment called Now, and as such has no concept of a future or a past. On this view what time it is never changes; what was once earlier than X will always be earlier, but that's all there is to say.

Importantly, according to the B-series – the time of physics – the past, even the skeletal past, does not exist. Although that might seem like a radical claim, it is actually common sense. If you asked the proverbial man in the street whether the past really exists, chances are he would say no. The Thirty Years' War does not exist, any more than the future exists; only Now exists.[32] So the past has ontological status only in the A-series, which is to say subjectively, as a matter of interpretation. Note that this does not mean the past is purely subjective, in the sense that one can just make something up and call it the past. Most of the intentional states that constitute interpretations of the past are shared and thus *inter*subjective, and even personal memories depend in part on social context, all of which helps stabilize the A-series past. Moreover, what the past can be is further constrained by the material traces – like a piece of paper called the Magna Carta – of what happened before. But the fact that only the A-series even has a concept of the past means that the question of whether it can be changed is one that can be posed only in the subjective register. Having said that, if the A-series is coupled with the classical assumption of temporal locality, as it usually is, then like the B-series it means the past cannot be changed, since even though it once existed as a now, it does not exist right now. All that can be changed is our interpretations of the past. It is that assumption which quantum theory calls into question.

The "hard problem" of time is how to reconcile these two conceptions of time, and especially how to interpret the A-series, for which the CCCP provides

---

[32] In the philosophy of time this view is called "presentism" (vs. "eternalism"). Although intuitive, presentism faces powerful objections, mostly centering on its supposed inability to justify truth claims about the past like "Germany lost WWII"; for a sense of the debate see Markosian (2004) and Mozersky (2011). While much concerned with the special theory of relativity, which poses a challenge to presentism, this debate to my knowledge makes no reference to quantum physics. My argument below I take it is a non-local presentism.

no place in nature. As we have seen, Atmanspacher and Primas' solution to this problem is to argue that the distinction between objective and subjective time emerges from a process of temporal symmetry-breaking from an underlying timeless reality.[33] The Schrödinger equation, which is used to describe wave functions, is deterministic and time-symmetric, meaning it can be solved either backwards or forwards. The Projection Postulate, in contrast, which is used to describe the transition from wave function to particle, is non-deterministic and time-asymmetric. It is in the collapse of the wave function, therefore, which I have associated with the emergence of consciousness, that the A-series is created.

So far that might appear just to give a quantum explanation for the Epistemological view of the past as something that cannot be changed, but this is where memory comes in. For the A-series to be a "series" and thus able to constitute a past, meanings must be remembered. A society that did not hand its collective memories down through the generations would have no past and so in effect have to reconstitute itself at every moment (if it could be called a "society" at all). Similarly, a person with advanced Alzheimer's has no past either, at least personally. Without memory there is no history.

Memory is built up out of experiences, which as the subjective aspect of wave function collapse take place in the classical (i.e. actual) world. As they occur experiences are automatically imprinted into memory, which over time builds up a classical history of our lives. The physical basis of memory is not well understood, but it has been known for some time that the storage of memories is not localized to specific neurons or even groups of neurons, but spread out across the brain. Quantum brain theory takes this finding to its logical conclusion, which is that memories are non-local in a quantum sense.[34] More specifically, "memory is printed to vacuum, i.e. minimum energy states of quantum fields that extend over macroscopic distances in the brain."[35] There memories exist in entangled states in the superposition that is our unconscious, from where they are available for subsequent recall.[36]

This mixed classical/quantum situation makes sense when you think about it. Take driving to the store. The drive consists of many choices – turning left at the first light, right at the second, left at the third, plus all the little choices in-between. As a quantum, free agent, what will become my past at each point is indeterminate (since I could choose an alternate route), but with each choice my wave function collapses into an experience which is recorded in memory as part of my history. However, those various memories gather meaning from

---

[33] See Atmanspacher (2003) and Primas (2003; 2007).
[34] See Stuart et al. (1978), Jibu and Yasue (1995), and Vitiello (2001); cf. Brainerd et al. (2013), who apply quantum probability theory to episodic memory, but take no position on quantum brain theory per se.
[35] See Franck (2004: 52).   [36] See Carminati and Martin (2008: 563).

their entanglement in the longer event of driving to the store – and so it is with that event in relation to every other event in my life to which it is connected. If we are walking wave functions, then even though our experiences at each moment are actualities, at the quantum level of the unconscious, "there are many histories that are there as potentialities."[37]

A crucial implication follows from this conception of memory: experiences of the past are present in the Now.[38] Not in the figurative sense that memory is *about* experiences in the past, as if the latter were now mere pictures or representations of what happened, but in the literal sense that *past experiences continue to exist today*. This is because the entanglement of past experiences with the wave function of unconscious memory implies a temporal non-locality, in which past and present are not fully separable. By virtue of this non-local aspect of memory, experience is preserved in time, enabling us to "re-live" it – and in so doing I shall argue also potentially to reconstitute it. Note that this non-separability does not mean the past is identical to the present, since recalling memories requires a new collapse of the wave function, i.e. a new experience of past experience.[39] However, it is only in virtue of the temporal non-locality of memory that we are able to recall past experiences at all. On this view, therefore, memory understood in quantum terms embodies a conception of time that is neither A-series nor B-series, since both series assume that the past no longer exists. If memory embodies temporal non-locality, then in some sense the past still exists, and as such might be changed.

More specifically, my idea for how the past can be changed begins by interpreting memory recall as a process of delayed choice. Recall Wheeler's Delayed-Choice Experiment, a variation on the Two-Slit Experiment, which is designed to capture temporally non-local effects (Chapter 2). In the Two-Slit case, a measuring apparatus is set up in which a photon can take two different paths to the recording device. If no effort is made to determine which path is taken, then a succession of photons entering the apparatus will build up an interference pattern on the recording device, indicating that each photon took both paths, which is to say remained in its wave form. However, if we try to measure which path is taken, then the photon's wave function will collapse and it will show up as having taken either one or the other path. In the delayed-choice variant this set-up is modified so that the determination of which path the photon follows takes place after it passes the two slits (but before hitting the recording device). Yet paradoxically the result is the same:

---

[37] Carminati and Martin (2008: 564). As Grove (2002: 577) puts it, "the past of our world may contain a great deal that did not happen, but in principle could have."

[38] Or more precisely "sub"-present, as Franck (2004) puts it, since past experiences remain in superposition as long as they are not recalled to present consciousness.

[39] For example, we can mentally *re*-enact what a past pain must have felt like, but we cannot have that particular pain again.

if we do not try to obtain which-path information then we get an interference pattern indicating both paths were taken, but if we do ask which path, then we find that just one was taken – even though we asked our question after the photon passed both slits. What this shows is that wave functions are non-local not only in space but in time, such that "there is a collapse of the wave function *on all the temporal duration* bounded by the moment the photon has been emitted by the source and the moment it has been detected."[40] In short, measurement creates a particular past that was indeterminate or "open" until that moment.[41]

So how might a delayed choice framework help us understand the possibility of changing our past? A recent paper by G. Galli Carminati and F. Martin addresses the question explicitly.[42] In their model the choice of what to observe is governed by free will; consciousness plays the role of the particle detector, registering what is observed; and "phenomena recorded by our unconscious [memories] persist as coherent superpositions of quantum states."[43] Now consider what happens when we recall a memory, making it present in consciousness. Just as the past of a photon can be created in different ways depending on where we put the mirror, as a superposition memory can be made to appear (collapse as a new experience) in different ways depending on the angle and context in which we look at it. The possibilities are not infinite, since superpositions are structured and as such make some outcomes more probable than others – my memory of driving to the store cannot be recalled as flying to Chicago. But within those constraints more than one past is possible. More specifically, there are two ways that the past can be influenced by delayed choices, which I will call the "Addition" and "Replacement" Effects.[44]

The Addition Effect is more intuitive, although still hard to make sense of from a classical standpoint. Sticking with the store example, now assume that because of an accident the next day playing football, I am never able to drive a car again. Because it was causally connected to the state vector we may call "my ability to drive," my choice to play football retroactively, non-locally, constitutes my experience of that drive to the store as my *last* drive, adding new content to the event which it did not have before. The Addition Effect fills the past in, making it more fine-grained, and so to that extent the "past is not finished."[45]

---

[40] Carminati and Martin (2008: 564); emphasis added.
[41] On the idea of an open past see Markosian (1995).
[42] See Carminati and Martin (2008); also see Yearsley and Pothos (2014), which appeared too late for me to incorporate into this text.
[43] See Carminati and Martin (2008: 563), and also Franck (2004: 55); cf. Brainerd et al. (2013).
[44] These terms are adapted from Bernecker (2004), though he does not consider the idea of changing the past.
[45] Sieroka (2007: 92), referring to Weyl's view.

The Replacement Effect goes further, suggesting that some aspects of the past can be literally changed. Recall Van Putten's war veteran turned peace activist.[46] Unlike the purely material happenings in his past, which are forever over and done with, his experiences in the past are preserved non-locally in the wave function of memory and as such still exist, not as representations *of* the past but as the past itself. There, in the present, they are available for reconstitution upon recall to consciousness. That will not change the veteran's past as he experienced it back in the war, but because that experience "lives on" it can in principle be changed by a new measurement in the present. Moreover, we do not have to rely just on hypotheticals here, for consider the well-documented phenomenon in psychoanalysis called an "après-coup."[47] The après-coup refers to a situation in which people have experiences that at the time they either repressed or did not view as troublesome, and then later came to see as problematic as a result of therapy, new information, or the introduction of new categories. An example is what happened to several thousand Jewish children who survived the Nazi occupation of Belgium by hiding with non-Jewish families.[48] Although many of these children understood something of their situation, most did not experience it as traumatic, and – in part because others did not constitute them as such – did not come to see themselves as survivors of the Holocaust. Decades later, however, after changes in collective understandings about the Holocaust, they became traumatized, in a "deferred retroactive effect" of their experience.

One might argue that this is just a re-interpretation of something that happened long ago and so is only an epistemological change. But that belies the very nature of the après-coup, which is to question just what *did* happen long ago. To be sure, the material aspects of the survivors' past did not change, nor could the experiences they had at that time. However, from a delayed choice perspective, memories are not separable from experiences in the past, but connected to them non-locally. So in recalling experiences in light of the "mirror" of new understandings, what those experiences *were* was changed – not causally but constitutively – which is an ontological change.[49] This gives new meaning to the idea of "mental time travel," which has recently been developed as a way of thinking about memory (and foresight).[50] Although its advocates treat the idea as only a metaphor, from a quantum perspective, there is

---

[46] Carminati and Martin's own example is one of mourning a father's death (2008: 569–572).
[47] See Birksted-Breen (2003) and Fohn and Heenen-Wolff (2011). The experience of being "born again" might have similar, though less traumatic, characteristics.
[48] For discussion of this case see Fohn and Heenen-Wolff (2011).
[49] The idea of a "quantum eraser" seems related to this effect, but I do not understand it well enough to discuss it here; for those who would like to follow up see Egg (2013).
[50] See Suddendorf and Corballis (2007), Suddendorf et al. (2009), and Gerrans and Sander (2014).

a sense in which people actually can travel to the past. Not by transporting their bodies, but by entangling their thoughts with past experiences that survive in memory.

Finally, taking the Addition and Replacement Effects together, we are led also to a broader conclusion. These effects operate not only on specific experiences, but on one's life as a whole, such that life itself may be seen as a delayed choice. It is only at the end of our lives that who we were, and what we did, is fully determined. Someone who did bad things when they were young can, within limits, change that past by good deeds when they are old. In short, quantum consciousness provides a physical basis for redemption, something which we all take for granted as a possibility, but which seems hard to explain if we are nothing but classical machines.

I have focused in this chapter on the experience of time by individuals and how it can enable changes in their past; by way of transition to the discussion of social structure in the rest of this book let me conclude by returning to the debate about changing the collective past, or history in a broader sense. Although I think the argument carries over, there is admittedly an important difference between the two cases, at least if we are talking about reconstituting events from which there are today no survivors. Individuals who were really there had direct experience and thus personal memories of events, whereas later generations have neither. So even if my argument is granted that individuals can change their past by reliving past experiences stored in memory, how could this be possible for future generations?[51]

In approaching this question let me first point out that even the access to historical events of individuals "who were there" is not as privileged as it may seem. First, no one who fought in WWII could have experienced "WWII," since they could not have been everywhere at once. They could not have even experienced the Battle of Stalingrad, but at most a small slice of it, such as the fight over the Tractor Factory – and even then, their experiences of that fight were all somewhat different. Second, WWII was not over until 1945, and so what individuals experienced before that depended in part on how it ended; had the Axis won, the nature of those experiences would have differed – for Germans, noble sacrifices for a thousand-year Reich instead of immoral fighting for a lost cause. (Or consider the Kiev Pocket in 1941, which was a major German victory at the time, but which turned out to have been such a strategic blunder that some historians think it cost Germany the war.) So, on both synchronic and diachronic counts even those who fought did not, as individuals, "own" the events that constitute WWII. The war was what it was only in virtue of the shared meanings that constituted their experiences.

---

[51] This question warrants a much more extensive treatment than I can give it here, so what follows is intended only as a suggestion.

Now extend this point to the case of World War *I*. Individual experiences of the war can no longer be relived directly, but the shared meanings that made those experiences what they were at the time have survived in the form of collective memory or history. That memory exists only in virtue of language, which I argue in Part IV establishes a semantic non-locality between individuals' minds. This non-locality does not enable us to literally *have* each other's experiences (language is not telepathy), but it does enable us to directly perceive as opposed to merely infer each other's experiences. A key implication of this analysis will be that minds are not fully separable, even though we inhabit separate bodies. By implication, given our semantic connection through collective memory to those who fought in WWI, we today are not fully separable from them either – their minds are in a sense part of ours. Thus, while we cannot personally relive their experiences, we can "re-enact" them, in R. G. Collingwood's terms.[52] To be sure, the greater the temporal distance from an event, and therefore the more minds involved in settling its meaning, the more difficult it may be to change. But while those who were there at the time might be said to have initialized the constitution of a past event, it then lives on non-locally in the memories of those who came after. As such, and as the lively debate today about WWI historiography attests, the Great War is also "our" event, and to that extent available to us for reconstitution.

---

[52] There is a large literature on the idea of re-enactment. See Stueber (2002) and Dharamsi (2011) for two accounts that seem particularly relevant to my argument here.

Part IV

Language, light, and other minds

Introduction

I began Part III by describing what human beings are if we imagine them under a strict classical physics constraint, which is to say without reference to consciousness and therefore intentional properties. The resulting image is one of a machine or zombie – material, well-defined, subject only to local causation, deterministic, and, in effect, dead.

I argued that a very different picture emerges if we imagine ourselves under a quantum constraint with a panpsychist ontology. Quantum Man is physical but not wholly material, conscious, in superposed rather than well-defined states, subject to and also a source of non-local causation, free, purposeful, and very much alive. In short, she is a subject rather than an object, and less an agent than an agen*cy*, someone always in a state of Becoming. Moreover, this agency is a process in and through which she is sovereign. She decides her present by how she collapses her wave function (Chapter 8); she decides her future by projecting herself forward in time and enforcing correlations backwards (Chapter 9), and to some extent she even decides her past, by adding to or replacing it in her practices (Chapter 10). Of course, these decisions are not unconstrained, both internally and externally, but within those constraints the quantum model of man posits an irreducible freedom to create who we are. It is what I take to be an existentialist picture, in which our lives are like works of art.[1]

I turn now in this and the next Part to the structural side of the agent–structure problem, which highlights not only the constraints but also the affordances in which human agencies are embedded. Some of these, like mountains and rivers, are purely material objects and as such external to human beings. To be sure, what those objects *mean* is interpreted and therefore can vary. To one person a mountain might be a sacred place to be revered, to another an obstacle to be blown up.[2] These different meanings implicate consciousness and as such could

---

[1] See Varga (2011) for reflections on this old theme.   [2] See Freudenburg et al. (1995).

profitably be analyzed from a quantum perspective. However, the mountain is also a brute fact independent of anyone's consciousness, which limits what people can do with it.[3] Social structures or institutional facts are different; while external to the individual, they are internal to human beings collectively. Although in practice many social structures mix brute and institutional facts – like states with river boundaries – in what follows I will bracket brute facts and focus on our relationships to each other that constitute institutional facts, or social structures in their elementary form.

Although a number of quantum concepts will figure in what follows, perhaps the most central one is the idea that, when entangled, quantum systems are not fully separable. A foundational assumption of the classical worldview,[4] separability means that "[t]he complete physical state of the world is determined by (supervenes on) the intrinsic physical states of each spacetime point (or each pointlike object) and the spatio-temporal relations between those points."[5]

It is this assumption that underwrites the reductionism of modern science, since it implies that everything can be decomposed into ever smaller parts that do not presuppose each other's existence.

In classical social science the separable, "pointlike objects" are individual human beings, whose intrinsic properties are constituted by our material states. This is a complicated way of saying something most people would find completely intuitive: that our skins form an impenetrable boundary between us, making us utterly distinct individuals.[6] Indeed, *so* intuitive is this idea that even if sub-atomic particles violate separability it is hard to see what this could mean at the social level – that you and I are the *same* person? Of course, human beings often experience "We-feelings," identifications that can be so strong that one may even sacrifice their life for another. But these feelings are states of mind that – as brain states – seem to be encased within our skins, and thus hardly make us numerically the same. If anything, separability seems to be their pre-condition, for what is a "We" if not a "you *and* I"? Given that, it is easy to see why for most social scientists an analysis of social structure must be based on an individualist ontology, which takes separability as its starting point.

It is the burden of my argument to show that despite its strong intuitive appeal, the separability assumption does not hold in social life. The burden only extends

---

[3] On brute facts see Searle (1995: 2 and passim); cf. the "rump materialism" of Wendt (1999: 109–113).
[4] As Kronz and Tiehen (2002: 332) put it, "there are no non-separable states in classical mechanics."
[5] Maudlin (2007: 51).
[6] See Farr (1997) on the significance of the skin as a boundary of identity, and Bentley (1941) for a critique.

so far, since I am not going to defend the opposite assumption, that human beings are completely *in*separable. This is not true even at the sub-atomic level, where entangled particles retain some individuality.[7] Rather, what characterizes people entangled in social structures is that they are not *fully* separable. As we will see that is still a radical claim, but at least it is not manifestly crazy, and where it leads is to a holist social ontology. Within social theory there are long-standing holist perspectives already on offer, but their arguments are qualitative rather than physical, appealing to features of intentional phenomena that seem inconsistent with human separability. Given the reductionism of modern science, that has kept social holism at the margins of contemporary thinking. By giving social holism a basis in quantum entanglement I hope not only to lend it legitimacy but also to shift the burden of proof onto advocates of the classical, individualist orthodoxy.

I develop this argument in two parts. Here in Part IV I examine the special case of language, which mediates all other social structures. Using recent scholarship on quantum semantics, in Chapter 11 I argue that linguistic meaning is irreducibly contextual and non-local. Then, building on an analysis of the nature of light, in Chapter 12 I address the Problem of Other Minds, making the case that, when understood quantum mechanically, language enables us to directly perceive each other's minds. In Part V I turn to the agent–structure problem itself, with a view toward showing how a quantum theory of consciousness and language provides a physical basis for an emergentist, holist, and vitalist conception of society.

---

[7] Though how much is a matter of debate; see for example Castellani, ed. (1998).

11 Quantum semantics and meaning holism

Language is the most fundamental institution of human society. It sets us apart from all other species,[1] and our other institutions would be impossible without it. However, to call language an "institution" is already to take sides in a debate about what language is, one which implicates the agent–structure problem. In recent years the dominant view in linguistics has been to think of language as being in the heads of individuals, whether as a "mental organ," "computational device," or even "instinct."[2] On this view linguistics is essentially a branch of cognitive science. A very different view is taken by those who emphasize the supra-individual character of language. Saussure, Wittgenstein, Davidson, and Searle – to drop just a few names – have all highlighted the ways in which our use of language is constituted and regulated by norms shared by a community of speakers. From this perspective language is less a cognitive than institutional fact.[3] Of course, institutions cannot exist apart from people, so "top-downers" would agree that what is in people's heads matters too, just as "bottom-uppers" would agree that language must be shared for communication to be possible. But, as in the agent–structure problem more generally, it is not obvious how to combine parts and whole.

Importantly, the interdependence between speakers and language communities means that an implicit subjective–objective polarity structures both perspectives.[4] For social scientists the most familiar example of this polarity is probably Saussure's distinction between *langue* and *parole*, the former referring to the structure of a language considered as an abstract system of signs, the

---

I am grateful to Karin Fierke, Ted Hopf, and Jennifer Mitzen for detailed comments on a draft of this chapter.

[1] Language in the broad sense is not unique to humans, but animal languages are obviously quite rudimentary in comparison.
[2] See Zlatev (2008: 37), who is summarizing the views of Noam Chomsky, Ray Jackendoff, and Steven Pinker respectively.
[3] See especially Harder (2003), who draws primarily on Searle (1995); for Davidsonian and Wittgensteinian approaches to the sociality of language see for example Williams (2000) and Verheggen (2006).
[4] See especially Cornejo (2004).

latter to actual speech. As an institutionalist Saussure neglects speech, whereas as a cognitivist Chomsky neglects institutions, but his distinction between "competence" and "performance" reflects the same polarity. A speaker is competent if she can use a language's generative grammar (a kind of "*langue* in the head"[5]) to communicate, while her performances actualize this potential in speech. These polarities map onto the difference within linguistics between semantics and pragmatics – the one concerned with how meaning is assigned by a language's grammar, or "meaning-in-itself," the other with how meaning is assigned in real-world situations, or "meaning-in-use." Semantics transcends particular agents and thus has an objective quality, whereas by virtue of being tied to speakers, pragmatics is (or tends toward the) subjective.[6] Moreover, and recalling my discussion in Chapter 1 about social structure, semantics is unobservable; what is observable is only utterances.[7] As we will see, an important problem in the philosophy of language is how to integrate these two poles under the same roof.

The pragmatic aspect of language raises another issue, however, which is that for interpretivists language is a fundamental barrier to a naturalistic social science. This is because the way language works has little in common with how things work in the material world. In the latter things happen through causal processes, in which a transmission of force or energy induces changes in material objects. These processes are fully objective, since their effects do not depend on how they are interpreted. In contrast, language makes things happen through performative processes that constitute phenomena with meaning. These processes are not fully objective, since their content depends on how they are understood. To be sure, social scientists can abstract away from these understandings, enabling them to do their work "as if" it were chemistry.[8] But they are a key source of the controversy between positivists and interpretivists that has bedeviled social science from the beginning. In this controversy I am with interpretivists in thinking that language is essentially different from chemistry, and that abstracting away from this difference is therefore to miss something essential about social life.

But what is language? According to the CCP everything in nature is ultimately physical, and so language must be physical too.[9] If it were not, then what (or where) else could it be? The alternative is either that it is a supernatural phenomenon, but that doesn't seem very plausible; or, as in substance dualism,

---

[5] Cornejo (2004: 9). See Chomsky (1995) for an overview of his views as they pertain to this chapter and Jackendoff (2002) for a magisterial synthesis of the generative tradition.
[6] I say "tends toward" because many philosophers deny that consciousness has any interesting role to play in linguistics; see below.
[7] See Itkonen (2008: 21–23).
[8] See Padgett et al. (2003) for a refreshingly explicit discussion to this effect.
[9] See Benioff (2002).

that it is a mental phenomenon with no connection to nature, which also seems problematic. So if language is part of the social world it must have some kind of physical basis. But *what* kind: classical (i.e. material) physical or quantum physical?

Until recently this question had not been asked, since all sides have implicitly assumed a classical answer. While a materialist explanation for consciousness – which the latter I argued is presupposed by intentional phenomena – has not yet been found, positivists are confident it will be, and that it will not undermine the unity of science. For their part, while interpretivists have not doubted that 'physical' means 'material,'[10] they seem confident that even if a materialist explanation for consciousness is found, it would not threaten at least the epistemological autonomy of the human sciences. Yet, in light of the mind–body problem there is no evidence a materialist explanation for consciousness will ever be found. While that might seem to pose the greater threat to positivists, it brings with it a threat of vitalism that even most interpretivists would probably be eager to avoid.

In this chapter I challenge the shared premise of this controversy: that language is a classical phenomenon. My discussion draws on the work of physicists who have begun applying quantum theory to concepts and linguistic meaning. Their analysis speaks to a long debate within linguistics about whether meaning is built up out of smaller semantic units with intrinsic content (the Compositional view) or depends irreducibly on local context (the Contextualist view). This is essentially a debate about the relative importance of semantics and pragmatics, and as we will see the quantum approach to language clearly favors Contextualism. In so doing, it justifies the interpretivist insight that, when it comes to epistemology, language *is* different from chemistry, and simultaneously the positivist insight that, when it comes to ontology, language is part of nature.

Composition versus context in meaning

The mainstream view about how linguistic meaning is created has long been Compositionalism, which social scientists would recognize as a species of reductionism: the meaning of a whole – a sentence, paragraph, or text – is a function of the meanings of its constituent parts (words) and the way in which they are combined. As Jerry Fodor puts it

It is extremely plausible that the productivity and systematicity of language and thought are both to be explained by appeal to the productivity and systematicity of mental representations, and that mental representations are systematic and productive because

---

[10] Though see Apel (1984) for an exception.

they are compositional. The idea is that mental representations are constructed by the application of a finite number of combinatorial principles to a finite basis of (relatively or absolutely) primitive concepts.[11]

By virtue of its reductionism and emphasis on primitive, separable parts this is clearly a classical theory of meaning. Such an approach puts a lot of pressure on our ability to define the elementary parts in which meaning bottoms out, namely words and especially concepts, which play a key role in helping people to interpret situations. This has proven difficult. It was originally thought that concepts could be defined in terms of necessary and sufficient conditions for their application, but it has become clear this is impossible. More flexible criteria have since been proposed, which invoke ideas like "prototypes," "exemplars," and "graded structure" to capture similarities in statistical evidence of how people actually use concepts.[12] The debate over the merits of these different models continues, though it remains unclear whether any can handle similarities that are not literal but figurative, as in metaphors and analogies.[13] But one way or the other, concepts need stable and well-defined properties if they are going to compose larger meanings.

Yet it turns out that defining concepts is the easy part; matters get quickly more difficult once concept-combinations and then whole sentences are taken into account. Some concept combinations are compositional, like "black cat,"[14] but consider "pet fish."[15] If you cue test subjects in a word association experiment with the concept 'pet' then 'guppy' is only rarely activated, and similarly with 'fish.' However, if you cue them with 'pet fish' then 'guppy' is activated with high frequency. If meaning is compositional, then why would a word that appears with low probability in association with two words considered separately suddenly become very salient when they are combined? This is not to say that the compositional approach has no resources for trying to explain the "guppy effect," though it is unclear whether any of them is up to the task.[16]

But now take the even harder – though still quite elementary – case of whether the meaning of a sentence can be decomposed into the meaning of its parts and how they are combined. An example much discussed in the literature is the story of "Pia and the Painted Leaves:"

Pia's Japanese maple has russet leaves; she paints them green. Addressing her neighbor, a photographer looking for a green subject, she says, apparently truly:

---

[11] The quote is from Busemeyer and Bruza (2012: 145).
[12] See Gabora et al. (2008) for a brief overview of the history of concepts research.
[13] See Thomas et al. (2012: 596), who argue that a connectionist variant of Compositionalism can deal with such cases.
[14] Busemeyer and Bruza (2012: 145).
[15] For a thorough discussion, albeit from a quantum point of view, see Gabora and Aerts (2002: 344–346).
[16] See Mitchell and Lapata (2010) for a sympathetic overview of such efforts.

(1) The[se] leaves are green.

Imagine now that Pia's botanist friend is interested in green leaves for her dissertation and that, in reply, Pia utters (1) again. This time, her utterance seems intuitively false.[17]

So, exact same sentence, yet completely different meaning, because the context of Pia's statement has changed.

The notion that context affects linguistic meaning goes back at least to Wittgenstein, but whereas in the past it was seen in the mainstream as being of only marginal interest, fueled by thought experiments like Pia's story, Contextualism has recently emerged as an important challenge to the compositional orthodoxy.[18] It comes in many forms, from moderate ones that might be synthesized with Compositionalism to Radical Contextualism, according to which there are no literal categories in cognition at all, just "temporary coalescences of dimensions of similarity, which are brought together by context."[19] But their common denominator is the view that pragmatics plays a crucial role, irreducible to semantics, in determining linguistic meaning. Pragmatic considerations can be almost anything, from the macroscopic structure behind a verbal exchange, to the interactional level of the intentions and/or knowledge of the interlocutors, to the way in which speech is inflected or sentences are organized, to the micro-level of concept combination ('pet' and 'fish' each forms a context for the other) – and all of which operate simultaneously.

Intuitively it seems obvious that context affects the meaning of what is said, and indeed this is recognized by Compositionalists, who alongside their primary emphasis on semantics accord a limited role to pragmatics as well.[20] However, the way in which they have defined and operationalized context – by identifying the values of fixed parameters like who is speaking, when, where, and so on – have not satisfied critics, who argue that at most this constitutes a "narrow" context, not the "wide" context in which speech actually takes place, where almost anything can affect meaning.[21] The complexity and potential subtlety of wide context is challenging in at least two respects. First, it creates doubt about the very possibility of a systematic semantics (read: science) of natural language. If the meaning of an utterance can change based on any number of contextual differences, then what hope is there of generalizing about

---

[17] Predelli (2005: 351). Hansen (2011) provides a good discussion of this and similar examples.
[18] See DeRose (2009) for a comprehensive statement of Contextualism. Semantic Contextualism is related to but not the same as Epistemic Contextualism, which I will not be discussing here; for a good survey of this literature see Rysiew (2011).
[19] Thomas et al. (2012: 595). See Recanati (2005) on the diversity of Contextualisms, which he juxtaposes to "Literalism."
[20] See Lasersohn (2012) for a good discussion of how the two traditions deal with contextual effects on meaning.
[21] See for example Recanati (2002: 110–112).

meaning?²² Judging from recent efforts to show that Compositionalism can deal with a fuller range of contextual effects, this threat seems to be taken seriously.²³ Second, Fodor has argued that computational models of the mind cannot cope with context-sensitivity even in principle, because they assume that the mind is a computer with a fixed representational and causal structure that processes information bit by bit.²⁴ Fodor's claim has been disputed,²⁵ but it raises a question for Contextualists as well, which is how can their wide theory of context be reconciled with cognitive science? What, in short, is the physical basis of contextual effects on meaning?

Quantum contextualism

I have neither the ability nor desire to adjudicate this debate on its own terms; rather, what I want to do is problematize its classical frame of reference. This is entirely implicit, since physics almost never comes up in the literature,²⁶ and even when a classical view of the mind is invoked, as by Fodor, this is not done in contrast to a quantum view. Thus, outside the quantum semantics literature I know of no references to quantum theory in the philosophy of language, and only one in experimental semantics, where Jeff Mitchell and Mirella Lapata quickly reject a quantum modeling option because its mathematical features "undermine not only [its] tractability in an artificial computational setting but also [its] plausibility as [a] model of human concept combination."²⁷ These complaints make sense if we assume the mind is classical, but that is precisely what is in question here.

Why make a quantum turn in linguistics? Well, first, languages are rooted in the brain, and so if the brain is a quantum computer, then that would be a compelling reason to think that language is quantum as well. Second, quantum theory is a contextual theory par excellence, in which precisely how a measurement is prepared makes all the difference to the outcome. As such, there is a "quite strong" analogy between quantum theory and language, such that the "exact same" modeling operations can be used in both.²⁸ One of the main contributors to quantum semantics, physicist Diederik Aerts, thinks the analogy is *so* robust that he has even used it in the other direction to propound

---

²² See Lasersohn (2012).
²³ See for example Pagin (1997), Predelli (2005), Hansen (2011), Lasersohn (2012), and Thomas et al. (2012).
²⁴ See Fodor (2000). ²⁵ See Thomas et al. (2012).
²⁶ The only exception of which I am aware is Chomsky (1995), who argues that we should approach the study of language in the same way we would any other physical system – by which he clearly means *classical* physical.
²⁷ See Mitchell and Lapata (2010: 1399–1400); quantum theorists Dalla Chiara et al. (2011: 85) are not impressed, calling the intractability charge a "common prejudice."
²⁸ Widdows (2004: 217); also see Neuman (2008).

a new interpretation of quantum theory. His argument is that whereas quantum processes are thought to be mysterious because they behave unlike anything we encounter in our daily lives, they actually behave just like concepts, with which we are all very familiar.[29] Aerts stops short of saying that concepts *really are* quantum mechanical, but his many examples are compelling, and suggest that it is a small leap to my stronger, realist claim.

Finally, like quantum mechanics, Contextualism is a holistic theory, in which "the meaning of the whole determines the meanings of its parts, but generally not the other way around."[30] Meaning (or semantic) holism comes in degrees, from radical – "[t]he meaning of an expression depends constitutively on its relations to *all* other expressions in the language . . . " – to moderate – "what a linguistic expression means depends on its relations to *many* or all other expressions within the same totality" – to almost benign – "all of the inferential properties of an expression constitute its meaning."[31] It is unclear to me which of these holisms quantum semantics implies, since the literature has not addressed the question directly. On the one hand, quantum theory is as holistic as holistic can be; on the other, as we saw in Chapter 8, the Hilbert space of the mind is partitioned into state vectors of regularly associated concepts, only some of which will be activated in a given context. So rather than try to sort this out here, let me just introduce quantum semantics, and then you can judge for yourself.

The starting point for the argument is that concepts typically have many meanings and as such lack definite properties in the abstract. Take 'suit,' which among other things can describe an item of clothing, a legal proceeding, or an aspect of a playing card – meanings that have little to do with each other.[32] This suggests that the "ground state" of a concept may be represented as a superposition of potential meanings, with each of the latter a distinct "vector" within its wave function. Being in superposition means these vectors are entangled quantum mechanically.[33] However, they will have different "weights," which can be identified experimentally through surveys asking respondents to rate how "typical" a meaning is of a given concept (a common practice in linguistics). These weights give the wave function structure, telling us how likely it is to collapse onto one actual meaning or "eigen-state"[34] vs. another, other things being equal. Since a concept cannot be potential/abstract and actual/concrete at the same time, Aerts argues the relationship between the two is one of Heisenberg uncertainty, just like our inability to know simultaneously the momentum

---

[29] See Aerts (2009; 2010).   [30] See Dalla Chiara et al. (2011: 85), and also their (2006).
[31] See Jorgensen (2009: 133–134), quoting Christopher Peacocke, Peter Pagin, and Michael Devitt respectively; emphases added. See Malpas (2002) and Pagin (2006) for good introductions to meaning holism.
[32] The example is from Bruza and Cole (2006: 12).   [33] Busemeyer and Bruza (2012: 151).
[34] 'Eigen' is German for actual or real.

and position of a particle.[35] Thus, "words could no more be said to 'possess' an intrinsic meaning that is independent of their use than, in Bohr's view, could an electron be said to 'possess' an intrinsic position or spin."[36]

In quantum mechanics measurement is what brings about a wave function's collapse, which is an inherently contextual process that involves first deciding what particular question to ask of nature and then preparing the experiment in such a way that it can be answered; if these steps are done differently, then a different result will be obtained. Similarly, in language what brings about a concept's collapse from potential meanings into an actual one is a speech act, which may be seen as a measurement that puts it into a context, with both other words and particular listeners.[37] This starts with a speaker's decision to try to communicate one meaning rather than another. But while communicative intent (analogous to the physicist's question and experimental design) structures outcomes in a certain way, the meaning that is actually produced (where the "particle" lands) depends also on the listener, whose understanding will depend on how what is said interacts with her memory of words and their associations (which may differ from the speaker's). So the guiding idea is that memory structures relate to concepts in the same way that measurement devices in physics relate to particles.[38] If that's right, then we should see quantum entanglement and interference manifested in actual language use, which is what quantum linguists seek to demonstrate. This is done primarily through the study of concept-combinations, such as "pet fish," in which the introduction of a new concept changes the context of the original one and thus the probability that a particular instance of the concept will be manifested.

Consider first the case of concept interference, for which Aerts develops a lengthy example drawing on experimental data about how people classify fruits and vegetables.[39] Subjects were asked to complete three tasks: (A) choose one item from a list of foods that they think is a good exemplar of "Fruits"; (B) choose one from the same list for "Vegetables"; and (C) again for "Fruits or Vegetables." The list contained twenty-four items, some of which were obvious (apples and broccoli), but most were not (mushrooms and mustard). The responses to (A) and (B) measure the "typicality" of each item as an instance of the concept, or in quantum terms, the weights for the different vectors in the superposition of Fruits and Vegetables. Regarding (C), when the two concepts are combined, if meaning is classical, then the expectation of someone

---

[35] See Aerts (2009: 388–389).   [36] Ford and Peat (1988: 1239).
[37] See Schneider (2005), who like Aerts actually makes the connection in the opposite direction, arguing that measurement *in physics* is a speech act.
[38] Aerts (2009: 371).
[39] Aerts (2009; 2010: 2954–2959); also see Busemeyer and Bruza (2012: 144–146), and the discussion in Chapter 8 of the "Linda" case.

choosing a given exemplar as an instance of "Fruits or Vegetables" would be the average of the answers to (A) and (B). However, this is not what the data from (C) show, which for each term deviate from expected values to varying degrees. After doing the math, Aerts shows that the deviation can be explained by interference, suggesting that the logic of concept combination is actually quantum.

Furthermore, Aerts shows that this experiment can be represented graphically as a direct analog of the Two-Slit Experiment in physics. Answers to (A) correspond to particles going through an open slit while the other slit is closed, while answers to (B) correspond to the reverse. The result in both the conceptual and physics experiments is a normal distribution, of choices and hits respectively, opposite the open slit. (C) corresponds to both slits being open, and instead of an average of the two distributions, we get the wavy line characteristic of an interference effect, just as in the Two-Slit Experiment. Although Aerts' purpose is to show that particles behave just like concepts, his mapping shows that the reverse is true as well. Moreover, in both cases we see a difference with classical material objects. This becomes clear if we consider that for any two concepts, say Furniture and Bird, their disjunction, "Furniture or Bird," will itself be a concept (even if one for which it might be hard to find exemplars); whereas the disjunction of two actual material objects, as this example shows, will not typically be an object.[40]

Now let's look at whether entanglement is found among concepts, which is tested for by deriving Bell inequalities and then seeing if they are violated experimentally; if they are then entanglement is present. To derive the inequalities Aerts considers the sentence "The Animal Acts" with its two concepts. He then considers "two couples of exemplars or states of the concept Animal, namely Horse, Bear and Tiger, Cat, and also two couples of exemplars or states of the concept Acts, namely Growls, Whinnies and Snorts, Meows."[41] In a series of four experiments subjects were asked which term in the first pair is a good exemplar of 'Animal' and again for the second pair, and then similarly for the latter two pairs for 'Acts.' These experiments provide weights for each exemplar of the concept that it instantiates, when considered in isolation from the other concept. Aerts then designed four more experiments in which the exemplars are paired in different combinations, like "The Tiger Meows," "The Cat Whinnies," and so on, and subjects are asked whether each is a good example of "The Animal Acts." In each experiment the expectation was that one combination, such as "The Cat Meows" would clearly dominate the others,

---

[40] Aerts (2010: 2965). Also see Sozzo (2014), which deals with borderline cases and also reproduces existing experimental results.
[41] See Aerts (2010: 2960), and for another example Aerts, Czachor and D'Hooghe (2006: 466–469).

which after assigning numerical values to different answers allowed Aerts to derive Bell inequalities for the entire set.

Although this research design lends itself to standard psychological testing, for data Aerts turned instead to an analysis of thousands of pages from the Web, which he argues can shed additional light on his quantum interpretation. The details are too much to go into here, but Aerts' results clearly show a violation of Bell inequalities – i.e. entanglement.

> This is because "The Animal Acts" is not only a combination of concepts, but a new concept on its own account. It is this new concept that determines the values attributed to weights of couples of exemplars, which will therefore be different from the values attributed if we consider only the products of weights determined by the constituent concepts... This shows that combining concepts in a natural and understandable way gives rise to entanglement, and it does this structurally in a completely analogous way as entanglement appears in quantum mechanics, namely by allowing all functions of joint variables of two entities to play a role as wave functions describing states of the joint entity consisting of those two entities.[42]

Relating this back to the debate in the philosophy of language, what this shows is that even for very simple combinations of concepts meaning is inherently contextual rather than compositional.

Further evidence of the entanglement of concepts is provided by word association experiments, in which subjects are cued with words and asked to list all the terms they associate with it. The idea here is that words vary in their implicit associations with other words – 'planet' is associated with 'Earth' and 'moon,' but not 'tiger' or 'chair' – so by mapping the number and connectivity of associations we can gain insight into the structure of a language. These experiments have established that words are stored in memory not as isolated entities, but as nodes in a network of related words. Note that this in itself does not imply entanglement, since networks can be classical (as in network theory, which is widely used in the social sciences). What points toward entanglement is how words are activated. The first hint was provided in a 2003 paper by Douglas Nelson, Cathy McEvoy, and Lisa Pointer, who tested two models of word activation.[43] According to the "Spreading Activation" model, "activation travels from the target to and among its associates and back to the target in a continuous chain" (p. 42). This is a classical picture, relying on local causal connections from each word to the next. In contrast, according to the "Activation at a Distance" model "the target activates its representation and the associates that comprise its network in parallel" (p. 42), or synchronously, without necessarily returning back to the target at all. After reviewing existing evidence and conducting two experiments, Nelson et al. concluded that the latter is the better predictor, and

---

[42] Aerts (2010: 2964).   [43] See Nelson, McEvoy, and Pointer (2003).

therefore that "the principle underlying memory activation is the synchrony of activation, not its spread" (p. 49).

This is not what they expected, perhaps in part because they did not theorize the Activation at a Distance model in explicitly quantum terms and so their findings were "counter-intuitive" and difficult to explain. However, Nelson et al. then teamed up with some quantum theorists, who were able to formalize the Activation at a Distance model in quantum terms, and thus provide an explanation for the observed results.[44] Whereas synchronous activation is an anomaly from a classical perspective, it is what we would expect if words are stored in memory as entangled superpositions and as such are not fully separable.[45] This means that in contrast to classical and specifically compositional approaches to our mental lexicon, the words that constitute the nodes of our semantic networks do not have distinct identities prior to their actualization. It is only with the introduction of context – in this case, the way in which associations are measured – that words take on specific identities, as a result of the network's "collapse." Concepts, in short, are not "objects" as in the orthodox view, but processes that unfold in time.[46]

Applications of quantum theory to language are still young, and as yet there has been no pushback from scholars invested in classical views of language. So it is not clear whether these arguments will carry the day. However, given that Compositionalists have struggled to give concepts precise definitions and also to explain the pervasive effects of context on meaning, the fact that quantum semantics predicts these difficulties is strongly suggestive. If the physical basis of the mind and therefore language really is quantum mechanical, then pragmatics is much more important in the production of meaning than is often thought, and the debate between Compositionalists and Contextualists should end with a clear victory for the latter.

By way of transition to the next chapter, I want to point to a curious lacuna throughout the philosophy of language: the role of consciousness. It makes sense that institutionalists would neglect consciousness, since they are uninterested in what is inside speakers' heads. Yet cognitivists too ignore it, conceiving of the mind instead in computational terms.[47] Indeed, the biggest role that consciousness seems to play in the literature is as an "illusion" *produced by language.*[48] Symptomatically, the *The Oxford Handbook of Philosophy of*

---

[44] See Bruza et al. (2009); for further discussion and examples of this formalization see Kitto et al. (2011) and Busemeyer and Bruza (2012: Chapter 7).
[45] Busemeyer and Bruza (2012: 200).
[46] This conception of concepts goes back to William James; see Larrain and Haye (2014).
[47] See for example Jackendoff (2002), who, when he finally does discuss consciousness (sic) (pp. 309–314), defines it in functional rather than experiential terms.
[48] See Dennett (1991).

Language (2006), a comprehensive, over-1000-page survey of the field, has no index entry for 'consciousness.'[49]

Quantum theorists of language have not thematized consciousness either, but my larger argument points to an important role for it in their approach. To recall, consciousness is produced in wave function collapse, conceived as a process of temporal symmetry-breaking initiated by free will (see Chapter 6). Will works through advanced action, enforcing correlations between the future and the present; experience complements that backward movement with retarded action, moving forward in time and in so doing restores temporal symmetry.

This suggests two things about language. First, the production of linguistic meaning is willful in the sense that it requires ongoing decisions to collapse the potential meanings of words into actual ones.[50] This means that although language as a whole is in a quantum coherent state, it is in *de*coherence that meaning is actually created.[51] Second, as an aspect of decoherence, it is only in the experience of language that meanings are realized.[52] Far from being an illusion produced by language, linguistic meaning *presupposes* consciousness. That points to a phenomenological view of language, which "counterposes the idea that speaking and listening, writing and reading, are unconscious, automatic roll-outs of thoughts and feelings formulated anterior to and outside of enactments of language."[53] Moreover, since consciousness originates in wave functions that are inherently non-local, insofar as these are shared – as they must be for language to be social – through experiences of language we might gain access to other minds.

[49] See Itkonen (2008), Zlatev (2008), and Ochs (2012) for critiques of this bias in the literature.
[50] On the purposeful character of language use see Zlatev (2008: 48–49) and Aerts (2009: 401).
[51] Cf. Aerts (2009: 371–372), who sees meaning rather as the analogue to *coherence*.
[52] See Itkonen (2008: 19) and Ochs (2012).
[53] Ochs (2012: 152); also see Robbins (2002) and Zlatev (2008: 49).

12   Direct perception and other minds

The quantum semantics literature has focused on individuals in isolation; I want now to extend it to the more realistic case of dialogue. Here it is not only neighboring words and sentences that form the context of speech acts, but the presence of other people with their own intentions. In shifting to interaction we run up against the Problem of Other Minds. Philosophers actually distinguish two such problems: an epistemological or "mind-reading" problem of knowing what others are thinking; and a deeper problem of knowing whether others have minds at all.[1] However, for social scientists the main issue is mind-reading, on which there is a substantial literature in psychology, and I shall limit my focus accordingly below.[2]

It is widely agreed that people are quite good at mind-reading; the question is how we do it, given the apparent privacy of consciousness. The dominant answer is that the mechanism is representational, inferential, and therefore "indirect." On this view, each of us has a "Theory of Mind" in our heads, which by analogy to a scientific theory represents others' minds well enough that we can infer their thoughts. Advocates of "direct perception" have challenged this picture, arguing that what we see in encounters with others is not representations of their minds in our heads, but their minds themselves in action. Both sides, however, assume that our minds are classical and therefore separable systems. This is particularly problematic for the direct perception account because it means that our contact with others' minds is subject to the constraints of local causality and as such cannot *literally* be direct.

I shall approach this problem indirectly, by way of how we visually perceive material objects. I do so for two reasons. First, the literature itself is preoccupied with the visual aspects of mind-reading, and as such curiously has not made language a central theme. There has been work on how learning language is important for developing children's mind-reading abilities,[3] and of course

---

[1] See Smith (2010a) and Gomes (2011).
[2] See Leudar and Costall (2004) for an historical but also skeptical view of psychology's concern with the problem of other minds.
[3] See for example Astington and Fillipova (2005).

much on expressive behavior, most of which is linguistic. But the specificity of dialogue as a route into other minds has been mostly neglected.[4] Second, "there are strong parallels to be drawn between the way in which the visual world is created and the way in which language is used to create our mental spaces."[5] In particular, I argue that by virtue of the non-locality of light, in vision we directly perceive objects in our environment. From there it will be easier to see that the semantic non-locality of language enables us to do the same with other minds. Language, in short, is like light.

The problem of perception

In a nutshell, the "problem" of perception is to explain how our senses hook onto external reality in such a way that we are usually able to navigate the world successfully. Most of the scholarship on this question focuses on visual perception. Vision is the primary sense through which humans interact with the world, and there seems to be a consensus that, notwithstanding the unique ways in which the other senses operate, the philosophical problems they pose are not essentially different.[6] The literature on vision addresses many issues,[7] but for my purposes the key question is whether perception is direct or indirect. My claim will be that the way this distinction is usually understood embodies a classical perspective that biases the discussion in favor of the indirect view.

Direct perception is generally considered the intuitive view of those untainted by philosophical training. Although there is no consensus on what precisely 'direct' means,[8] the basic idea is that in visual perception what we see are literally objects themselves rather than representations of them in our brains. This view, sometimes called "naïve realism," has at least three things going for it. First, it accords with what we experience. Objects *do* appear to be out there, in the world, not in our heads. Second, when we open our eyes objects are immediately present to consciousness, with no hint of concept mediation or mental inference. Finally, it extends naturally to lower organisms, for which talk of mental "representations" seems strained.

Despite the intuitive appeal of direct perception, most philosophers and vision scientists today think perception is indirect, or inferential. On this view, which is closely tied to the computational model of the mind, in perception

---

[4] For exceptions see Gallese (2008), Iacoboni (2008: Chapter 3), and Fusaroli et al. (2014).
[5] Ford and Peat (1988: 1235).
[6] See for example Velmans (2000: Chapter 6), Matthiessen (2010), and Crane (2014: 14–15).
[7] Crane (2014) provides an excellent overview of contemporary philosophy of perception, though addressing only its mainstream varieties; James Gibson's ecological view does not appear, for example, much less any quantum perspectives.
[8] McDermid (2001) identifies five primary meanings. My own usage below is closest to Warren (2005).

we are in contact with representations[9] of objects, not objects per se – "no perception without representation."[10] In part this is driven by philosophical arguments about hallucinations and illusions, but the crux of the matter is that vision science has shown there is a large gap between the information in the light striking our eyes and what we actually see – a gap filled in by unconscious inferences made in the brain. This makes theories of indirect perception "constructivist," since the brain must integrate or "construct" the output of its retinal sensors into perceptions.[11] And while many might eschew that term, most social scientists also think that perception is indirect. The belief that all observation is "theory-laden" is a rare point of agreement among positivists and interpretivists, and post-structuralists might go even further, arguing that perception is theory-*determined*.

Yet despite the claim that efforts to explain perception without positing representations have "failed in systematic and massive ways,"[12] the debate goes on, largely because the phenomenology of perception is hard to square with the idea that what we actually see in vision are representations of objects, not objects themselves.[13] This has kept "ecological" theories in the field, which take visual perception at least partly out of the head and try to embed or situate it in relation to the environment.[14] Moreover, a relatively neglected feature of the phenomenology of vision that will be important to my own argument is its projected quality. Given that one way or another, information processing is clearly involved, how does the brain project its internal "inferences" into an experience of objects out in the world? Or for that matter, why is a pain in your foot experienced in your foot, rather than in your brain, where the information is being processed?[15] Given the mind–body problem, it is not surprising that this would be puzzling. If we can't explain any experience, then we won't be able to explain visual projection either. But the difficulty is also empirical in that the information present on the retina is two-dimensional, whereas what is perceived is three-dimensional, and as we will see it is not clear how a classical computational brain could convert the one into the other.

As usual quantum theory almost never comes up in this debate, which suggests that both sides implicitly share a classical worldview. This manifests itself in various ways, but in particular in an assumption that the brain/mind is fully

---

[9] What used to be called "sense-data," but this is just one version of representational theorizing, most of which today disavows the term.
[10] Warren (2005: 337).
[11] See Paternoster (2007). The canonical text here is Marr (1982); also see Palmer (1999).
[12] See Burge (2005: 20).
[13] Though for a defense of the representational view on this score see Millar (2014).
[14] See Gibson (1979) and Orlandi (2013) for a recent example; cf. Hudson (2000) and Brewer (2007). Although critical of indirect perception, Orlandi's article is particularly good on why it initially seems so plausible.
[15] See Velmans (2000: 116).

separable from the world around it. A classical approach to perception is a kind of dualism, therefore, though one of subject vs. object rather than the traditional one of brain vs. mind.

Separability has a crucial implication that stacks the deck against direct perception, which is that the relationship of world to mind must be causal and local. Causality is implicit in the view that light is something that travels through space and initiates the perceptual process only when it makes contact with our retinal sensors.[16] As Michael Sollberger puts it, "entering into causal chains seems to be the *sole* way for entities to become *epistemically salient* to us perceivers."[17] Causality implies locality, or the classical assumption that no influence can propagate faster than the speed of light. From this Tyler Burge derives a "Proximity Principle," according to which "the effects of distal causes are entirely exhausted by their effects on proximal causes"[18] – which is to say, in the end what matters is what is going on inside our heads. Burge goes too far in claiming this principle is basic to science, since it does not hold in quantum physics. But he could still note, with Sollberger, that "empirical researchers and philosophers of perception alike agree that *direct action at a distance has to be banned from the macrophysical realm of perception.*"[19] This leads to a seemingly decisive "Time-Lag Argument" against direct perception: what we can directly perceive is happening right now, and since even at light speed it takes time for visual information to reach our retinas, what we directly perceive must therefore be "mind-dependent proxies" of the world rather than the world itself.[20]

Although they bias the debate in favor of indirect perception, the classical assumptions of separability, locality, and causality are so taken for granted in the literature that even advocates of direct perception have not called them into question. In a recent defense of direct perception, for example, Julie Zahle divides perception into two stages:

> In the first stage, the environment causes, via the light it reflects and emits, the activation of the retinal cells. In the following, I do not go further into this stage of the process. Instead, I concentrate on the second stage. It begins when the retinal cells are stimulated, and it ends with the formation of a perceptual belief. I refer to this second stage as the perceptual process.[21]

However, this reduces the Great Perception Debate to what is going on inside the brain, which is the home turf of the inferential view and begs the question against the Time-Lag Argument. Some ecological psychologists sense the

---

[16] This idea goes back to Galileo; see Reed (1983: 88).
[17] See Sollberger (2012: 590); emphasis in the original.   [18] See Burge (2005: 22).
[19] See Sollberger (2012: 587), emphasis added, and Burge (2005: 24–25); also see Sollberger (2008).
[20] The quote is from Warren (2005: 337), but I am drawing primarily here on Power (2010).
[21] Zahle (2014: 506).

danger here, that the "crucial problem [for their direct perception view] is action-at-a-distance."[22] But while gesturing in a quantum direction they have not yet developed a full-blown solution.

To my knowledge no one has studied empirically whether there are quantum effects in human vision.[23] However, as we saw in Chapter 7, plants, birds, and several other organisms are known to use quantum processes in perceiving their environments; and as we saw in Chapter 8, there is also considerable evidence that human beings perceive probabilities in quantum terms. These findings point toward a definition of direct perception genuinely distinct from indirect perception – i.e. as an intrinsically non-local phenomenon rather than as a different view about what happens in the brain. Problematizing the taken-for-granted classical bias in the literature would level the playing field and thereby give advocates of direct perception a much stronger hand.

However, in order to show that human perception really is non-local it is not enough that our "receivers" (brains) be quantum mechanical; we also need a change of perspective on the sending side. The reason is the standard view of light as something that travels from objects to our retinas, which justifies the separability assumption and motivates the Time-Lag Argument. So first we need to see that this is only half the story of light.

The dual nature of light

In the standard view, light consists of tiny particles (photons) traveling through space at a speed of 186,000 miles per second. Although relativity theory tells us that nothing can travel faster than light, according to the orthodoxy light is not special in any further, metaphysical sense.

Yet philosophers and physicists who have thought about the nature of light are not so sure. The physics of light have been puzzling ever since Michelson and Morley discovered in 1887 that light always has the same speed, regardless of one's perspective, and thus is never at rest. In this respect light is unlike everything else in nature, the measured speed of which can vary and is always relative to an observer. Relative to someone on the Earth's surface a car might be traveling at 60 mph, but relative to an observer in space, where the car is riding on Earth as it hurtles through the heavens, it would appear to be traveling at thousands of miles per hour. And likewise for every other moving object: its speed all depends on your distance from it and the angle and velocity of your own motion. Light is fundamentally different. No matter from what frame of reference it is measured, its speed is always the same. This "puzzle of light

---

[22] Kadar and Effken (1994: 322); also see Kadar and Shaw (2000: 167).
[23] Though there has been work on the theory side; see Woolf and Hameroff (2001), Flanagan (2001; 2007), Rahnama et al. (2009), and Khoshbin-e-Khoshnazar and Pizzi (2014).

speed constancy" was one of the key factors that led to Einstein's theory of relativity, and still provokes discussion among physicists today.[24]

Beyond its physics, it has long been noted that light is also unique in at least three, more philosophical respects, which David Grandy has ably summarized in a thought-provoking series of articles and books.[25] First, we never see light itself, only lighted surfaces or things. Grandy provides the example of a movie projector in a theater. Although we take it for granted that there is a beam of light from projector to the screen, unless there are dust particles in the air the beam itself will be invisible.[26] Or consider why outer space is dark despite the billions of stars around: as a vacuum, there is nothing *in* space for light radiating from stars to reflect off of and thereby reveal itself. The fact that light is only seen in conjunction with other objects means that it is impossible to objectify it, suggesting that light is not an object at all, but a *principle by which we see*, an "unframed window" within which the rest of the world is situated.[27]

Second, "time does not exist in the world of light."[28] We all know from *Star Trek* that as one accelerates toward the speed of light time slows down. What tends not to be noted, however, is that if we could actually travel at light speed, time would stand still. This calls into question our usual way of speaking about inter-stellar distances in terms of "light years," the years it takes light to get from a star to Earth. Relative to *our* frame of reference that is how it appears, but from the point of view of light itself – and in a panpsychist universe photons have such a thing! – it takes no time at all to travel from point A to point B, no matter how far apart they are.

Finally, if time has no meaning for light then in what sense could space? It seems strange to say that a phenomenon that, from its point of view, can get from Andromeda to Earth instantaneously is "moving" at all. And indeed, light moving is not something that can literally be seen at all, but an *inference* from what we can see.[29] To be sure, light has a "speed," but this is in its particle aspect from our perspective and thus only a partial truth; in its wave aspect light is inherently non-local.[30] In effect, the speed of light is not a normal part of our space-time regime at all, but, as Einstein thought, an exogenously given value that regulates the space-time properties of material bodies in that regime.[31]

It is considerations like these that have led some to think of light as not "just another kind of particle,"[32] but as metaphysically special, even if the

---

[24] See Grandy (2012) for an overview.
[25] See Grandy (2001; 2002; 2009; 2012); Heidegger and Merleau-Ponty are the two philosophers who appear most frequently in contemporary philosophical reflection on light.
[26] Grandy (2012: 542).
[27] The quote is from Grandy (2001: 11); also see Young (1976: 11) and Rosen (2008: 164).
[28] Young (1976: 24); also see Germine (2008: 153).   [29] Grandy (2012: 544).
[30] See Healey (2013: 50–53) for discussion of physicists' views on light as wave vs. particle.
[31] Grandy (2012: 542).   [32] Young (1976: 11).

idea is counter-intuitive. Grandy locates the intuitive difficulty in the modern – and one might add, classical and materialist – tendency to think of light as a "free standing" phenomenon, as something that exists on its own like any other material object. Yet as physicist Mendel Sachs has asked, "[w]hat is 'it' that propagates from an emitter of light, such as the sun, to an absorber of light, such as one's eye? Is 'it' truly a thing on its own, or is it a manifestation of the coupling of an emitter to an absorber?"[33] In other words, *light breaks down the separability of subject and object*. To develop the implications of this holistic view for vision and ultimately dialogue let me now turn back to the receiver side of the equation.

Holographic projection and visual perception

As we saw above, a key problem in making sense of visual perception is explaining its projection into the world, that we experience objects as out there rather than on our retinas. The difficulty here lies in understanding how two-dimensional information on the retina is transformed into three-dimensional perceptions of objects.

A promising analogy for thinking about this process has been suggested by Max Velmans, who argues that perceptual projection is holographic.[34] To produce a holographic image in the lab, laser light is split into two beams or waves. The "reference wave" proceeds from the origin of the split to a holographic film; the "object wave" proceeds to an object, envelops it, and then proceeds to the film, where it is reunited with the reference wave. Together they record an interference pattern on the film, like crossing ripples from stones dropped in a pond. Looking at the film, one would see just ripples. But if a third wave of the same frequency but reverse phase as the reference wave – a "reconstructive" or "phase-conjugate" wave – is directed at the film then a ghostly three-dimensional object will appear. Apart from the striking visual effect, holograms have three other features that distinguish them from traditional photographs.

First, they are holistic in the sense that all of the information about the object is recorded on each pixel of the film ('hologram' = "to write the whole"). In a photograph there is a 1:1 correspondence between points in the image and on the object itself, so that cutting the film in half cuts the picture in half. In contrast, cutting holographic film merely makes the overall image slightly less clear; with a hologram it is possible to reproduce the whole image even from

---

[33] The quote is in Rosen (2008: 164). Also see Flanagan (2007), who identifies perceptual fields with photon fields.
[34] See Velmans (2000: 114–127; 2008); on holographic perception also see Gillett (1989), and see Talbot (1991) for a popular introduction to holographic thinking more generally.

small pieces of film. This suggests a participatory rather than compositional relationship between parts and whole: the whole is present *in* the parts, not made up *of* them.[35] Second, by changing the frequency of the reference wave a single holographic film can record multiple images. This enables holographic film to store vastly more information than can simple photographs. Finally, holography is all about projecting a virtual, or simulated, image to a place where the real image (which is on the film) is not located. By analogy to perception, the "real" image is two-dimensional information on our retinas, but what we see is a three-dimensional object out in the world.

As a philosopher Velmans suggests only that there is an interesting analogy here; however, several scientists have gone further and speculated that the brain is *actually* a holographic projector. The first to make this claim was Karl Pribram in the 1960s,[36] and his work was later taken in an explicitly quantum direction by Peter Marcer and Walter Schempp.[37] In their model, neural surfaces function as holographic planes (the film), and the brain as a holographic projector continually emitting reference waves (millions per second). Object waves pour in from light reflected off material objects in the world. The resulting interference pattern is encoded on the neural surfaces and then decoded through a reverse process of "phase conjugate adaptive resonance" or "pcar," the brain's reconstructive wave. Marcer and Schempp argue that it is the ongoing process of phase conjugation which enables us to perceive objects as out in the world.

A further element has recently been added to this model that I believe completes the picture and also ties in nicely with my discussion in previous chapters. Mitja Perus and Rajat Pradhan have independently suggested that the pcar approach is best understood in conjunction with Cramer's Transactional Interpretation of quantum theory.[38] Recall that Cramer's is a time-symmetric framework that uses two kinds of waves, retarded waves moving forward in time and advanced waves propagating backwards; wave function collapse (and thus on my account, consciousness) occurs when the two waves meet and their "transaction" is completed. Applying this to vision, the idea is that the holographic reconstructive waves emanating from the brain are in fact advanced waves. This responds directly to the Time-Lag Argument, which assumes perception is a local, causal process. Local causation is still present in the transactional approach, in the form of retarded waves, but "advanced waves are the vehicles of mental perception through backward ray-tracing along a straight

---

[35] See Bortoft (1985: 282–283).
[36] See Pribram (1971; 1986); however, Robbins (2006) argues that the truly first holographic view of the mind was due to Bergson, albeit *avant la lettre*.
[37] See Marcer (1995) and Marcer and Schempp (1997; 1998).
[38] See Perus (2001: 583–584) and Pradhan (2012: 635–637).

line of the signals received by the brain through the knowledge sequence."[39] In short, by virtue of their backward propagation in time, advanced waves exploit light's non-locality, enabling us to directly "touch" objects quantum mechanically.[40]

This model is admittedly speculative but three considerations argue for it. First, it assumes that the brain is a quantum rather than classical computational device, for which I have argued earlier there is considerable evidence. Second, it provides a coherent account of how 2-D information on the retina is projected into 3-D experiences of the world. Advocates of indirect perception have suggested that this is accomplished by "heuristics" and "biasing principles" stored in the brain,[41] but absent a materialist basis for consciousness this amounts to hand waving. Pradhan thinks that projection is "impossible to comprehend" without advanced waves, and for Perus, "neural networks... without having electro-magnetic or quantum embedding, can definitely not back-project their images into the external space on their own."[42] Finally, recent work in cosmology suggests that holographic processes are not limited to the brain, but are universal.[43] If the "universe is a system of holographic surfaces within surfaces,"[44] then it would be surprising if organisms did not embody this principle as well.

Semantic non-locality and intersubjectivity

I return now to the debate about mind-reading, all sides in which I will argue at least implicitly buy into the classical worldview. As I note in passing, because the debate has been preoccupied with visual perception, the quantum account of vision developed above offers a distinct perspective in its own right. However, my main purpose is to leverage that account for thinking about the connection between quantum semantics and mind-reading, and specifically the role of semantic non-locality.

The theory of mind debate

Recall the "problem" in the Problem of Other Minds. On the one hand, it is widely assumed that mental states are exclusively intra-cranial phenomena and

---

[39] Pradhan (2012: 635); also see Perus (2001: 583).
[40] Perus (2001: 584); also see Manzotti (2006: 27–28).   [41] See Burge (2005: 10–18).
[42] See Pradhan (2012: 636) and Perus (2001: 583); also see Mitchell and Staretz (2011: 939).
[43] See for example Bousso (2002) and Bekenstein (2003). Long before these discoveries, David Bohm (1980) used holographic reasoning in a more qualitative way to conceptualize the relationship between the implicate order and explicate order.
[44] Germine (2008: 152).

as such intrinsically unobservable.[45] Thus, all we have to go on in dealing with others is their verbal and non-verbal behavior, which could be deceptive or admit various interpretations. On the other hand, in real world dialogue people generally know what others are thinking, since we communicate and coordinate our activities effortlessly most of the time. So how do we manage to read minds so well despite their unobservability?

A widely held answer is that human beings carry around a "theory of mind" in their heads which, when applied to others' behavior, enables us to know their mental states. Within this orthodoxy however there has been a long-running debate about two different views of how this theory of mind works. According to the "Theory-Theory," mind-reading is like scientific theorizing, in which others' behavior serves as observational data, and then from these our minds draw an inference to the best explanation about the mental states that cause them. According to the "Simulation Theory," in contrast, mind-reading is more akin to empathy, in which we interpret others' behavior by reference to what *we* would be thinking if we were behaving that way, and then project that simulation onto their mental states. The debate continues,[46] but whereas the two theories used to be seen as opposed, today there is more recognition of their respective limits, and a corresponding growth of hybrid arguments. This is possible because ultimately they agree on fundamentals: (1) minds reside entirely within brains; (2) perception is therefore indirect, mediated by representations in our minds; (3) in perceiving each other we are in a third-person, "spectator" mode;[47] and (4) communication therefore follows a signaling model, in which a sender emits signals to a receiver, who processes them in light of her own representations, and then sends signals in return. While never stated explicitly, this is clearly a classical picture premised on the separability of minds.

While the two theory of mind perspectives remain dominant in the literature, recently they have come under attack from a theoretically diverse group of scholars who reject its foundational assumption that mental states are actually unobservable. On this alternative view our perception of other minds is *direct* and thus does not require a "theory" at all.[48] That might seem counter-intuitive

---

[45] Krueger (2012: 149) calls this the "unobservability principle" underlying social cognition research; Gallagher and Varga (2014: 185) call it the "imperceptibility principle." For an argument that this principle is actually more nuanced than is usually assumed see Bohl and Gangopadhyay (2014).

[46] In part this is due to the discovery of "mirror neurons" in the brain, which have given a boost to Simulation Theory; see for example Gallese and Goldman (1998) and Iacoboni (2009). For more skeptical, Theory Theory perspectives on mirror neurons see Jacob (2008) and Spaulding (2012).

[47] See Hutto (2004).

[48] See for example Hutto (2004), Zahavi (2005), Gallagher (2008a; 2008b), and De Jaegher (2009).

given our lack of telepathic powers, but its advocates point to intuitions of their own to make their case.

First, in real life we are always dealing with concrete Others in particular contexts for particular purposes, which sits uneasily with the idea that we are "spectators" to each other's behavior, watching abstractly from afar. Social cognition is a participatory, "I-You" relationship, in other words, not a detached, "I-She" one,[49] which suggests that the *second* person rather than the third is the appropriate stance from which to think about mind-reading.[50] Such a perspective accords a special role to the space *in-between* us as a locus of intersubjectivity, as thinkers as diverse as Martin Buber, Georg Simmel, and Hannah Arendt have emphasized.[51]

Second, within these situated encounters there is little evidence at the conscious level that, to understand each other, human beings routinely engage in either inference or simulation. As Wittgenstein puts it:

In general I do not surmise fear in him – I *see* it. I do not feel that I am deducing the probable existence of something inside from something outside; rather it is as if the human face were in a way translucent and that I were seeing it not in reflected light but rather its own.

"We *see* emotion." – As opposed to what? – We do not see facial contortions and *make the inference* that he is feeling joy, grief, boredom. We describe a face immediately as sad, radiant, bored, even when we are unable to give any other description of the features. Grief, one would like to say, is personified in the face.[52]

The claim here is that mental states are expressed in behavior, not hidden away inside our brains causing it. So when we see other people behaving what we are experiencing is literally their *minds in action*.[53] In unusual circumstances yes, when we have no clue what others are thinking, we may be forced to theorize consciously about their minds, but most of the time perception is "smart" and so we are able to see their intentions directly.[54]

The battle between these rival approaches has now definitely been joined. Defenders of indirect perception have largely conceded that mind-reading is rarely conscious, but mostly operates on a sub-personal level. There the critics have continued to press their attack, arguing that events deep inside the brain

---

[49] See Reddy and Morris (2004) and Stawarska (2008).
[50] See Schilbach et al. (2013) for a discussion of what this would mean for neuroscience and Pauen (2012) for an interpretation more centered on intersubjectivity.
[51] See Fuchs and De Jaegher (2009: 476–477) and Bertau (2014a); on the "in-between" in Buber and Simmel see Stawarska (2009) and Pyyhtinen (2009) respectively.
[52] Quoted in Gallagher (2008b: 538); emphases in the original.
[53] See Zahavi (2008) and Krueger (2012); cf. Smith (2010b).
[54] On "smart" perception see Gallagher (2008a; 2008b).

hardly qualify as "theorizing" or "simulation."[55] Yet the orthodoxy has also launched attacks of its own. In particular, if mental states are enacted in behavior, then they cannot be perceived independent of the physical and cultural context in which they appear, which suggests a need for the kind of perceptual mechanisms postulated especially by the Theory-Theory – and that direct perception therefore cannot truly be a rival to a theory of mind approach.[56] However, rather than weigh in on this debate on its own terms, I want to highlight two absences from it, which motivate an alternative, quantum approach.

One is that neither side has considered that mind-reading could involve anything but the causal transmission of information from one actor to another. Since there is a time lag in any such process, this means that actors' separability – and thus a classical frame of reference – has not been problematized. That makes sense for theory of mind advocates, who think perception is indirect, but not for the other side. Gallagher recognizes that his view stands or falls on the definition of 'direct,' and distances himself somewhat from Gibson's view,[57] but he does not question that there is a local causal process underlying "smart" perception. So despite breaking with the Cartesian orthodoxy in thinking that minds spill out of brains into behavior, advocates of direct perception remain Cartesians with respect to the separability of minds. There is a lingering individualism here, in other words, which undermines the holistic thrust of their approach.[58] This is where a quantum theory of vision could help, since it suggests there is a *non*-local connection between minds-in-action, and as such that in mind-reading our minds are not fully separable in the first place.

To really make that point, however, we need to attend to a second absence in the mind-reading literature, which is the role of language.[59] What we see obviously matters, but communication depends at least as much on what we hear. That involves unique processes of sharing and comprehending linguistic meanings which cannot be reduced to visual perception, and when understood in quantum terms support the direct perception view of other minds.

Semantic non-locality and other minds

Two features of language stand out as the basis for its unique contribution to mind-reading. First, unlike visual perception, language use is intrinsically

---

[55] See Herschbach (2008) for a defense of theory of mind at this level, and Zahavi and Gallagher (2008) for a response.
[56] See for example Jacob (2011) and Lavelle (2012).
[57] See Gallagher (2008b: 537 footnote 2).
[58] See De Jaegher (2009), and also Chapter 13 below.
[59] The most significant exception of which I am aware is Hutto (2008), who argues that direct perception of other minds depends on constructing narratives about them.

dialogical, performative, and dynamic. Its main purpose is to do things together, and it does this over time, not typically by one-off perceptions of others' states of mind. Second, to be dialogical language must be shared. The effect of an actor's speech on how someone else perceives them will depend on whether or not they speak the same tongue; what is meaningful behavior if they do will be gobbledygook if they don't. In what follows I take up the sharing point first, and then turn to dynamics.

The classical assumption that our minds are fully separable means that sharing a language amounts to nothing more than that Emma's brain is wired to speak English and Otto's brain is too. In a material sense – and given the CCCP, ultimately that's all there is – nothing about these two facts presupposes the other, or exists in-between. It is as if Emma and Otto both had blue eyes.

From a quantum perspective sharing a language is something more than that. Recall first that in contrast to the classical view of language as a set of well-defined meanings and rules of combination, in the quantum view the language in each of our heads is a superposition of potential meanings that are actualized only by its collapse in speech acts. So what happens if we put two quantum minds together and they start trying to communicate? Well, if one is from Vietnam on holiday in Denmark and the other is a local merchant, neither of whom speaks the other's native tongue, then initially their linguistic competencies will be separable, and their dialogue limited to gestures and other visual cues. However, if one ventures to say "English?" and the other says "yes," then suddenly a new superposition will be created in which the meaning of their potential English speech acts will be entangled with their meaning in the other's mind. Their linguistic competencies are now no longer fully separable, but correlated non-locally through an over-arching system of meaning *between* them. And insofar as language is constitutive of mind, this means not just that shared languages are not fully separable, but their associated minds are not either.[60]

Now consider what happens when we move from non-locality at the semantic level, which is static, to the pragmatics of actual dialogue. In the classical view, dialogue is a causal signaling process, in which Emma tries to convey meanings in her head by moving her vocal chords, the vibrations from which travel through the air, strike Otto's ears and are then processed by his brain into meanings in his head, on the basis of which he responds in turn. Call this the "transport" or "conduit" model of how meanings are communicated,[61] which posits a chain reaction of local triggers. *Truly* direct perception of linguistic

---

[60] Cf. "distributed language theory" (e.g. Steffensen, 2009), which especially in Steffensen and Cowley (2010) flirts explicitly with the idea that language is non-local. However, they make no reference to quantum semantics and in the end seem ambivalent about how far to push the quantum connection (see Linell, 2013: 170).

[61] See respectively Ford and Peat (1988: 1235) and Lipari (2014: 506).

meaning, by which I mean perception that does not reduce to a causal, time-lagged process, is therefore impossible. Moreover, the meanings that senders and receivers of linguistic signals produce together cannot be shared in any deeper sense than sharing blue eyes. The meanings each assigns to sound waves remain locked inside their heads.

The starting point for a quantum alternative would be speakers' co-presence in a concrete situation, which means that each forms part of the context for the other's speech. As we have seen contextuality is intrinsic to quantum theory, which is why quantum semantics handles the contextual features of individual word meanings so naturally. Putting that analysis into a dialogical setting expands the context in two ways. One, at least in the case of a face-to-face encounter, is to bring on board the non-local properties of visual perception that I described above. The other, which I shall focus on here, follows Jean Schneider in likening a measurement in quantum physics to a speech act.[62] An important consequence of this will be that communication through speech is not primarily causal, at least not in the classical, efficient sense.

Measurement in quantum physics can be broken down into two steps. In Step One, the experimental design selects a "preferred basis" for the wave function that determines the probabilities associated with its collapse. This selection does not cause a change in the wave function's structure, because there has not yet been an interaction with "it"; if there had then it would have already collapsed. Rather, what Step One does is create an entanglement with the observer that constitutes a specific context for the measurement. In Step Two, the physicist does the measurement and the particle's wave function collapses into an observed outcome. Two aspects of this process are important to emphasize. First, in actually performing the measurement (vs. taking a lunch break) she is also collapsing her own wave function through a teleological act of will.[63] Second, her action does not cause the observed outcome, since wave functions are not material objects upon which one can exert force. Rather, triggered by the decoherence inducing measurement the particle's wave function causes *itself* to collapse, though – like the experimenter – in a teleological sense of causation. With these two steps in mind we can now see why measurement is like a speech act. 'I now pronounce you husband and wife' "does not describe a situation independent of itself, it creates what at the same time it describes."[64] Schneider is careful to make clear that in physics, the 'create' here does not mean the experimenter's consciousness causes the outcome. Her point is that what is being registered is not a state of the world independent of the observer,

---

[62] See Schneider (2005).
[63] Also see Chapter 8, p. 163, where decision-making is treated as measuring one's own preferences.
[64] Schneider (2005: 349).

but a state of consciousness in relation to a world with which it is entangled. In that sense "the measurement act... is an act of attribution, a declarative act,"[65] as if to say, "this is what has happened."

Now consider Emma and Otto's conversation, with their quantum minds. The initial "preferred basis" is constituted by (a) the fact that each knows the other speaks English and so their linguistic repertoires are already entangled at a general level, and (b) the immediate context that prompts them to start talking – such as their parents bringing home a set of Lego building blocks to play with. That event reduces their superposed minds to the particular state vector associated with "Legos," and thus changes the probabilities that the first words out of their mouths (ideally after "thank you, Mom and Dad") will concern Legos rather than, say, what to watch on TV. Note that the appearance of the Legos did not *cause* this change in probabilities, because the kids' minds are still in superposition with respect to the Legos; what has happened so far is merely that, through visual contact, their minds have adjusted non-locally to the new context.

Emma now opens the dialogue. Assuming she prefers to play with Legos, within that state vector she still has various (free) choices to make, from "let's build something together" to "let's build separately" to "let's make a mess by throwing the Legos around the room." Say she opts for "build together." In willing this speech act she makes two measurements: one on herself, collapsing her previously superposed preferences into one preference, and the other on Otto – though for the latter with a twist. For unlike measurements in physics, where the decision to actually run an experiment or break for lunch does not affect the context for the particle's eventual choice (which has been fixed by the research design), in this case Emma's statement has precisely that effect. It changes the preferred basis for Otto's response rather than immediately inducing a collapse of his wave function in a speech act of his own, which is to say it happens in Step *One* above, not Step Two. So while Emma's words affect the context and thus probabilities for Otto's reply, this is a non-local change made possible by their entanglement, not a causal one. And similarly then when Otto starts an argument by collapsing his own wave function with "no, I want to build my own structure," which changes the preferred basis for Emma's next statement, but the actual choice of which is still up to her. In short, in dialogue there *is* no Step Two, because whereas the particle is trapped in a stable context, every time people speak it changes the context for those who are listening. Put in different terms, as the literature on order effects shows (Chapter 8), speech acts *interfere* with trying to measure the true state of our interlocutors. The only unmediated measurements we can take in social life are on ourselves, not each other.

[65] Ibid.

In sum, the classical model of dialogue assumes that our minds are separable information-processing machines in well-defined states. Given that assumption, the process of understanding someone's meaning will be necessarily inferential and indirect: verbal signals come in from Sender and then Receiver's brain crunches the data to infer their meaning. The quantum model assumes that our minds are entangled through language and context and thus not fully separable, and that our brains are quantum computers in superposed states. On this view, there is normally no need to infer a speaker's meaning, since it is contained right there in her words and their context, which are picked up non-locally, i.e. directly, rather than "transported" to the listener's mind. This is not to say that the meaning communicated will always match the intended meaning, since a speaker may have expressed himself poorly, and the actual associations a listener makes will also reflect the state of her own quantum mind. And nor is it to say that inferences to meaning never take place. Sometimes it is unclear what someone is trying to say, in which case conscious reflection may be needed to figure it out. But given how effortless everyday communication is, such cases are likely to be the exception rather than the norm.[66]

Three objections considered

In concluding this discussion of dialogue and semantic non-locality I want to address three potential objections. The three are related, but each offers a distinct window on what I am trying to say, and as such my hope is that even if it creates some redundancy, responding separately to them will help pull my argument together.

The first is that what my Lego example portrays is more like the "parallel play" of toddlers than genuine dialogue, since there is no real *interaction* between Emma and Otto, just a series of speech acts that have no direct impact on each other's minds. This might seem problematic on the grounds both that it is an unrealistic description of adult conversation, and that after my complaints about the classical view of shared meanings reducing to sharing blue eyes, it appears that shared meanings are not possible at all.

If by 'interaction' we mean causal interaction, then my claim is indeed that when we are talking with each other adults are actually not interacting. It is a measure of how deeply the classical worldview has affected contemporary social theory that most readers will probably find this idea counter-intuitive, but it follows directly from quantum theory, which is *a*causal.[67] In experiments physicists do not "interact" with particles, since until they perform their

---

[66] Although he does not mention quantum theory, Recanati (2002) makes a strong case for an anti-inferential view of linguistic communication that resonates with my argument here.

[67] It is also a salient feature of Leibniz's metaphysics, in which monads do not interact causally; see Bobro and Clatterbaugh (1996), Piro (1997), and Puryear (2010), and see Nakagomi (2003a: 16) for a quantum rendition of Leibniz's conception here.

measurements there are no particles there to be interacted *with*, just wave functions, and as such no prospect of causally affecting their behavior. And if not just particles but people are quantum systems, then the same is true in dialogue, on both sides of the "measurement": when it is my turn to speak others will be in a superposition rather than well-defined states, and I will be so myself prior to collapsing my wave function in a speech act. The concept of interaction presupposes entities with well-defined states, in other words, which does not hold in the quantum world. "Intra-action" is more like it,[68] according to which it is in entangled speech acts that we actualize ourselves as separable beings.

However, this is not to say that we cannot *influence* each other through speech acts, which clearly we do. The issue is how this influence comes about. In my account, it occurs indirectly, through the effects our speech acts have on the shared *context* of our conversation. These are not causal effects, since a dialogical context is an intentional object rather than a material one. But by adding words to the context and thereby changing the preferred basis for subsequent speech acts we can influence the probabilities of what others will say, even if the actual cause (in the teleological sense) of what they *do* say comes entirely from them. The difference between dialogue and parallel play, therefore, is not that we are interacting and the toddlers are not, but that they are not attending to the same context and as such have less opportunity to influence each other's actions, whereas we typically do and therefore can.

What about the worry that if through dialogue we cannot have any causal effect on each other's minds, then the prospects for creating shared meanings – and thus social cooperation – look rather grim? The answer depends on what is meant by 'shared.' If it means *exact same*, then yes, my argument suggests such meanings will be difficult if not impossible to create. In my panpsychist reading of quantum theory, every quantum system from human beings on down is its own coherent, impenetrable monad, situated in a distinct time and place, and as such will have a unique subjective perspective on the world around it.[69] However, if "shared" language refers to a *superposition* state with which participants are entangled and thus is available non-locally to all, then as long as that state does not admit too many conflicting meanings, cooperation will be possible. And indeed, narrowing down potential meanings – not to just one, but to something like an overlapping consensus – is one of the primary functions of dialogue. We do not all have to agree in the privacy of our own minds on what "the US Constitution" means, for example, for us to work together to uphold it.

---

[68] See Barad (2003; 2007 passim), as well as my discussion of quantum game theory in Chapter 8.
[69] See Jorgensen (2009) on the challenge this poses to holism and communication about meaning, and a non-quantum attempt to meet it.

The second objection is that it is a well-established principle of quantum theory that non-locality cannot be exploited to communicate between entangled entities; meaningful signals can only be exchanged classically. Given my claim that shared language entangles people through semantic non-locality, this ban on super-luminal signaling might seem to preclude us from communicating at all!

However, dialogue has not only a quantum dimension, but also a classical one, in two respects. One is the words that are actually said. Like the particles that emerge from collapsing wave functions, speech acts are actualizations of potentials, and once "out there" in consciousness remain in memory and cannot be erased (though their meaning may change after the fact; see Chapter 10). These words constitute classical events, and it is to them that a listener ultimately responds. But that does not reduce the quantum theory of dialogue to a classical one, because the meaning of words does not stand alone, but is given only in context. What semantic non-locality adds is a physically grounded, holistic notion of that context, which predisposes the listener to interpret spoken words in certain ways rather than others. In that way semantic non-locality makes classical signaling possible.

The other classical aspect of dialogue is what the participants are talking about. The "minimal communicative situation" always involves three elements, not just the two on which I have focused so far: Emma, Otto, *and Legos, the shared object of their conversation.*[70] In my example this was an actual material object, but often it will be an intentional object, like the state or what to watch on TV. Intentional objects are not material and as such in my view cannot be explained by the classical worldview. However, when we are thinking or talking about them they are objects of which we are conscious, and consciousness takes place in the classical world of actuality rather the quantum world of potentiality. This means that intentional objects are *real*, but not in the usual sense. Real in what sense, then? Building on my account of visual perception above, in Part V I argue that shared intentional objects are holographic – projections of conscious minds of "virtual" objects into the space between individuals, which exist only as long as people are conscious of them. That argument will take some time to unfold, so suffice it to say here that the presence of such virtual objects in dialogue makes communication possible by giving participants a common referent, beyond their semantic entanglement, to which their speech acts are directed.

The last objection I want to address goes to the heart of my thesis that in dialogue people can directly perceive other minds. Recall that an important criticism of all theories of direct perception is the Time-Lag Argument: that even if information about objects or others' behavior is picked up directly from

---

[70] See Cornejo (2008: 174).

the light striking our eyes, the light itself still takes time to travel from there to here, and therefore perception must necessarily be indirect. In the case of dialogue, the problem is compounded by the fact that sound waves travel much more slowly than light, and so unlike my non-local theorization of light above, there is no sense in which one could argue that, from sound's "point of view," time stands still. Interestingly, the direct perception model is today one of the leading accounts of speech comprehension, according to which what we hear in speech are not acoustic cues per se but the vocal gestures that produced those cues.[71] Yet this is explicitly understood as a causal process,[72] which challenges my argument because causation is local and classical – and therefore in my view not *really* direct.

My response to this objection is two-part. The first is to point out that speech perception is not an additive and linear process but a holistic and dynamic one. We do not grasp the meaning of someone's speech word-by-word or even sentence-by-sentence, only achieving comprehension at the end of their soliloquy. Rather, we do so through *gestalts*, which sub-consciously relate what is being said backwards to what they have already said in the past and also by anticipating what they are going to say in the future. As Francesco Ferretti and Erica Cosentino put it,

understanding each sentence is not enough to understand the sense of what is said; the comprehension of speech flux implies that the listener is actively listening for a flux of speech and continuously checking for coherence in what is being said as well as predicting what the speaker is about to say.[73]

This makes sense in light of the importance of context in determining meaning. The context of speech comprehension is not only synchronic, with words sitting side-by-side in one moment of time, but also diachronic, with words gaining their specific meaning in virtue of how they fit into a moving, temporal whole. The boundaries of this whole are set not only by the speaker's decision to start and stop talking at certain points, but also by the listener's recognition of what they are trying to do together through dialogue.[74] If semantic context is a non-local, quantum phenomenon, then the dynamics of speech comprehension suggest that this non-locality can be temporal as well as spatial.

The second part of my response builds on this point by invoking the idea of "mental time travel," "the faculty that allows us to mentally project into the past

---

[71] This is due especially to Carol Fowler's work (e.g. 1986; 1996), which also highlights that speech perception is not intrinsically different than other forms of perception; also see Worgan and Moore (2010). For a recent overview of this field see Samuel (2011).

[72] Fowler (1996: 1732).

[73] Ferretti and Cosentino (2013: 28); also see Cornejo (2008), Pickering and Garrod (2013), and Bertau (2014b).

[74] See Togeby (2000).

and the future in order to relive or anticipate events."[75] A key aspect of "MTT" is "temporal perspective taking," or projecting oneself into different temporal contexts, which enables "individuals to extend their consciousness beyond the 'here and now'."[76] In Chapter 10 I used this idea to argue that it is possible in certain respects to change one's past. In the case of dialogue MTT means being able to project oneself into the temporal perspective of someone else, whose communicative intention is expressed in her words taken all together, which are spread out both backwards and forwards in time. Advocates of the concept of MTT have not so far linked it to quantum theory, but given everything that has been said above, making such a link would be quite natural. On this view, what listeners are doing in projecting themselves temporally into others' speech is taking advantage of semantic and now also temporal non-locality to gain access to their minds.[77] Because it is non-local this projection is not subject to the time-lag involved in the causal transmission of sound waves from speaker to listener, and as such it supports a truly direct account of speech perception. At the same time, because it is sub-conscious, this projection also means that there is still a role for that causal process, which takes place in the classical world where conscious speech takes place.[78]

In sum, I have tried to do three things in Part IV. The first was simply to review the emerging literature on quantum semantics, which not only shows that quantum theory provides a compelling way to model the production of linguistic meaning, but in doing so provides leverage on the debate between compositional and contextual approaches to language. The second was to weigh in on the visual aspect of the Problem of Other Minds, where I argued that the non-local quality of light not only sets up a much more demanding test for theories of direct perception (i.e. that it be non-causal), but also helps them meet that test. Finally, I suggested that the idea of semantic non-locality allows much the same reasoning to be applied to dialogue, which has been relatively neglected in the mind-reading debate. Far from being locked away inside our brains, in dialogue our minds spill out into the world, where they can be directly perceived by others. From this standpoint the default question is less "how can we ever know what others are thinking?" than "how is it that sometimes we make mistakes?"

---

[75] See Ferretti and Cosentino (2013: 24–25), and more generally Suddendorf and Corballis (2007) and Suddendorf et al. (2009).

[76] Ferretti and Cosentino (2013: 39); also see Gerrans and Sander (2014).

[77] See Velmans' (2000: 118–119) discussion of projected auditory sensations, although he does not make the point in quantum terms.

[78] It would be interesting to extend this argument from speech perception to the case of writing/reading, which would allow for non-local communication over longer temporal distances, including with the unborn (from an author's point of view traveling forward in time) and with the dead (from a reader's traveling backward), but I won't try to do so here. See Togeby (2000) and Tylén et al. (2010: 5–8), however, for good places to start.

In developing these arguments Part IV is the first of two installments on my larger, holistic claim that human beings are not fully separable. Indeed, what I aim to suggest with my quantum account of direct perception is that in our "intra-actions," there is a literal sense in which – in Daniel Kolak's enigmatic formulation – "I am You."[79] I am You insofar as we are entangled in a linguistic wave function, which is to say at the unconscious level. And I am You when we collapse our wave functions together in intra-action, which is to say even in consciousness. To be sure, this does not mean we are numerically identical, any more than entangled particles are. By virtue of being organisms constituted and sustained by quantum coherence you and I must also live our own lives, which will end separately. Thus, I am You only *in potentia*, as a shared superposition, and even though this potential is actualized in intra-action, that is only for the moment, as long as we are co-present. But the non-locality of vision and language still affords a deeper conception of "We" than the classical I *and* You. It is I think captured well by Kolak's idea of "Open Individualism," according to which the boundaries between individuals are blurred in much the way as the boundaries between oceans. The North Pacific is not the same as the South Atlantic, but at the level of the shared unconscious we are all part of *the* (one) ocean, and in intra-action we form temporary unities of consciousness, even if experienced from separate points of view.

[79] See Kolak (2004); for a useful introduction to this very difficult text see Zovko (2008).

Part V

The agent–structure problem redux

Introduction

The problem of how to understand language and its relationship to speakers is an instance of the more general problem of how to understand the relationship between social structures and agents. Languages per se do not constitute social structures – one needs particular forms of language, such as discourses about "capitalism" or "marriage," to do that – but they make such social structures possible in the first place.[1] In this Part, I push closer to such discourses by re-considering, from a quantum perspective, the ontology of social structures and their relationship to agents in general.

The agent–structure problem is concerned with understanding the relationship between intentional agents and the structured social systems or societies in which they are embedded. Typically the agents in question are assumed to be individuals, but in some disciplines, notably IR, they are corporate or group-level phenomena like the state, to which scholars often attribute agentic properties. Since I have already discussed at length how individual agents should be conceptualized in quantum terms (Part III), it makes sense to retain that focus on individuals as I attend to the concept of social structure here, although in Chapter 14 I take up the question of state agency as a way of illustrating the potential purchase of my approach for my own field of IR.

In the social sciences the concept of social structure is defined in a bewildering variety of ways.[2] What precisely a quantum view of structure entails will become clearer below; first, following Anthony Giddens, I want to distinguish it from the concept of social system (or society).[3] The latter refers to behavioral regularities that our ET friends might discover simply by looking through their surveillance cameras. Social 'structure,' in contrast, refers to mind-dependent, relatively enduring relationships that explain those regularities. As such, to speak of the agent–structure rather than –*system* (or –society) problem is to

---

[1] See Elder-Vass (2010b) for a good discussion of this distinction.
[2] See Porpora (1989) for a useful taxonomy, and Wight (2006: Chapter 4) for a further elaboration and extended discussion with reference to IR scholarship.
[3] See Giddens (1979: 61–66).

highlight (unobservable) cause (sic) rather than (observable) effect. That makes sense in the present context, since identifying the physical location of social structures was one of the puzzles with which I motivated this book. However, although my focus will be on the ontology of social structure, it should be clear that this cannot be completely divorced from the ontology of social systems. Without enduring relationships there can be no system, and without behavioral regularities to instantiate those regularities there can be no structure.

The "problem" of agents and social structures is that although all sides agree that they are related, it is not clear how this relationship should be understood. Following Philip Pettit, I shall distinguish two questions in the debate, which he calls vertical and horizontal.[4] These are often conflated, but given my eventual argument in this chapter, it is important to separate them. To each question Pettit identifies two main answers, which yields a 2×2 matrix of positions.

The vertical question is whether social structures are reducible to agents and their interactions or emergent from them – or, put another way, whether social reality is stratified into distinct "levels," the macro-level of structure and the micro-level of agents.[5] The debate here pits what Pettit calls collectivists against individualists. Going back at least to Durkheim, the former advocate hierarchical ontologies, in which structures are seen as emergent phenomena that cannot be reduced to agents. Durkheim's own view of structure is widely seen as suffering from intractable problems of reification and as such is not a significant position today. However, more sophisticated forms of collectivism have enjoyed considerable attention of late: in sociology, where emergence is a central theme of critical realism and neo-Durkheimian views; in analytical philosophy, where the idea of irreducible collective intentions has gained widespread currency; and in IR, where "levels-talk" is routine.[6] Individualists in contrast defend "flat" ontologies in which social structures are seen as nothing over and above the properties and interactions of agents.[7] This is the ontology behind the push for micro-foundations in economics, political science, and elsewhere, with game theory leading the way.

The horizontal question has been less widely thematized. Rather than ask whether social structures are reducible to individuals, it concerns the relationship among individuals themselves, in particular what makes them as

---

[4] See Pettit (1993a: 111 and passim).
[5] Pettit sees the question somewhat differently, as one of whether social regularities can undermine our picture of human beings as intentional agents, which is more about social systems than structures.
[6] Though in IR it is often meant in only an analytical rather than ontological sense; see Temby (2013) on the state of this art.
[7] The term "flat" ontology has been popularized by post-structuralists like Deleuze and Latour, whose work differs in many ways from those who would describe themselves as individualists. However, the two groups share an opposition to hierarchical ontologies, and "flat" certainly seems an apt description of the ontology of mainstream individualism.

"individuals" in the first place. The debate here is between what Pettit calls atomists and holists. Atomists take the view that it is our material properties as organisms that constitute us as individuals, which are completely encased within our skins.[8] Note that this is not to deny that our mental states are causally shaped by interaction with other people; it is to claim that, *constitutionally*, those mental states are what they are solely in virtue of the material states of our brains. The upshot is that individuals are ontologically prior to society.[9] For their part, holists of course do not deny that our bodies have skins, but argue that the content of our minds presupposes relationships to other individuals. Again, the claim here is not the causal one that mental states are formed through interaction with other people, but a constitutive one – that our thoughts cannot even be defined except in relation to others.[10] On this view the body may be atomistic but the mind is not, which means that, considered in their totality, agents are not ontologically prior to social structure. While that might seem counter-intuitive, as we will see it is a widely held view in philosophy and I will argue follows from a quantum perspective as well.

When combined, these dimensions and distinctions generate four types of social ontology.[11] However, judging from the complete absence of engagement with quantum theory, all sides in the debate implicitly assume a classical worldview.[12] To be sure, this judgment is complicated by the fact that, unlike part–whole debates in chemistry or biology, here the parts are intentional agents, and the wholes are collective intentions. If my claim in Chapter 1 is correct that the classical worldview can never make room for intentional phenomena, then since all sides in the debate appeal to such phenomena one might equally well argue they implicitly assume a quantum worldview instead. Nonetheless, as we will see classical thinking still plays a key role in structuring the debate, particularly in assigning burdens of proof, which have made it difficult especially for emergentism and holism to gain traction and thereby kept them at the margins of social theory.

The following chapter is organized into four sections. I begin with the emergentist-reductionist debate in philosophy – the vertical axis of the problem – and show that a classical worldview has privileged individualism as a

---

[8] See Farr (1997) on the significance of the skin as a boundary of identity, and Bentley (1941) for a critique.
[9] Note that the issue here is not whether people are considered individuals or actors in a *social* sense, which is arguably a modern invention (see Meyer and Jepperson [2000]).
[10] On the distinction between causation and constitution see for example Wendt (1998) and Ylikoski (2013).
[11] Pettit (1993a: 172–173) does a nice job showing that the two "mixed" positions – atomist collectivism and individualist holism – are intellectually coherent.
[12] The only exceptions of which I am aware are Kessler (2007), who is responding to my (2006); Lawson (2012: 355–356), who briefly invokes quantum field theory; and Pratten (2013), who develops a process-theoretic ontology of social life.

starting point for the corresponding debate in social theory. In the second section I focus on efforts to articulate an emergentist conception of social structure based on the concept of supervenience. After arguing that such efforts are doomed to failure, I bring in the horizontal axis to critique the supervenience approach from a holist, and specifically "externalist" perspective. While that dislodges supervenience thinking from its privileged position, I argue that externalism is itself incompatible with the classical worldview. In the third section I show how the unique nature of emergence in quantum systems enables us to recast the entire debate. The result is a social ontology that is at once holistic and emergentist, yet flat. With this framework in place, I illustrate its ability to help us solve a key problem for all forms of emergentism, which is how to understand the "downward" causal powers of social structure. As we will see in Chapter 14, the result is a holographic or monadological model of society in which each of us is a "pixel," entangled in social structures that both enable our agency and give it potentially far-reaching, non-local effects.

13   An emergent, holistic but flat ontology

The heyday of emergentist thinking was in the 1920s, when it was seen as a middle position in the reductionist–vitalist debate.[1] However, as a result of ambiguities in its original formulation and the rise on the larger philosophical scene of logical positivism, with its strong reductionist impulse, emergentism fell rapidly out of favor. However, since the 1970s it has undergone a revival, due on the one hand to the failure of reductionism to solve the problems of life and mind, and on the other to the invention of dynamical systems and complexity theories, which seem to cry out for an emergentist interpretation. Emergentism is by no means now orthodoxy, especially in the practice of science, where reductionism is still the default method, but in philosophy it is again taken quite seriously.[2]

The situation for emergentism is if anything even better in the social sciences, where the issue (if not the term itself) was first taken up in the debate about methodological individualism in the 1950s. Although that debate was inconclusive, the importation of positivist philosophy of science in the 1960s and then the spread of rational choice theory from economics to the other social sciences gave the momentum to reductionism. Yet, in sociology macro-level theorizing never lost its premier position, and critical realists there like Roy Bhaskar and Margaret Archer were among the first to develop explicitly emergentist social theories;[3] in economics it has become apparent that macro-theory cannot be reduced to micro; and in political science the rise of rational choice theory has been balanced by historical institutionalism, constructivism, and other non-reductionist social theories. So in the social sciences emergentism has always been a live option.

Emergentism and reductionism share at least one key assumption: a materialist ontology.[4] This is not surprising for reductionism, which is all about showing

---

[1] See McLaughlin (1992) for a good overview of this work.
[2] See Kim (2006), Corradini and O'Connor, eds. (2010) and O'Connor and Wong (2012) for good overviews of the philosophical debate.
[3] See Bhaskar (1979; 1982), Archer (1995), Wight (2006), and Elder-Vass (2010a) for the contemporary state of the critical realist art.
[4] See for example El-Hani and Emmeche (2000: 241) and Kim (2006: 549–550).

how, not just macroscopic physical objects, but seemingly non-material phenomena such as life, consciousness, and social structures are all ultimately made of the material stuff ostensibly described by physics. In emergentism the materialist premise is less obvious but no less important. By definition, an emergent phenomenon emerges *from* something else, and does so not capriciously, but necessarily as a result of the organization of its elements, which emergentists have also assumed are material. If there were no such connection to a material base, as in Cartesian dualism, then the higher level would not be "emergent" at all, but just *there*, untethered to physical reality. So while emergentists tend not to privilege physics in the way that reductionists do, they too accept the CCP.

Despite this common starting point, however, the two theories differ fundamentally on whether anything is truly emergent, which comes down to three main emergentist theses. The first is irreducibility. Although there is debate about what reductionism actually requires,[5] the intuitive idea of irreducibility is straightforward: "the whole is more than the sum of its parts." The second thesis is that emergent phenomena have novel properties not possessed by their parts. Again there is debate,[6] but it is widely agreed that the novelty must appear to be qualitative (e.g. life or mind relative to matter) rather than merely quantitative (a ten-pound bag of sugar vs. ten one-pound bags). This makes emergentists property-dualists even as they reject substance dualism. The third thesis is that emergent phenomena exhibit "downward causation," in which the whole affects the properties and/or behavior of the parts. This is the most controversial claim of emergentism, since it raises a question of circularity, or how a whole can cause its own parts, which critics think is incoherent and inconsistent with the CCP.[7] Even some emergentists are wary of downward causation, but in the view of many it is an essential part of the doctrine, since if emergent properties lack causal powers then they are epiphenomenal and thus redundant. A key question here is whether emergence is taken to be diachronic or synchronic. The former is much the easier argument for emergentists to make because it allows for causal interaction between parts and whole, but for that reason it is also less interesting philosophically.[8]

How the debate about emergence plays out depends crucially on whether one is talking about ontology or epistemology, or what sometimes is called "strong" vs. "weak" emergence.[9] Both address the relationship between levels, but the things being related are different – in the former it is entities, events,

---

[5] See for example Silberstein (2002: 82–89) and Wimsatt (2006).
[6] For discussion see Francescotti (2007).
[7] See for example Kim (2006). I discuss this issue at length below.
[8] E.g. Humphreys (2008), and see Elder-Vass (2007) for a good discussion of how this distinction relates to emergentism in critical realism.
[9] See, for example, Clayton (2006), McIntyre (2007), and O'Connor and Wong (2012).

and properties (i.e. elements of reality), in the latter it is theories, concepts, and models (elements of descriptions or explanations).[10] Importantly, positions on the two issues need not be the same: one can be a reductionist about ontology but an emergentist about epistemology (but not vice-versa).[11] Although in the social sciences the distinction is often blurred, most of the discussion has been about epistemological emergence. This is not surprising, since unlike chemistry or biology, where different levels of material reality are manifest, in social science all we can literally see are people. From a materialist perspective, what appear to be inter-level relationships are actually intra-level ones, among individuals.[12]

To be sure, some social scientists, notably critical realists, think that, even though invisible, emergent social structures are *real*. However, as I argued in Chapter 1, in the classical worldview everything is ultimately material, so if social structures are real then why can't we see them? The realist reply is that we can infer them from their effects, but (a) this is unlike other classical phenomena, and (b) it assumes the reality of the intentional states that ostensibly generate those effects. Because those states imply consciousness, the existence of which materialism puts in question, this puts realists on shaky ground: if consciousness is an illusion, then why aren't social structures too?[13] And indeed, as avowed materialists, realists too are wary of reifying social structures. So realists are caught on the horns of a dilemma: they either can accept that social structures are ultimately material and therefore not really emergent; or accept a non-materialist, dualist ontology.[14] In short, *if* one accepts the constraints of the classical worldview, it seems impossible to defend social emergentism in an *ontological* sense. This may explain why, unlike the long debate about epistemological and methodological individualism,[15] there has been almost none about ontological individualism, which outside critical realism seems to be taken for granted on all sides.[16] Nonetheless, if the latter were to fail, then epistemological individualism would fail with it.

With this dialectical situation in mind, and since this is a book about ontology, I am going to focus on the ontological sense of emergence, which is the hard

---

[10] Silberstein (2002: 90).
[11] Though see Hüttemann (2005), who argues that the reverse is true in quantum physics.
[12] Le Boutillier (2013: 214).
[13] Although his starting point is different, see Harré (2002) for a provocative argument to this effect.
[14] This may help explain the ambiguities of Bhaskar's emergentism, on which see Kaidesoja (2009). Sawyer (2005: 80–85) reads Bhaskar as committed to a supervenience (and thus ontologically individualist) view of social structure, whereas Le Boutillier (2013) thinks he ultimately reifies structure.
[15] See for example O'Neill, ed. (1973), Pettit (1993a), Sawyer (2005), and Greve (2012).
[16] Within the analytical tradition Epstein (2009) is the only explicit critique of ontological individualism of which I am aware. Even Durkheim, who is routinely criticized for reifying social structure, conceded that only individuals are real; see Sawyer (2002: 241–242).

case for the social sciences. (So, unless otherwise noted, when I use the term 'individualism' below I mean its ontological form.) In doing so, I will not challenge the view that only individuals are really – i.e. classically – real, and as such what follows is not a defense of ontological collectivism. Rather, what I will be defending is social holism. Holism is compatible with a flat ontology that on the surface might look individualist.[17] However, I argue that the only way to justify holism metaphysically is in quantum terms, and that the result is incompatible with individualism. This will pave the way for a new solution to the agent–structure problem in which both sides of the equation are emergent, but only in a quantum sense.

Supervenience meets externalism

These days, attempts to defend epistemological emergence are almost always based on the ontologically individualist concept of supervenience.[18] Supervenience is an asymmetric relationship of non-causal dependency between a lower-order "subvening" base and a higher-order "supervening" structure, in which once all the properties of the base are fixed then so are the properties of the structure. It comes in various forms – weak, strong, local, global, even "super-duper"(!) – which vary in the strength and scope of their dependency relations.[19] But they are all synchronic rather than diachronic, which is another way of saying that supervenience is a constitutive rather than causal relationship between levels (sic). The concept is highly general and has found uses in many domains, from aesthetics (where the beauty of a painting supervenes on how its colors are arranged on the canvas) to philosophy of mind (where "non-reductive materialists" have used it to try to explain consciousness). And whatever its attractions elsewhere, it seems particularly apt in social science. As Pettit puts it with respect to the latter, "Individualism insists on the supervenience claim that if we replicate how things are with and between individuals, then we will replicate all the social realities that obtain in their midst: there are no social properties or powers that will be left out."[20] This in turn suggests a straight-forward definition of ontological emergentism: "The individual-level facts do not fully determine the social facts; i.e. there can be possible worlds that are identical with respect to all individual-level facts but different with

---

[17] My approach bears a closer relation to the non-individualist flat ontologies of Manuel DeLanda (2002) and Bruno Latour (2005), although as materialists of a sort they neither thematize consciousness nor make an explicit link to quantum physics (cf. Jones [2014] on Whitehead's flat ontology). It would be instructive to compare the argument that follows to theirs, but given that my target in this chapter is individualism, I shall not undertake that formidable task here.
[18] See for example Currie (1984), Pettit (1993a), Sawyer (2005), and List and Spiekermann (2013).
[19] For an overview of different forms of supervenience see Kim (1990).
[20] The quote is in Epstein (2009: 188).

respect to some social facts."[21] My guess is that most social scientists today would find such a view implausible.[22]

Although hardly ever mentioned in social theory, the ontology of supervenience is physicalist in the classical sense.[23] As we saw in Chapter 1, the "physical" in 'physicalism' is doubly ambiguous: whether it refers to classical or quantum physics, and regarding the latter whether it implies No Fundamental Mentality (materialist interpretations of quantum theory) or Mentality All the Way Down (panpsychist ones). However, in the case of supervenience it is clear that physicalism means classical materialism, since in physics it is widely agreed that *quantum systems entail a failure of supervenience*.[24] This is because at the heart of all forms of supervenience is the assumption that the elements of the subvening base are separable.[25] That means the properties of the subvening parts must be intrinsic and non-relational (which is violated by entanglement), and also compatible or non-disjunctive (which is violated by the Uncertainty Principle). In short, supervenience assumes that the supervening structure is ultimately a function of "local matters of particular fact."[26] As such, it is not surprising that materialist philosophers of mind were the first to make widespread use of the concept of supervenience, and due especially to Keith Sawyer some of the debate about individualism has been in explicit analogy to their work.[27] The classical nature of supervenience theory will play a crucial role below.

One of the key attractions of the concept of supervenience for social theorists is that it creates space for epistemological emergence within an individualist ontology, which can be seen if we consider its implications for change. Ontologically, changes in the supervening structure imply changes in the subvening base, but because the relation is asymmetric, the converse does not hold: it may be possible for various configurations of elements to realize the same macro-level structure. If the number of possibilities is low this might not preclude explanatory reduction, but if the degree of multiple realizability is high – what Sawyer calls a "wild disjunction" between base and superstructure – then even though social structure is not ontologically autonomous from agents,

---

[21] List and Spiekermann (2013: 633). Note that List and Spiekermann think that they are here defining *holism* rather than emergence, but as will become clear below this conflates one with the other.
[22] Though see Wight (2006: 116), who rejects the supervenience approach. My argument below is that the only way to ground this naturalistically is through quantum theory.
[23] See List and Spiekermann (2013) for a particularly clear illustration to this effect.
[24] Teller (1986) was the first to make this point, which to my knowledge has never been directly challenged; see Esfeld (2001: 245–256), Belousek (2003), Karakostas (2009), and Darby (2012) for further discussion of the quantum challenge to supervenience.
[25] On the definition of separability see the Introduction to Part IV, p. 208.
[26] See Esfeld (2001: 247–248).
[27] See Sawyer (2005), Greve (2012), and List and Spiekermann (2013).

epistemologically it will be impossible to reduce it to its base.[28] This justifies a separation of macro- from micro-theory and with it the autonomy of the social sciences from psychology.

There is considerable controversy about whether the concept of supervenience can bear the weight even of epistemological or weak emergentism.[29] Of particular interest are doubts about its ability to accommodate downward causation. However, rather than get into that right now, I want to focus on an aspect of the *ontology* of supervenience that, in the social sciences at least, is widely believed to be unproblematic. Namely, how to specify the properties of the base – individuals – in a way that respects a classical physics constraint?

Unlike structures in the physical and biological sciences, most of the properties that subvene social structures are intentional states rather than material ones: norms and institutions, identities, beliefs and so on. To be sure, people must act on these states for social structures to be actualized, and those practices have a material aspect, but without intentionality there would be no practices, just behavior. This dependency of social structure on intentional states creates two problems. Since I have already discussed the first – that intentional states have no place in the materialist worldview – let me focus on the second, which is a widely held view in the philosophy of mind known as externalism.[30] Unlike supervenience, externalism justifies an emergentist view of social structure in an ontological sense. However, this is true only if it is quantized; in its current, implicitly classical form I argue externalism lacks a physical basis and as such is inconsistent with the CCP.

If the ontology of the agent–structure relationship is to be conceived in terms of supervenience, then in specifying the intentional properties of agents it is essential that they not presuppose the very structures they are supposed to constitute, which would be circular and violate the asymmetry of the supervenience relation. The key to satisfying this requirement, and to individualism more generally, is that the states attributed to individuals be *intrinsic*, which means people could have them all by themselves.[31] Note that this is a synchronic, constitutive condition; intrinsic properties are perfectly consistent with having been acquired through a diachronic, causal process like socialization. The point is only that, once acquired, their existence at any given moment is independent of other individuals.

So how are intentional states constituted, or, in philosophers' jargon, "individuated"? Intuitively it might be thought the answer is obvious: by brain

---

[28] See Sawyer (2005: 67–69), and also Wendt (1999: 152–156).
[29] See for example Humphreys (1997b), Heil (1998), and Greve (2012).
[30] Brian Epstein (2009: 188) argues that even on the purely material side ontological individualism is not true, since "social properties are often determined by physical ones that cannot plausibly be taken to be individualistic properties of persons." Although very interesting, Epstein's argument is orthogonal to my purposes and so I set it aside here.
[31] See Esfeld (2004: 626).

states. Common sense suggests that our thoughts do not depend intrinsically on anything beyond our skins, and materialist philosophy of mind reinforces that belief with its view that mental states are constituted by brain states. This view of intentional states is known as "internalism," and it clearly supports an individualist, supervenience-based ontology of social structure, since on it "[t]hought is logically prior to society."[32]

Given its intuitive force, it may therefore surprise social scientists (or at least it did me the first time I heard it) that internalism is a minority view in the philosophy of mind.[33] *Externalism* holds that the content of at least some mental states is constituted by conditions external to the mind/body. On this view, thinking is dependent upon society, not in the causal sense that we wouldn't have the thoughts we do if we had not learned a language and been socialized to a culture (though that is of course also true), but in the constitutive sense that our thoughts depend for their very content on external context.[34] More surprising still, this view dominates not just in continental scholarship influenced by Hegel or Wittgenstein, but in analytic philosophy, where it gained traction as a result of so-called "Twin Earth" thought experiments in the 1970s by Hilary Putnam and Tyler Burge.[35]

Consider an example of Burge's.[36] $Jones_1$ has various correct beliefs about arthritis – that he has it in his ankle, that his father had it, that it is painful, and so on – as well as the incorrect belief that it can afflict the thigh. Concerned about recent pains, $Jones_1$ tells his doctor that he fears his arthritis has spread to his thigh. His doctor says that is impossible because arthritis is an inflammation of the joints. Relieved, Jones changes his belief. Now imagine a counterfactual ("twin") world in which $Jones_2$ is in every way identical – same beliefs, same physical history – but in this world the term 'arthritis' *is* applied to thigh pain. Hence, upon complaint, $Jones_2$' doctor treats him for "arthritis." Burge concludes that the content or meaning of $Jones_2$' belief is different than in the first case, even though his mental state is the same. The difference is due to his social context.

What this example and others like it purport to show is that the content of mental states is individuated by what "conceptual grid" is used,[37] and those grids, which correspond to the social norms and rules pertaining to the mental

---

[32] Gilbert (1989: 58).
[33] According to a recent survey of philosophers, 20 percent are internalists about mental content, 51 percent are externalists, and the rest are various types of "other"; see Bourget and Chalmers (2014: 495). For an excellent overview of externalism see Lau and Deutsch (2014).
[34] Currie (1984: 354), Burge (1986: 16), Pettit (1993a: 170), Esfeld (1998: 367) and others all emphasize that the point here is constitutive rather than causal.
[35] See Putnam (1975) and Burge (1979). Esfeld (2001) provides an excellent overview of an alternative path to externalism through Wittgenstein's problem of rule-following.
[36] This paragraph is reproduced from Wendt (1999: 174); for further examples from IR see ibid: 176–178.
[37] Bhargava (1992: 223).

states in question, are not the possession of the individual but of the community. Note that this is not to say people don't have private thoughts in their heads. The point rather is that the conditions that make those thoughts what they are are not up to the individual. If I think I saw a Bigfoot and the community in which I live denies Bigfoot's existence, then my thought is a "hallucination" or "crazy"; in an otherwise identical society that recognized Bigfoot's existence the same thought would have a different content. In this way thinking depends intrinsically on social relations. And since the latter are linguistic phenomena, this means that language does not merely mediate thought, but makes it possible in the first place.[38] In short, as Putnam summarizes externalism, "meanings ain't in the head."[39]

Externalism is an explicitly holistic doctrine and as such widely seen as inconsistent with a narrow or "local" form of supervenience, according to which an individual's mental states supervene on just her brain states.[40] However, what about *global* supervenience, in which the subvening base includes all individuals, their interactions, and the physical environment? Here the consensus seems to be the opposite, that externalism is no threat to individualism. Pettit for example argues that the external constitution of intentional states implicates only the relationships of individuals to each other, not emergent social structures. Once the properties of the (socially constituted) minds of all individuals are set, the supervenience argument can proceed from there – a perspective he calls "holistic individualism."[41] Similarly, in response to Burge, Gregory Currie argues that even if $Jones_1$ and $Jones_2$ are in identical mental states, their communities are different because *other individuals* behave differently in the face of 'arthritis,' which he claims secures the ontological priority of the subvening base.[42]

However, to my knowledge no advocate of externalism has considered the possibility that materialism cannot explain consciousness or that the mind is a quantum system. Thus, they are at least implicitly assuming that the ontology of the supervenience base for social life is classical and materialist. This is problematic in several respects for the compatibility of externalism and supervenience.

First, classical materialism requires that the parts of the subvening base – individuals' mental states – are constituted by intrinsic properties, such that ontologically, the supervening structure is nothing but "local matters of particular fact." Even globally formulated, it is unclear how this is consistent with

---

[38] See Pettit (1993a: 169) and Searle (1995: 59–78). This applies only to higher-order, human thought; in my view animals also have thoughts, just not linguistically constituted ones.
[39] See Putnam (1975).
[40] See for example Currie (1984), Bhargava (1992), Esfeld (2001: 157), and Howell (2009: 84).
[41] See Pettit (1993a).   [42] See Currie (1984: 354–355).

externalism. After all, the content of other individuals' mental states are *themselves* socially constituted, so at what point does supervenience bottom out in the intrinsic, non-relational properties of separable individuals per se? "Holistic individualism" seems circular in this regard. Second, the social constitution of mental states is non-local and non-causal. How does this square with a classical physics constraint? It is telling that Currie reduces the differences between the two communities first to what is said or written by individuals and then from there to "bodily movements."[43] This preserves the materialist basis of supervenience theory, but at the cost of stripping mental states of any genuine role. Finally, as I argued in Chapter 1, if materialism cannot explain consciousness, social scientists are forced either to treat mental states on an "as if" basis and accept a tacit vitalism, or abandon them as illusions and thus also the structures they subvene. If these criticisms have force, then the classical worldview is not compatible with externalism, and as such Pettit's holistic individualism lacks a naturalistic basis.

In the face of this problem one could embrace internalism instead, which is more clearly compatible with individualism and not vulnerable to at least some of the objections above. However, such a retreat is both unattractive and unnecessary: unattractive because the externalist critique of internalism still stands, and unnecessary because there is an alternative. If the mind and language are quantum systems, then externalism is the logical consequence. Its basis is semantic non-locality: by participating in a shared language, the content of individuals' thoughts is entangled with other minds and thus irreducibly contextual. However, this means we have to give up supervenience as the ontology of social life, since supervenience fails in quantum systems. The resulting ontology is holistic and flat, but one in which emergence plays a central role.

Agents, structures, and quantum emergence

So far I have tried to establish that (1) an ontologically emergentist conception of social structure is incompatible with the classical worldview, which leads to a supervenience ontology that treats individuals and their properties as constitutionally exogenous to social life; (2) this individualist ontology is incompatible with the holism of externalist philosophy of mind; and (3) the latter is itself incompatible with the classical worldview. These contradictions set the stage for a quantum intervention. When quantized, externalism in my view satisfies all three of the criteria laid down above for ontological emergence: the social constitution of intentional states is not reducible to the intrinsic properties of

---

[43] Also see Esfeld's (2001: 157) emphasis on practices as the source of meaning, which likewise are either nothing but bodily movements or presuppose collective intentions.

individuals; it introduces qualitative novelty; and it can make sense of downward causation. In this section I first review the literature on emergence in quantum mechanics and then recast the agent–structure problem in that light.

While there is considerable debate about the status of ontological emergence within the classical worldview, there is little disagreement that it exists in the quantum world. Michael Silberstein and John McGeever argue that quantum mechanics "provides the most conclusive evidence for the existence of ontological emergence,"[44] and starting with a seminal article by Paul Humphreys, a small but coherent body of scholarship has grown up exploring how best to understand it.[45] Humphreys sees quantum entanglement as the source of emergence, which subsequent scholarship follows and I address in this section. This is a synchronic view of emergence, which is the truly hard case for ontological emergentism. In the next section I take up a more diachronic form of quantum emergence, the collapse of the wave function, which speaks to downward causation.

Humphreys' argument is motivated by what is known as the "exclusion argument" against mental causation.[46] Namely if, as non-reductive materialists would have it, mental states supervene ontologically on brain states, then it would seem that all the causal work in the mind is done by brain states. This leaves no room for mental states to have causal powers, and since phenomena without causal powers are usually excluded from scientific ontologies this raises the specter of epiphenomenalism with respect to the mind. In response Humphreys first develops an abstract argument that if there were a "fusion" of basal properties, such that they no longer had separate identities, then the effects of their fusion "[could] not be correctly represented in terms of the separate causal effects of its constituents."[47] He then argues that quantum entanglement fits this bill. In entanglement, only the composite system is in a "pure state," while the components lose their identity as fully separable elements with intrinsic properties. In this situation "the state of the whole determines the parts rather than the other way around."[48] Crucially, this is the *exact opposite of supervenience*. Although Humphreys does not return to the issue of mental causation, his argument seems an apt description of quantum coherence, the breakdown of which quantum brain theory sees as the physical basis of consciousness.

---

[44] Silberstein and McGeever (1999: 187).
[45] See Humphreys (1997a), Kronz and Tiehen (2002), Hüttemann (2005), Wong (2006), Bitbol (2007; 2012) and Prosser (2012). Bitbol is the skeptic in this group, although his approach if anything pushes beyond ontological emergence in the sense that it rejects the notion of an emergence "base" altogether.
[46] On this argument see Kim (1998: 150).   [47] Wong (2006: 352).
[48] Bitbol (2007: 299); Humphreys (1997a: 15).

Subsequent work has challenged an important detail of Humphreys' argument, that in fusion the parts lose their identity *completely*.[49] This "basal loss" is seen as unmotivated and overstated, and as leading to the unwelcome consequence that there could be mental states with no neural states underneath at all. Either way, in social science the more qualified view of Humphreys' critics seems apt. While there has been a long debate among philosophers of physics about whether entangled particles retain any individuality,[50] in social life the "particles" are biological individuals whose bodies cannot fuse even in principle. As for our minds, although I have argued that shared language gives rise to entangled mental states, here too fusion faces a limit in the physical structures protecting quantum coherence within each of our brains.[51] Importantly, however, despite their more qualified view of basal loss, Humphreys' critics agree with him that entanglement involves ontological emergence. By the same token, I will argue that the fact that individuals cannot completely fuse their identities is compatible with the ontological emergence of social structures. In both cases we are dealing with wholes that are not reducible ontologically to parts, because the identity of the latter is not separable from the whole.

Before proceeding there is one important wrinkle to add. The discussion around Humphreys' argument has been asymmetric in the sense that it has assumed that what is emergent is the entangled state, which is to say the whole. However, as Bitbol argues, we can see entangled parts as emergent as well.[52] Drawing on quantum field theory, he points out that sub-atomic particles are not substantial individuals in themselves, but particle-*like* processes or vibrations in the universal quantum flux. As such, "there is no essential difference between the alleged 'basic' level and the emergent level,"[53] and thus no "ground" upon which one might build a stratified ontology of levels. So when particles become entangled they too acquire new properties, namely relational properties to the whole.[54] In entanglement therefore, parts and whole are "*co*-emergent," rather than only the latter emerging from an ontologically prior base of the former. This symmetry of emergent entanglement plays a key role in what follows.

The CCP tells us that everything in the world, including individuals and society, is physical (vv. material). In the case of people the referent of that physicality is clear – the body/brain/mind complex. However, I have argued

---

[49] See Kronz and Tiehen (2002), Wong (2006), and Bitbol (2007); though see French and Ladyman (2003).
[50] See for example Castellani, ed. (1998) and Winsberg and Fine (2003).
[51] Cf. Swann et al. (2009).
[52] See Bitbol (2007: 302–303); note that Bitbol's general approach to quantum theory is epistemological (see Chapter 4), and as such he rejects the idea that there is ontological emergence in entanglement. However, I don't think that precludes adapting his argument to my purposes here.
[53] Ibid: 303; also see Campbell and Bickhard (2011). [54] Also see Francescotti (2007).

that the physicality of social structures is puzzling, because unlike other macroscopic objects (sic) they are unobservable. So let me begin by defining social structure in qualitative terms and then translating it into physical ones.

I won't try to review the numerous definitions of social structure in the sociological literature; for my purposes it is sufficient to distinguish two basic types of phenomena that often fall under that description.[55] The first is exemplified by a demographic structure.[56] While demographic patterns are "structural" in the sense that they constrain human agency, they are not particularly "social," given that as distributions of objective attributes they would exist even if people were completely unaware of them. Since I do not think we need quantum theory to understand demography, I shall set this first kind of social structure aside. That leaves the second, which encompasses phenomena like norms, rules, culture, institutions, and so on. What distinguishes these from a pattern of objective attributes is that they presuppose a discourse on the basis of which people act.[57] Social structures here are mind-dependent or intentional objects, and specifically collective intentions in Searle's sense, involving shared mental states and language.[58] Specific social structures like the institution of marriage, the market economy, or the state are instantiations of this general category, constituted by the discourses peculiar to their purposes and functioning.

If the physical basis of the mind and language is quantum mechanical, then, given this definition, that is true of social structures as well. Which is to say, *what social structures actually are, physically, are superpositions of shared mental states – social wave functions*. I want to highlight here four implications of this ontology, two of which are at odds with critical realism, but two of which realists should find congenial.

First, against the realist view of social structures as "real but unobservable entities," as superposition states social structures are pure potentialities and as such not classically or really real (much less entities), any more than the wave functions of sub-atomic particles are real. This explains why social structures are unobservable, since one cannot observe a wave function, only the result of its collapse into a particle. Second, and by the same token, against realism's stratified ontology, social structures do not lie on their own level of reality above (or below) individual agents. In the *real* world there are just people and their practices, which imply a flat ontology more like Giddens' structuration theory than critical realism.[59]

---

[55] This distinction is inspired by Hodgson (2002: 167–168); see Porpora (1989) for a wider view.
[56] See Archer (1995: 174–175), and for critical discussion see Elder-Vass (2007).
[57] Note that this is not to say that people necessarily understand what they are doing when they act upon such discourses; they might have no idea.
[58] See Searle (1995).
[59] See Giddens (1979; 1984), and Hodgson (2002: 161–166) for a good overview of the similarities and differences between critical realism and structuration theory.

Third, despite this flat ontology, conceiving of social structures as superpositions of shared mental states means that they are ontologically emergent. Not in the classical sense of an autonomous level of reality, but in the quantum sense of entanglement among the agents who constitute them. This provides a more plausible physical basis for critical realists' commitment to emergence than the classical discourse of levels. Finally, the superposition approach implies a holistic social ontology, in which the agents participating in a structure are connected non-locally to each other. This affirms Bhaskar's "relational" view of structure as a set of internal relations, as well as the more process-theoretic relationalism of Mustafa Emirbayer.[60]

I will flesh out some substantive implications of this view below, but first I want to consider a potential individualist objection to it, which from a classical standpoint may seem all very mysterious. Doing so will complete the circle by bringing quantum agents back in. Esfeld, who sees holism as an ontology of relations rather than things, summarizes the suspicion well:

A metaphysics of relations is often dismissed out of hand, for it seems to be paradoxical. It seems that (a) relations require relata, that is, things which stand in the relations, and that (b) these things have to be something in themselves, that is, must have intrinsic properties over and above the relations in which they stand.[61]

On this view, supervenience is the only way to understand part/whole relations, which when applied to social life – where the "relata" are human agents – implies individualism. As such, any genuinely holistic model of social structure is a non-starter.

Given that externalism – a form of holism – is widely accepted in philosophy of mind I have argued we already have some reasons to resist this conclusion, and the tensions between externalism and supervenience come even more into view if the elements of social structures are quantum rather than classical agents. In that case, the relata of social relations are *themselves* superpositions – "walking wave functions" – and as such do not have intrinsic properties in the first place, but only have the properties they do by virtue of their entanglement with other agents. Indeed, not only do quantum agents lack intrinsic properties, but if quantum decision theorists are right their properties are often disjunctive or "incompatible," which the ontology of supervenience rules out.[62] This puts in a different light Pettit's claim that those who reject supervenience think that "individual-level facts do not fully determine the social facts." For what quantum emergentism implies is that there *are* no "individual-level facts"

---

[60] See Bhaskar (1979: Chapter 2, passim) and Emirbayer (1997).
[61] Esfeld (2004: 626); also see his (1998). See French and Ladyman (2003) for a defense of a strong version of relations without relata within the context of quantum field theory.
[62] See Chapter 8.

apart from the social ones that constitute them. They are co-emergent, because individuals' minds are not fully separable.

In sum, a quantum social ontology suggests – as structuration theorists and critical realists alike have long argued – that agents and social structures are "mutually constitutive." I should emphasize that this does not mean "reciprocal causation" or "co-determination," with which "mutual constitution" is often conflated in social theory. As quantum entanglement, the relationship of agents and social structures is not a process of causal interaction over time, but a non-local, synchronic state from which both are emergent.[63] And it is also not co-determination, since until they collapse in practices neither agents' minds nor social structures are in determinate (i.e. actual) states, only potential ones. Granting these stipulations, however, given that Giddens came up with the "mutual constitution" formula in 1979, it might fairly be asked what the value added of a quantum perspective actually is. I would say three things.

First, it provides a physical basis for a thesis that is otherwise untenable. Given the classical worldview, there are in my view irresistible arguments to favor an individualist ontology, so that thinking about agents and structures must be in terms of supervenience. Because supervenience is asymmetric it rules out even mutual constitution, much less emergence. Quantizing the debate removes this a priori constraint, which puts a burden of proof on individualism's advocates to justify their assumption of agent separability, and enables emergentists, externalists, and holists to ground physically arguments that, qualitatively at least, seem quite plausible. Second, in a more forward-looking vein, there is the prospect of using the quantum formalism in new areas. Quantum decision theory has already proven its worth in thinking about individual agents, and quantum game theory might do the same for interacting agents. However, to my knowledge no one has used the formalism to model social structure, which could have the widest impact of all. Finally, the quantum approach also offers a way to deal with the heretofore most difficult question for social emergentism, the problem of downward causation, to which I turn now.

Downward causation in social structures

For emergent properties in any domain to be real rather than just explanatory contrivances, then on pain of epiphenomenalism they must have causal powers, and specifically the "downward" (sic) power to affect the parts from which they emerge.[64] Although the language of downward causation is not widely used in

---

[63] Cf. Archer (1995) and Wight (2006: 117).

[64] Note that there is nothing mysterious about large objects affecting small ones if the latter are not *part of* the former, so we are only talking here about what Kim (2000: 311) calls "reflexive" downward causation.

An emergent, holistic but flat ontology

the social sciences,[65] the question of how to think about the causal powers of social structure (and/or discourse) is an old one, and continues today to exercise social scientists of many different theoretical persuasions. This is particularly true of those committed to macro-level theorizing, which includes not just critical realists and other structuralists, but non-reductive individualists as well.[66] In this section I first use the philosophical debate to explain why downward causation is a "problem," then suggest how a quantum approach offers a solution, and conclude by applying this framework to the case of social structure.

Empirically it seems that downward causation is all around us.[67] If Jones is thinking happy thoughts and then starts to worry about losing his job, this mental change will have measurable effects on his blood pressure, anxiety level, and other biological functions. Similarly, it is well known in biology that what happens inside individual cells is affected by macroscopic changes within organs and the body as a whole. Yet the idea of downward causation has proven very difficult to reconcile with naturalism, and while philosophers have paid considerable attention to it lately, the dominant view seems to be that it is an incoherent notion.[68]

The reason stems from a key principle of the classical worldview: causal exclusion, which says that if an event has a complete micro-level cause – and in the classical worldview micro-level causes ultimately rule – then there can be no "over-determination" of that event by further micro- or macro-level causes.[69] Given this principle, if downward causation is defined synchronically then a vicious circularity ensues, since the putative causal powers of wholes only exist in virtue of their parts, and so how could they simultaneously cause the latter? This circularity can be removed by defining downward causation diachronically, as an interaction between parts and wholes over time.[70] However, that strips the concept of much of its interest. Since wholes are made up of parts, then by causal exclusion their putative effects at T2 are *really* just the effects of their parts at T1. In short, there is "no room at the bottom" for downward causation, making it only an "illusion."[71] Robert van Gulick sums up the problem nicely:

---

[65] Hodgson (2002) and Elder-Vass (2010a: 58–62) are important exceptions. The phrase was first introduced by Donald Campbell in his (1974).
[66] For recent efforts in this vein, see Hodgson (2002), Sawyer (2005), Wight (2006), Elder-Vass (2010a; 2010b), and List and Spiekermann (2013).
[67] Bitbol (2012: 233).   [68] See Hulswit (2006) for an excellent overview of the debate.
[69] See Kim (1999; 2000), and for other skeptical treatments of downward causation Robinson (2005) and Davies (2006).
[70] This is the strategy of Hodgson (2002) and Elder-Vass (2010a: 60–61).
[71] See Davies (2006: 46) and Robinson (2005: 133). Elder-Vass (2010a: 60) rejects this conclusion as "pure ontological prejudice" on the grounds that the *organization* of parts plays an irreducible causal role. While I am sympathetic to his complaint, as we saw in the discussion of supervenience the classical worldview does not support emergentism, and as such 'organization' could have at most an explanatory, not ontological status.

The challenge of those who wish to combine physicalism with a robustly causal version of emergence is to find a way in which higher-order properties can be causally significant without violating the causal laws that operate at lower physical levels. On one hand, if they override the micro-physical laws, they threaten physicalism. On the other hand, if the higher-level laws are merely convenient ways of summarizing complex micro-patterns that arise in special contexts, then whatever practical cognitive value such laws may have, they seem to leave the higher-order properties without any real causal work to do.[72]

Critics of downward causation routinely assume that 'causation' means efficient causation, in which causes precede and are therefore distinct from their effects, and there must be a transfer of energy from one to the other. Defenders will often concede that given such a stipulation, the idea is incoherent, especially in its synchronic form.[73] However, the door is still open to more pluralistic views of causation, like Aristotle's four-fold typology of efficient, material, formal, and final causality, which has been the principal resource for those trying to make sense of the idea.[74] One option in this vein, "medium" downward causation, is to view the higher level as setting constraints or boundary conditions on activities at the lower level.[75] While using all four Aristotelian causes, the logic here is primarily functionalist (a modern version of final causality), in which the whole controls the parts by selecting activities consistent with its survival.[76] Then there is "weak" downward causation, in which the structure of the whole describes the form or arrangement of the parts. This view has the virtue of being synchronic, which is the hard case for downward causation. And it is also attractive from the standpoint of the agent–structure problem because in emphasizing the arrangement of parts it does not commit us to the reality of the whole, which in the case of social structure is problematic. Still, this approach is more descriptive than explanatory, and, as its name suggests, it is the farthest removed from the everyday sense of causation.

Whether an Aristotelian approach to causality can make sense of downward causation is unclear. Ignored by defenders of the orthodoxy, even some who are sympathetic have questions about what precisely it amounts to or if it is anything more than a useful heuristic.[77] However, what I want to emphasize here is that, as Bitbol makes clear, the literature on downward causation, both pro and con, has almost without exception assumed a classical CCP.[78] That

---

[72] The quote is from Tabaczek (2013:390).
[73] See for example Emmeche et al.'s (2000) rejection of "strong" downward causation.
[74] On neo-Aristotelian approaches to downward causation, pro and con, see Emmeche et al. (2000), Moreno and Umerez (2000), de Souza Vieira and El-Hani (2008), and Tabaczek (2013); cf. Craver and Bechtel (2007), who argue that the role of the whole should not be seen as causal at all, but as constitutive.
[75] See Emmeche et al. (2000: 24–25).   [76] Also see Meyering (2000: 194–196).
[77] See for example Hulswit (2006) and Bitbol (2012).
[78] See Bitbol (2012). This is true even of Davies' (2006) article on the physics of downward causation, which addresses quantum theory only at the end. Unfortunately the literature on quantum emergence has to date not taken up the issue.

would be fine if the world in question really was classical, but most of the apparent examples of downward causation that motivate interest in it concern life and mind, which I have argued are quantum mechanical. If that's right, then the presuppositions of the debate – that physicalism means materialism, the latter is about substances, causation is local, and so on – are all wrong.

In introducing a quantum approach to downward causation it is useful to recall that the idea of emergence as entanglement is synchronic. Parts and wholes are co-emergent simply by virtue of being entangled. This affects the distribution of probabilities for the parts' behavior,[79] but that in itself does not cause anything to happen because what are entangled are superpositions. While this is nonetheless important given that the static case is the hard one for emergentism, one would be hard pressed to call it "causation" in anything but a formal causality sense. However, there is another kind of emergence in quantum mechanics that has received less attention: the collapse of the wave function. Collapse is a dynamic process that makes things happen, but I will argue still in a sense synchronic. Let me take up this issue first with reference to the individual in isolation and then turn to the case of multiple agents entangled in a social structure.

In Chapter 6 I argued that collapse is a process of temporal symmetry-breaking in which two phenomena result. One is a material particle that is experienced as moving forward in time from the past to the future, and the other is a force of sub-atomic will moving backward from what will become the future to the past. Both are irreducible to the wave function from which they ensued (since collapse is non-deterministic), and both exhibit novel properties relative to its characteristics, satisfying two of the three criteria for ontological emergence.

What about the third criterion, downward causation? The "force" that collapses the wave function is Will. Will is purposive and as such instantiates teleological or final causation in a strong sense. It is what animates sub-atomic particles and, more to the point here, all life forms. Since unlike particles the latter have persisting minds, we may say that Will is the locus of mental causation, the ability of the mind to direct the behavior of the body, which is often invoked as a paradigmatic example of downward causation.[80]

Now, as we have seen, the case for genuine downward causation turns heavily on its being synchronic, since if it is only diachronic, then the counter-argument can be made that really what is going on is parts at T1 are affecting parts at T2. Because wave function collapse is a process that brings forth actuality from potentiality, it might seem as if it is subject to this criticism. But I don't think so.

---

[79] Prosser (2012: 37).
[80] Or perhaps more precisely, downward "self"-causation given that it is willful; see Bitbol (2012: 251–252).

Collapse is a process, yes, but it does not take place *in* time, since it happens instantaneously. Rather, it *creates* time, by breaking the temporal symmetry of the wave function. This may seem very abstract, but I submit that it conforms to our own experience of Will. When I move my arm I do not first have the will to move and then act on it, I just *do* it. This is not to deny that a hugely complex set of material processes are entrained by and necessary for my movement, but all of these processes are governed by and instantiate the effect of my will on my body. The downward causation of will is not diachronic, in other words, but nor is it static, since it brings consciousness and movement into being. The difficulty here lies in the synchronic/diachronic dichotomy itself, which presupposes a classical ontology of substances exerting causal powers *in* time. Quantum theory in contrast points to a process ontology, in which processes rather than substances are primary and time itself is an effect.[81]

In contrast to downward causation within the individual, within a social structure it is distributed across many different individuals, each of whose free decisions to collapse the structure's social wave function make it a material reality in that moment. This means that downward causation in social structures always happens locally, in concrete practices in particular contexts, and once those practices are over the structure that enabled them disappears back into its wave function. In effect, social structures are continuously popping in and out of existence with the practices through which they are instantiated.[82]

This perspective has many resonances with the "practice turn" in social theory, the advocates of which also reject substantialism and argue that the agent-structure problem can be solved by setting agents and social structures aside and focusing instead on practices – on the process of what people do.[83] In taking this line, however, practice theorists might reject altogether the idea of "downward" causation by social structures, with its connotation of distinct levels of reality in which a higher level exerts causal powers on a lower one. I share their rejection of levels discourse, and so I agree that the downward metaphor is misleading. What is going on here is more accurately described as structures being *pulled* out of the quantum world of potentiality into the classical world of actuality by agents. But that does not mean social structures have no causal powers, for several reasons.

First, a social wave function constitutes a different probability distribution for agents' actions than would exist in its absence. Being entangled in a social structure makes certain practices more likely than others, which I take to involve

---

[81] For process-theoretic critiques of substantialist assumptions in the debate about emergence and downward causation see Bitbol (2007; 2012), Campbell and Bickhard (2011), and Pratten (2013).

[82] Cf. Schatzki (2006).

[83] See for example Bourdieu (1990), Schatzki et al., eds. (2001), and Schatzki (2002); for a succinct and particularly clear overview of practice thinking see Adler and Pouliot (2011).

formal causation. Second, as we saw in the discussions of quantum semantics and externalism above, the intentional states that accompany agents' practices are not fully separable from the collective intentions which make them possible, since the latter define the context in which agents are acting.[84] Third, given my quantum reading of the performative model of man, whenever agents engage in practices they are actualizing not only a social structure, but also a determinate identity for themselves. And since human action is purposive, which I have argued involves final causation, this means that when people act in light of a social structure they are expressing its teleological purpose – they are agents *of* the structure as much as they are the agents who pull it "down." Thus, while practices might be where we can actually "see" social structures, by themselves they tell only part of the story.

In sum, social emergentists have been looking for ontological emergence in the wrong place, in a vertical relationship between a lower level of agents and a higher level of social structure. There *is* no higher level in social life above that of individuals: the reality of social life is flat. Note that this is not to say that social structures do not vary in *scale*, since some are quite localized (the structure of Amish society, say) while others are global (the international system). The point is rather that scale should be seen as a "horizontal measure of 'scope' or 'extensiveness'" rather than as a "level – a vertically imagined, 'nested hierarchical ordering of space.'"[85]

However, a flat ontology does not mean that social emergentism's historical rival, reductionism or individualism, is true. The crux of my critique was that, given its classical worldview, individualism assumes that individuals are fully separable entities in well-defined determinate states. Externalist philosophy of mind provides qualitative arguments against this assumption, which are backed up by the physical arguments of quantum decision theory, quantum semantics and quantum emergence. This view finds emergence not in vertical relationships between levels, but in holistic horizontal ones among agents, whose states are constituted by non-local entanglements mediated by language. As superpositions social structures are only potentialities rather than actualities, but this is equally the case for agents. Their superposed states are co-emergent, and if they become real realities they do so *together* in localized practices, which themselves are emergent from the dynamic process of wave function collapse.[86] Perhaps paradoxically, therefore, a quantum approach suggests that

---

[84] This might be seen as analogous to material causation in the Aristotelian scheme, though there is nothing "material" about it except its strictly behavioral aspect.

[85] See Marston et al. (2005: 420) and the "scale debate" in geography which their work has inspired, and especially El-Khoury (2015), who in a comprehensive review links this debate explicitly to quantum ideas about the social.

[86] Theodore Schatzki's (2002; 2005) concept of a "site ontology" seems like a productive "site," as it were, for further unpacking this suggestion in social theoretic terms; also see Woodward et al.

to understand emergence properly we need a horizontal rather than vertical worldview.[87]

Throughout this chapter my focus has been on emergence in an ontological sense. Given that ontological individualism is compatible with non-reductionism in an epistemological or methodological sense, this might make the argument seem far removed from the actual explanatory practices of social scientists. However, if ontological individualism falls, then the supervenience approach that supports epistemological non-reductionism falls as well, and as such will mischaracterize the explanatory role of both agents and structures. What is going on in situations where it seems that irreducible structures need to be invoked is not that multiple realizability means there is an excess of explanatory leverage, beyond what the intrinsic properties and interactions of agents can tell us. What is going on is that agents are themselves emergent from interaction, and that the causal power of the structures in which they are embedded is not that of efficient causation, but final causation or collective purposiveness.[88]

(2012), who while writing from a New Materialist perspective, develop Schatzki's idea in interesting ways that relate to a number of my own concerns.

[87] Cf. van Dijk and Withagen (2014), who argue that Wittgenstein's later work embodies such a worldview.

[88] I hope to develop the teleological powers of social structures in future work; for a preliminary attempt to do so outside of a quantum framework see Wendt (2003).

14 Toward a quantum vitalist sociology

The dominant model of man in naturalistic social science today is materialist, ontologically deterministic, and mechanistic. At least it is no longer behaviorist too; although behavior*al*ism remains strong, almost all social scientists agree that human behavior is affected by unobservable intentional states that must be dealt with as best we can. Yet while it is recognized that people have minds, the fact that our minds are *conscious* plays little role in mainstream scholarship, where we are modeled as either machines or zombies, and thus, in effect, as dead. If this is true of human beings, in turn, then it must be all the more so of society. And indeed, in contemporary sociology[1] materialism is unquestioned, error terms are chalked up to complexity or poor data rather than free will, and causal mechanisms are the gold standard of explanation. At the end of the day, social systems are just matter in motion – complex, even intelligent matter, but dead just the same.

In this book I have laid the groundwork for a different, albeit still naturalist, sociology. Building on quantum consciousness theory, in Parts II and III I argued that human beings are conscious, free, and purposive in a teleological sense – in short, very much alive. I suggested this amounts to a genuinely vitalist ontology – not the ersatz vitalism of New Materialism, but a phenomenological vitalism in which subjectivity is constituted by a physical but non-material and unobservable life force: quantum coherence. In subsequent chapters I extended this framework to social structures, which by virtue of the quantum character of language entangle individuals with and enable them to act non-locally upon each other. Like the minds of individuals, social structures are in superposition and thus also exhibit quantum coherence. The question that arises, therefore, is whether my phenomenological vitalism extends to social systems, which is to say, is society itself an organism, with subjectivity and consciousness? The logic of my argument I think demands an affirmative answer, though it is such a controversial claim that I hesitate to bring it up – both because it risks undermining all my hard work, and because I cannot

---

[1] I mean this term in the broad rather than narrow, disciplinary sense, as encompassing all macro-level social sciences.

give it the theoretical development it deserves here. Yet I can hardly avoid the question either, since if it is implied by my argument then critics will surely bring it up for me. So while what follows is only a gesture in this direction, let me put the idea of a vitalist sociology on the table.[2]

I shall do so using the state as an example. Unlike the decentralized social structures that I addressed in the previous chapter, the state has a centralized structure that gives it a capacity for corporate agency. That makes it an "easy" case for the thesis that social systems are organisms, so if the argument does not fly here then it won't for less centralized systems either (though currently I'm unsure what to make of the latter in any case). I unfold the argument in three parts, the first filling in a missing piece of the agent–structure relationship, the second exploring some contours of the state as organism idea, and the third taking up the question of collective consciousness. Since I'm almost out of gas and over my word limit, and have addressed the first and second pieces in previous work,[3] I will concentrate my energies on the third.

The holographic state

The state is a social system constituted, on the one hand, by a social structure organized around particular forms of language (of citizenship, territoriality, sovereignty, and so on), and on the other, by the myriad practices of those who participate in this discourse (both citizens and outsiders). Conceptualized in quantum terms, as a structure the state is a wave function shared non-locally across both time and space by millions of people, but as such it is only a potential reality, not an actual one. As a practice, in turn, the state is an actual but local phenomenon, materializing momentarily as people collapse its wave function in their daily affairs such as voting, paying taxes, and going to war, and then disappearing again. Neither aspect captures our ability to "see" the state, the former because wave functions are not really "there," the latter because practices are not the state as a whole. I argue in a moment that what is missing from the model is the state's holographic character. But first let me say a bit more about the individuals who make up the state.

According to the flat ontology I defended above, only individuals and their practices are really real. As such, in a classical sense, states are *nothing but* individuals and their interactions; there is no higher reality of the state existing above them. However, unlike classical agents, quantum ones are endowed with superposed minds entangled through language, which means that they enfold socially shared wave functions within their subjectivities, one of which concerns the state. So the members of states are not fully separable from its

---

[2] I hope to develop these ideas further in future work.
[3] See Wendt (2004) and (2010) respectively.

wave function either, but constituted in irreducibly relational terms by their co-participation in this discursive form.

The picture that suggests itself here is one of individuals as Leibnizian monads, who mirror the social whole within their minds and reflect it back in their practices. Leibniz's monads were "windowless" and thus needed God to ensure their harmony with the whole. In contrast, quantum monads have windows that enable us to directly perceive the world, including other minds. While in this substitution we lose the pre-established harmony provided by God, we gain the ability to work toward harmony through naturalistic processes of learning and socialization, which evidently have enabled us to create relatively durable societies.[4] Still, as quantum monads we do not mirror the social whole all of the time or equally, which suggests a need for three distinctions.

First, since an individual cannot act simultaneously on the many different wave functions which they embody, following Teruaki Nakagomi we can distinguish between "active" and "passive" monads,[5] which describe two modes of relating to our social entanglements. When we are in active mode we are thinking about the potentials of a given wave function and collapsing it to actualize a desired reality; in passive mode we are doing something else. Considering the myriad practices in which we all engage, this means that with respect to most of our entanglements we will be in passive mode at any given moment. The switch from passive to active mode is therefore crucial, which is accomplished by attention.[6] Attention to a particular wave function will often result from measurements by other people, which focus the individual on that set of potentials rather than others. Thus, even though "the state" is always present as a potentiality within its subjects, most of the time we are not thinking about it, and as such it only has "occasional relevance" for us.[7]

Consider for example the War in Iraq between 2003 and 2011. The active monads were the leaders, combatants, and supporters on all sides, during those times of the day when their decisions were actually making war; the passive monads were everyone else, including the previously active monads when they were eating breakfast or sleeping. This means that most Americans most of the time were not *actually* at war with Iraq. Our imagery should be the horizontal one of warring ant colonies rather than the vertical one of dueling Leviathans: viewed by aliens in space, the war was nothing but particular "ants" living in one colony flying a long way to fight with those in another. As casualties rose new individuals arrived to take their place, but at no point were all Americans actually making war. Importantly, this is not to say that U.S. citizens as a whole

---

[4] On quantum monadology see especially Nakagomi (2003a; 2003b); cf. Tarde (1895/2012) and Lash (2005) on Georg Simmel's vitalism, which in its phenomenological, intersubjective spirit seems relatively close to my own.
[5] Nakagomi (2003a: 19) refers to the latter as "null" monads.
[6] See Schwartz et al. (2005: 1322–1323).   [7] See Coulter (2001: 36–38).

were not party to the war, since when those who made munitions or MREs were on the job they were active monads enabling the troops to fight, and even as passive monads their acceptance of the war made it possible. But at the end of the day the war was fought by those who actually lived it, not by the rest of us.

Speaking of decisions to make war, a second distinction should be made between individuals who have the authority to speak for a state as a whole – its leaders – and those who do not. Following Leibniz we might call leaders "dominant" monads, which contain within themselves the reasons for the collective actions of their members.[8] Other monads defer to the dominant, giving the latter "first mover" status in collapsing the state's wave function and by implication giving up their own right to act against the chosen path (at least at that moment). In quantum terms this may be understood as a system of entangled particles in which, by virtue of its internal structure, when measurements are made on the system by the environment the choice of how to respond is not made locally by the particles on the spot, but centrally by the leader.

This has at least two interesting implications. One is that because the state's wave function has many potential outcomes, the intentions and character of leaders are crucial in determining which policies are realized. Even in highly constrained situations, small differences in leaders can make big differences in what actually happens (think of Al Gore being President in 2003 rather than Bush), so there is a reason here to "bring leaders back in." The other is that when a dominant monad collapses a state's potentialities into an actual choice, it has non-local consequences for everyone else in the group, and even beyond. Just as Socrates' death instantaneously constituted Xantippe as a widow, Bush's decision for war was a Cambridge Change for all Americans and Iraqis,[9] altering our status from peoples at peace to ones at war. Importantly, this was not a causal change. To be sure, just as a chain of classical events was required for Xantippe to learn of Socrates' death, implementing Bush's choice required millions of other choices, each of which also collapsed the state's wave function into classical events in the world. But by virtue of being the dominant monad, Bush's decision non-locally changed the probabilities that all those other choices would be made, and once they were made, what gave them all the meaning of "going to war" was the new socially shared superposition that Bush had created. This kind of causation might be seen as the basis for a quantum conceptualization of structural power, according to which dominant monads, by virtue of their position within a social wave function, can affect others through action *at a distance*.[10]

The third distinction reflects the fact that individuals vary hugely in the extent to which they possess the knowledge that constitutes a given social wave

---

[8] See Look (2002), Nachtomy (2007).
[9] On Cambridge Changes see Chapter 10, p. 195 and passim.
[10] Cf. Barnett and Duvall (2005).

function. In medieval Europe, many peasants probably had no idea what polity they were "members" of, much less what its policies were, and even today not every monad carries a perfect image of the social whole within. This suggests that people who are unaware of their state and/or what is being done in its name are not subjects of the state, only objects. By this I mean that, on the one hand, since leaders are empowered to act on behalf of all their citizens, we are all potential *objects* of manipulation by leaders; on the other hand, however, only those with an awareness of what is going on – the "attentive public" – can be subjects in/of the state, in the sense of being able to act purposefully on its potentials. Note that being an object of the state is different than being a passive monad. Whereas the latter can become active monads if they choose, the latter cannot even in principle until they acquire the relevant knowledge. Everyone is an active monad somewhere, but to do so you have to be "in the know."

Conceptualizing individuals as monads is half of the story, but what is the whole – in this case the state – that we are mirroring in our practices and thoughts? Recall that practices and thoughts are the classical effects of measurements of quantum phenomena, and as such what they mirror can't be the state as social structure, which exists only as a potentiality. So when we observe a policeman arresting a drunk driver, although the state's wave function is collapsing before our very eyes, we are not actually seeing the structure that makes the collapse possible. Similarly when we imagine the state in our thoughts – we are not "seeing" it as a superposition, but as an intentional *object*, which is to say what its *decoherence* looks like in our consciousness. So what is the ontological status of this object, which we all know is "there" even though we can't see it?

The answer is that the state is a kind of hologram.[11] This hologram is different from those created artificially by scientists in the lab, and also from the holographic projection that I argued in Chapter 11 enables us to see ordinary material objects, since in these cases there is something there visible to the naked eye. However, if, as some physicists have argued, the whole universe is a hologram, then there is no reason to demand that holograms be visually perceptible (which would be anthropocentric to boot). What matters is whether the holographic principle is operative in social contexts, where three considerations suggest an affirmative answer.

First, in any hologram, the information that generates the whole is encoded in each pixel rather than, as in a photograph, distributed across pixels that stand in a 1:1 correspondence to points in the image. This is precisely what a monadological view of individuals entails: that the whole (here the state) is present *in* the parts, not made up *of* them. Granted, I just suggested that in social holograms it is rarely the case that *every* part contains the whole, but

---

[11] Cf. Bradley's (2000) holographic analysis of social collectives, and see Milovanovic (2014) for a comprehensive overview of holographic thinking for social science.

this is more a mark of the imperfection of social holograms than a difference in kind. Thus, just as one could destroy most of a holographic plate and still recover the overall image (albeit fuzzier) from what is left, so a state could lose most of its population in a natural disaster and yet be able to reconstruct its core institutions from those who survived.[12] So there is a dual implication here. On the one hand, a holographic perspective suggests that even in "state-centric" fields like IR, individuals matter much more than is typically assumed. On the other, however, it also makes individuals massively redundant, since with the partial exception of leaders, the unique qualities of a state's members will mostly wash out in the practices that realize it. This "democratic" quality enables the state to project a stable identity over time, but at the cost of most individuals not mattering *qua* individuals most of the time.

Second, it is not just what monads are statically, but what they do dynamically – how they behave – which encodes holographic information. As Edgar Mitchell and Robert Staretz put it, "quantum emissions from any material entity carry information non-locally about the event history (e.g. an evolving record of everything that has happened) of the quantum states of the emitting matter."[13] In the case of the policeman, the emissions are the words, "you are under arrest," and by virtue of his entanglement with society the history of those words is not just his alone, but the history of the shared quantum state that defines what putting someone under arrest means. The policeman's practices enfold the history of the whole state, in other words, rather than being a purely local and one-off phenomenon.[14]

Finally, the essence of any holographic process is "wave front reconstruction,"[15] without which there is no perception of the object in question. Recall that holography involves three kinds of waves: object waves emitted by objects in the world form an interference pattern with reference waves emitted from a holographic projector (in this case the brain), the object hidden in which is then deciphered by reconstructive waves on the same frequency as the reference waves. In the policeman case, the object waves are the sights and sounds of "you are under arrest," while the reference waves are coming continuously from our own visual and auditory senses. The key is the reconstructive waves in our brain, which by virtue of our entanglement in the state and English language are on the same frequency as the object waves coming from the policeman. The resulting reconstruction enables us to understand immediately what is going on. Note that this does not constitute seeing "the state," per se; what we are seeing is just an arrest, so the state is only implicit. However, if we step back and think

---

[12] In the ideal case, "the whole information of current states of all active monads can be obtained from that of a single active monad"; Nakagomi (2003a: 21).
[13] Mitchell and Staretz (2011: 942).   [14] Cf. Schmidt (2007: 145–146).
[15] Robbins (2006: 367).

about what makes the arrest possible, then we will be conscious of "the state" as an intentional *object* or Idea.[16] So even though we cannot literally see the state, because it is enfolded holographically within our minds and practices, by attending to it we can nonetheless *perceive* it.[17] In short, the state is like a rainbow – it only exists when someone is looking at it.[18]

The state as an organism

With this holographic model in place we can turn now to the idea of the state as an organism. Organicism can be based on either a materialist or vitalist ontology, and as such it is not the same thing as vitalism; indeed within biology today materialist organicism is a perfectly respectable (if still minority) view, whereas vitalism is not.[19] So while I will be combining the two, in principle it is possible to accept the argument in this section while rejecting that in the next.

The idea that society and/or the state are organisms was widely held in the nineteenth century.[20] Yet, as a result of the genetic revolution, the fact that critics often conflated organicism with vitalism, and the perceived association of both with fascism, social organicism fell into disrepute by the mid-twentieth century. Recently, however, it has regained some of its currency, especially in biology of all places, where research on insect colonies has led to a substantial revival of the concept of the "superorganism," "a collection of single creatures that together possess the functional organization implicit in the formal definition of organism."[21] David Sloan Wilson has done more than anyone to take the logical next step of applying this concept to human society, where it has been picked up by some others.[22] For reasons that will become apparent in a moment, at this stage it is not possible to say anything definitive about what this revival entails for the nature of the state. But it does raise two big questions that might help structure our thinking about it.[23]

First, is the state an organism or superorganism? The literature on social insects has focused on the latter, on the assumption that the concept of an organism is well understood and clearly not exemplified by colonies. Yet, in fact, biologists have no more idea how to define 'organism' than they do 'life.'[24]

---

[16] On the Idea of the state see Buzan (1991: 65–66) and Wendt (1999: 218–219).
[17] Cf. Ittelson (2007).   [18] On rainbows and perception see Manzotti (2006: 10–14).
[19] See for example Gilbert and Sarkar (2000).
[20] For overviews see for example Levine (1995) and Cheah (2003).
[21] Wilson and Sober (1989: 341); see Hölldobler and Wilson (2009) for a good introduction to the superorganism literature.
[22] See Wilson (2002), Wendt (2004), Heylighen (2007), Keseber (2012), and Hoffecker (2013).
[23] See Mainville (2015) for an effort to grapple with these questions in an IR context.
[24] Pepper and Herron (2008) is a good overview of various definitions, and on the current debate see the special issue on organisms in *History and Philosophy of the Life Sciences* (2010) and also Bouchard and Huneman, eds. (2013).

Part of the problem is that organisms take a mind-boggling variety of forms, over which it has proven impossible to define a set of necessary and sufficient conditions. But recently this difficulty has been compounded by recognition that "paradigmatic" organisms like vertebrates contain millions of other organisms that exist in symbiosis with their host, which makes human beings look a lot like *super*organisms.[25] Matt Haber calls this the "problem of the paradigm," which takes two forms – assuming that an organism is the same thing as an individual, and assuming there is a paradigmatic organism at all.[26] Haber's own suggestion is to dispense with all organism concepts, and conceive of colonies instead as individuals; Samir Okasha's "rank-free" approach, in contrast, would treat colonies as a kind of organism.[27] Point being, while we can probably agree that states are not vertebrates,[28] what form of life they are is by no means clear.

Second, and related, how should we conceptualize the boundaries of social organisms? This question is especially pressing in the human case, since the answer would depend on whether we are talking about languages, states, societies, the international system, and/or world society (to name just some of the possibilities), and therefore how many social organisms there are. Here too biological theory offers little guidance, except that the issue is related to the equally murky question of what defines an individual, to which various answers have been proposed.[29] If quantum coherence is constitutive of life, however, then the need to protect it would lend support to the "immunological" approach to individuality. In recent years it has become clear that all organisms (and colonies) possess an immune system, which monitors the environment for threats and generates an immune response when they appear.[30] It is easy to see such a process at work in the state (the securitization of immigration anyone?), although it might figure in decentralized social organisms as well. Either way, given that the boundaries of human societies are constantly evolving, there is an opportunity here for social scientists to learn from the biological literature on the evolution of individuality.[31]

Radical as it is in some ways, social organicism does not in itself entail a vitalist sociology, which its advocates would surely want no truck with. As I suggested in Chapter 7, this is because biologists take a materialist ontology for granted, and as such, although some may think that even primitive organisms have minds, by virtue of their materialism their conception of mind does not thematize consciousness. Thus, it comes as no surprise that the question of

---

[25] See for example Gilbert et al. (2012). [26] Haber (2013: 198).
[27] See Okasha (2011). [28] The cover of Hobbes' *Leviathan* notwithstanding...
[29] See Pradeu (2010) for a good overview.
[30] See for example Tauber (1994) and Pradeu (2010). This approach also resonates with the theory of autopoiesis, which has already found its way into the social sciences.
[31] See Buss (1987) and Bouchard and Huneman, eds. (2013); cf. Wendt (2003).

whether superorganisms are conscious has to my knowledge never been asked. I don't have that luxury, given that I have argued that quantum coherence is the physical basis not only of life but of consciousness. So here goes . . . !

The state and collective consciousness

For many years it was assumed that groups could not have minds even in the truncated, materialist sense simply because they do not have brains. However, while groups have not acquired brains since then, today multiple lines of scholarship argue they do have the functional equivalent. In philosophy there has been substantial work on the concept of collective intentionality, which suggests that the attribution of intentional states and agency to groups is not just a useful fiction, but – depending on whom you read – justified in a realist sense.[32] And then there are the even larger and more empirically grounded literatures on distributed, extended, and group cognition, all of which suggests that cognition takes place not just in the head but also outside in transactions with the world.[33] While by no means uncontested,[34] these ideas are well within the mainstream of contemporary thinking about social cognition, and have also found resonance among biologists, who coined the phrase "swarm intelligence" to describe cognition in insect colonies.

These literatures offer rich resources for theorizing about states as organisms, but I will not review them here,[35] because although they imply that states have minds, their contributors almost without exception reject the possibility of group *consciousness*. In this they have common sense on their side, since even to my biased ear "what is it like to be a state?" is a much stranger question than "what is it like to be a bat?" and experiments have shown that people are in fact much more willing to attribute cognitive properties to groups than they are conscious and/or emotional ones.[36] The main reason seems to be that our normal experience of consciousness (sic) is indivisible and private. Thus, while it is not hard to conceive of the members of a group sharing the cognitive labor involved in working toward a common end (you write that section of the paper, I'll write this one . . .), it is difficult even to imagine what shared consciousness would be like. Telepathic powers or a unitary, supra-individual like the Borg on *Star Trek* come to mind, but you'll be reassured to hear that I won't be going

---

[32] For the major points of departure in this debate see Gilbert (1989), Bratman (1993), Searle (1995), and most recently Pettit and List (2011).
[33] These literatures are expanding rapidly; for seminal statements see Hutchins (1995), Clark and Chalmers (1998), and Wilson (2001) respectively, and Theiner et al. (2010) and Walter (2014b) for recent integrative perspectives.
[34] See for example Adams and Aizawa (2008) and Rupert (2009).
[35] See Wendt (2004) for a preliminary effort.
[36] See Huebner et al. (2010), which shows, however, that this also varies by culture.

there.[37] In short, group "consciousness" seems reducible to the consciousness of individuals, in which case the state is quite literally a "zombie."[38]

Beyond its disturbing imagery (yikes!), I take this to be a significant challenge to my conception of a vitalist sociology, since unlike neo-vitalists like Bennett and Latour who are already operating at the sociological level, I have made consciousness a constitutive feature of life. Indeed, without a physical basis for collective consciousness, even treating the state as an organism would on my view be problematic in anything but a metaphorical sense. Despite its counter-intuitive character, however, with quantum help I believe such a concept can be fashioned, which I shall do in three steps.

The first is to free consciousness from the confines of the skull and get it out into the world. Although that might itself seem counter-intuitive, it is a core principle of enactivism, which is a relatively new but now well-established position in the philosophy of mind.[39] To set up the enactivist intervention it makes sense to start with the "extended mind" hypothesis put forward by Andy Clark and David Chalmers.[40] Their idea is that if a mind is "reliably coupled" with devices that aid thinking, like calculators, notebooks, or for that matter, other people, then the material "vehicle" of that mind includes parts of its environment, since there is no principled reason to distinguish cognitive operations inside the brain from those in "processing loops" with prostheses outside.[41] The claim is controversial,[42] but it supports my larger argument and so let me assume it is true, since my more immediate problem is that most of its advocates, including Clark, reject the idea of extended *consciousness*. The crux of his critique is that information processing coupled with the environment is vastly slower than in the brain alone, and only the latter has enough "band-width" to generate experience.[43] The upshot is externalism about cognition, but internalism about consciousness.

This is a curious critique, considering that materialists like Clark actually have no idea what generates experience, much less how much "band-width" it requires. However, the real issue here is an assumption about what consciousness *is*.[44] If materialists are right that consciousness supervenes on brain states trapped inside of a skull, then yes, we should reject the idea of extended consciousness, but it is precisely this belief that enactivists question. In their view,

---

[37] See Mathiesen (2005: 237) and Szanto (2014: 109).
[38] See Szanto (2014), and also Huebner (2011) for an argument to similar effect.
[39] On enactivism see for example Hurley (1998), Noë (2004), and Thompson (2007).
[40] See Clark and Chalmers (1998).
[41] Note that this "vehicle externalism" is different than the "content" externalism discussed in Chapter 13.
[42] See the special issue of *Cognitive Systems Research* in 2010 devoted to the debate.
[43] See Clark (2009: 984–985).
[44] The following discussion draws especially on Ward (2012) and Laughlin (2013); also see Manzotti (2006).

consciousness is a transaction between the mind and its environment, understood not just as a causal relationship in which the world affects consciousness, but as a constitutive relationship in which experience is intrinsically world-involving. The vehicle not just of cognition but of consciousness too is the brain *plus* the world.

Enactivism, I submit, conforms to our actual experience: when we open our eyes the world is right there, rather than something from which it seems we have to wait for input. From a classical standpoint it may be objected that light takes time to get from objects to our eyes, which means that mind and world are fully separable and the latter's role must therefore be causal. However, as I argued in Chapter 12, a quantum view of light buttresses the enactivist case. If light enables direct, non-local perception of the world, then the time-lag objection is moot and consciousness could indeed be intrinsically world-involving.

A quantum interpretation of extended consciousness takes us part way toward collective consciousness, but only part, because even extended consciousness is still centered in individual brains and thus solipsistic. A plausible second step therefore would be to invoke the concept of 'We-feeling,' which seems to get at something like 'collective consciousness,' and is not only widely used by philosophers of collective intentionality, but has been studied empirically by social psychologists as well. Unfortunately, the philosophers almost never mention collective consciousness (except to emphasize that it is *not* what they are talking about);[45] and while the social psychologists have shown that We-feelings are routine and ubiquitous, they almost all consider it an individual-level emotion.[46] A provocative exception is Jonathan Mercer, who argues that since it is well known that emotions track identities, and that identities can be emergent, group-level phenomena, then "feeling like a state" is emergent too.[47] But while open to the possibility of collective consciousness,[48] Mercer perhaps wisely does not bring it up, and ultimately it is unclear in his account what the ontological status of group-level feeling is, since he is as wary of reification as anyone. Indeed so am I, having defended a flat ontology in which even collective cognitions are not a "level" of reality above the individual.[49]

We can nevertheless move forward on the basis of Hans Bernhard Schmid's phenomenological analysis of the "sense of us," which he sees as being presupposed by collective intentions.[50] Schmid argues that the dominant understandings of this sense – social identity theory and plural subject theory – are logically

---

[45] Mathiesen (2005) and Midgley (2006) are the only exceptions of which I am aware.
[46] See Stephan et al. (2014) for a review and helpful typology of approaches to emotions "beyond the body."
[47] See Mercer (2014), which also provides an excellent overview and analysis of relevant social psychological scholarship on group emotions.
[48] Personal communication.   [49] Cf. Wendt (1999: Chapter 4).
[50] See Schmid (2014); all page references in this and the next two paragraphs are to this article.

problematic. According to the former, the sense of us refers to an object – the group – around which actors' beliefs coalesce; "social unities are 'the participants' consciousness of that unity'" (p. 10). In Schmid's view this is circular, because to constitute a social unity it is not enough that members just happen to have the same beliefs, they have to form a unity *together*, which implies a collective intention that the sense of us is supposed to clarify. According to the latter, in turn, the sense of us refers to a (plural) subject, in which the relevant beliefs are not yours or mine, but "ours." In Schmid's view this leads to an infinite regress, because forming a plural subject is something that – once again – can only be done together, which implies a preexisting plural subject, and so on. Schmid's own proposal is to define the sense of us not as an object or a subject but as a *mode* of relating to each other, a "plural self-awareness" "in which members experience the world from a shared perspective" (p. 11).[51] For my purposes a key feature of his view is that, unlike other accounts of collective intentionality, which deny that there is anything "it is like" for a group to be in a mental state, Schmid's "plural self-awareness" is explicitly about "what it is like *for us*" (p. 14, emphasis added). He does not actually come out and say "collective consciousness," but invoking Nagel's famous phrase brings him tantalizingly close.

Schmid makes his argument in qualitative terms, with no thought to what it implies physically, but if we were to ask that question it seems clear that this is implicitly a quantum model. For example, he emphasizes that the individual self does not exist prior to self-awareness, which agrees with the argument of Chapter 8 that the mind is a superposition that is only actualized in the collapse of its wave function, i.e. in experience. Based on his experiential conception of the self, in turn, Schmid argues that plural self-awareness is just what the plural self *is* (p. 18). In his own argument, this serves the purpose of enabling him to avoid the logical problems that attend thinking of the sense of us as an object or a subject, but it is what a quantum perspective would suggest as well. We-feeling does not reduce to the consciousness of separable individuals, but neither does it presuppose a collective object or subject existing over and above them. What it presupposes is only a social wave function, which could come into being either from co-presence in a concrete situation (before they say a word, strangers meeting on a desert island would become entangled simply by virtue of visual perception) or, more often, from prior socialization (being taught in school that we are all citizens of X). However, as a potentiality a social wave function alone does not constitute We-feeling; it is only when individuals actualize it in *collapse* that they experience "us." As with any experience, in so doing they constitute themselves as separable individuals, but by virtue of entanglement this experience is also non-locally connected to others, whether concrete or imagined, and as such it is more than just what We-feeling is like

---

[51] Cf. Mathiesen (2005: 247–248), which takes a simulationist view of collective consciousness that I read as more individualistic than Schmid's.

for me, but what it is like *for us*. So the object and subject of collective intentions are not prior to collective consciousness, but emergent from it, in a quantum sense.

This argument, I think, brings us much closer to a physical basis for collective consciousness, but it might now be objected that the whole idea of collective consciousness is a contradiction in terms, since consciousness is normally indivisible and unitary – an I-feeling rather than a We-feeling. Schmid himself argues that groups do not have an authoritative point of view like individuals do, and as such there is no "I, the state," only "We, the People" (p. 23). So let me take a third step, which is to try to turn the tables by arguing that even the consciousness of individuals is collective.

What is the primary unit of life? Intuitively one might think the organism, which lives and dies as a whole, but in biology the dominant view has actually long been the opposite, that *cells* are primary.[52] Rather than seeing multi-cellular organisms in top-down or organicist terms as a whole to which the parts are subservient, "cell theory" does so in bottom-up terms as a symbiosis of autonomous living beings. The evidence for this view is multi-faceted, including the fact that single-celled organisms are the smallest unambiguous life forms, that despite being highly specialized every cell has a copy of its organism's DNA, that cells can be extracted from organisms and kept alive, that cells have behavioral autonomy within the organism, and so on.[53] This is not to deny that multi-cellular organisms are in some ways special, but from a cell theoretic perspective they are superorganisms, massive colonies of cells with no real ontological status of their own.[54]

In considering the relevance of cell theory for the possibility of collective consciousness two potential objections stand out. The first is that cell theory is atomistic and reductionist, and as such might seem to conflict with the holism of quantum biology. And indeed, in suggesting above that the state is an organism, I myself invoked not cell theory but organicism to help make the case. However, if emergence is understood in quantum terms and, as I have argued, quantum coherence is a condition of possibility for life, then the two theories can be reconciled without reducing one to the other. Cells would constitute emergent, coherent states of their own elements (microtubules and the rest), and then in virtue of quantum entanglement with other cells a macroscopic coherent state would emerge at the level of the whole organism. As one would expect from a quantum perspective on any part/whole relation, with respect to cells, organisms and what is alive, the answer is both/and, not either/or.

---

[52] As Baluska et al. (2004: 9) put it, "[t]he cell doctrine is firmly embedded in all biological disciplines and acts as a general paradigm of organismal and tissue construction and function."

[53] See for example Sitte (1992: S1–2) and Reynolds (2010: 198–199), and Reynolds (2007) for a good overview of the historical debate over the cell theory.

[54] Nicholson (2010: 205). Interestingly, the most serious challenges to cell theory concern not vertebrates, but plants; see Baluska et al. (2004).

The second objection is that if cells are the basic units of life, and if life is co-extensive with subjectivity, then that implies that each cell in our bodies is independently conscious. Most modern cell theorists would want none of that.[55] For them cells are just machines – though then again for them even organisms are just machines, which seems equally counter-intuitive. However, outside biology the idea of cell consciousness has been taken seriously by some eminent philosophers, most notably Whitehead, who conceptualized the individual as a "society" of elementary conscious units, and also Schrödinger, who saw the individual as a "republic of cells" the consciousness of which is not separable from the universe as a whole.[56] So why then is our experience unitary rather than a cacophony of squawking cells? The qualitative answer would be that organisms[57] are hierarchically structured such that it is only a dominant monad which experiences the whole. I recognize that this points toward a potentially problematic homunculus view of subjectivity, but if so then the physical answer at least suggests a new understanding of what that would entail. Traditionally the homunculus has been understood in classical terms as a physically separable seat of consciousness within the brain. In contrast, the quantum coherence of life means that even if our experience is that of a dominant monad, that monad is a superposition entangled non-locally with all the other cells in the organism, and as such is not separable from them. So when it decoheres into experiences, what it is instantiating is the consciousness of a *collective*, not that of a single cell.

As a social scientist I can't help but take pleasure in the fact that this argument suggests a "pan-social" ontology, in which, rather than starting with individuals and building up from there, sociology instead goes all the way *down* (at least within life).[58] Indeed, sociological metaphors played a key role in the development of cell theory in the nineteenth century,[59] and today there are indications that more than a metaphor is involved here, if cells communicate with each other through a language that is isomorphic with human language.[60]

Of more immediate relevance however, is what this inversion of our usual way of thinking does to skepticism about the possibility of collective consciousness. If the seemingly unitary consciousness of individuals is itself collective, then

---

[55] Though there are exceptions; see for example Margulis (2001), Edwards (2005), and Sevush (2006); for a less radical but not unrelated argument see Zeki (2003) on the "disunity of consciousness."

[56] On Whitehead in this respect see Hartshorne (1972) and Griffin (1998: 185–198), and on Schrödinger see Poser (1992: 160).

[57] Or at least animals; in nineteenth-century cell theory plants were routinely treated as having a more "egalitarian" structure and thus would not have a unitary consciousness.

[58] See d'Hombres and Mehdaoui (2012) on Alfred Espinas' "sociologization" of biology.

[59] See Reynolds (2008).

[60] See Ji (1997) on this specific argument, and more generally Clark (2010), Baslow (2011) and Marijuan et al. (2013) on inter-cellular communication.

collective consciousness at the sociological level, and in particular in states and other centralized organizations, is no longer an oxymoron. As with the individual, in the state there is a dominant monad that experiences and speaks for the whole, namely the leader, who, contra Schmid, can legitimately say, "*I*, the State," and by virtue of his structural position does have an authoritative point of view on what the state feels. To be sure, anyone else who is a subject in the state's hologram can "feel like a state" too – like when children pledge allegiance to the flag or soldiers take an oath to defend their country – although because they are not dominant monads their experience is more "what it is like for *us*" than the leader's "... for *me*." Still, this democratic quality of collective consciousness might suggest a difference from individual consciousness, for which – even if it is collective – there is no evidence that cells in my spleen can experience what it is like to be me, and (fortunately!) I in turn have no experience of their experiences. Yet, perhaps the two cases are not so dissimilar. After all, we have no way of knowing what our cells experience, and one should also allow for the fact that individuals are far more autonomous and complex than cells, and as such capable of a much wider range of experiences. And just as with me and my spleen, leaders have no idea when their citizens are feeling like a state. So while what I have offered here is far from a thorough examination of the concept, I hope my basic strategy for giving collective consciousness a physical basis is sound enough to pursue further.

The politics of vitalist sociology

Social organicism, collective consciousness, the cell state embodied by the leader: what next, the *Führerprinzip*? I grant it could all sound rather sinister, so would a quantum sociology – on my rendition at least – lead inexorably toward fascism? Well, it is true that the Nazis did invoke Driesch's vitalism to help justify conquering "less vital" peoples – although they had to fire Professor Driesch as a result, who objected to this use of his ideas.[61] On the other hand, Jakob von Uexküll, another thinker associated with vitalism and a father of biosemiotics, did not object, having written a book on the state as an organism that conceived of foreigners as "parasites" and was about "as harsh a model of othering as one could possibly come up with."[62] And then of course most famously there is Carl Schmitt, vitalist and erstwhile Nazi Party member. However, Schmitt's vitalism was "*not* taken from a biological but from a theological, philosophical, or political-ideological register,"[63] and as

---

[61] See Bennett (2010: 69).
[62] Drechsler (2009: 90). For further discussion of von Uexküll in relation to fascism see Harrington (1996) and Stella and Kleisner (2010), and on Oka Asajiro's similar ideas in pre-war Japan see Sullivan (2011).
[63] See Braun (2012: 4), emphasis in the original.

such is very different from the naturalist vitalism on offer here. And while other nineteenth-century reactionaries sometimes made use of vitalist ideas, anti-colonial and progressive movements did so as well,[64] and today vitalism is associated less with fascism than with the New Ageism of a post-modern culture trying to "re-enchant" the world.[65] In sum, Bennett I think is right that whatever link there is between vitalism and violence is contingent.[66]

Indeed, if there is any inherent politics to quantum vitalist sociology, then I would argue that it is *voluntarist*. Against determinists, voluntarists emphasize the creativity and freedom of individuals to resist and overcome structural constraints. To be sure, even more than vitalism voluntarism was central to Nazi ideology,[67] and some of the philosophers to whom the Nazis looked for voluntarist inspiration, such as Leibniz and Schopenhauer, have inspired my own argument. Yet, as with vitalists, voluntarists worked for the good guys too, like the many existentialists who fought in the French Resistance.

Consider in contrast the political slope of materialism, with its picture of a deterministic reality in which people and society are just machines and consciousness and freedom have no place. Whereas vitalism privileges life, as Schopenhauer and more recently Hans Jonas argued materialism is a philosophy that privileges death. For materialists dead matter is the norm, the baseline in terms of which life must be explained; the living, in effect, are nothing but assemblages of the dead.[68] Not only does that blur the distinction between life and non-life philosophically, it also raises questions about what happens when such a view permeates society, like it does today, such that it becomes "under the ontological dominance of death."[69] For while it is difficult for moderns to conceive, the ancients thought of life as the normal state, not death, and as such there is an implicit societal choice here, one that might plausibly shape attitudes not just toward nature but toward each other. While in a vitalist ontology there is no guarantee that individuals will put their agency to progressive causes, at least there they have the option, to experience and make of society what they will.

[64] See Schwartz (1992), Reill (2005), and Jones (2010).
[65] A serious manifesto for which might be Berman (1981).
[66] Bennett (2010: 90). [67] See Strehle (2011), and also Braun (2012).
[68] For a good discussion of Jonas (1966) on this issue see Wolters (2001).
[69] See Wolters (2001: 91), quoting Jonas.

Conclusion

In this book I have addressed the physical basis of social life. Within the social sciences the de facto ontology is dualism. While most social scientists would probably consider themselves materialists, and virtually everyone at least implicitly accepts the CCP, positivists and interpretivists alike routinely reference intentional phenomena in their social theorizing. This is problematic because such phenomena presuppose consciousness, and no progress has been made on integrating consciousness with the materialist worldview. Philosophers seem increasingly inclined to think that consciousness must therefore be an illusion, but that would leave social scientists in a tough spot. Either we become behaviorists, eschewing reference to intentional phenomena altogether, or we retain them in our explanations and become dualists and tacit vitalists.

The source of this dualism is an assumption that the relevant causal closure constraints on solving the mind–body problem are those of classical mechanics, which describes a purely material world of matter and energy. But since the 1930s we have known that the causal closure principles in the universe as a whole are quantum mechanical rather than classical, where the physical constraints on explanation are radically different. In particular, quantum theory admits a neutral monist/panpsychist interpretation in which 'physical' does not equal 'material,' and instead sees the material world described by classical physics and the mental world of consciousness as joint effects of an underlying reality that is neither. The question then is whether an ontology in which consciousness goes "all the way down" can scale up to the human and specifically sociological level. While there are a priori reasons to doubt it, there is growing experimental evidence that human behavior in fact follows quantum principles. If that evidence continues to mount, it would confirm a key prediction of quantum consciousness theory, according to which our subjectivity is a macroscopic quantum mechanical phenomenon – that we are walking wave functions. That would constitute a basis for solving the mind–body problem, and in so doing unifying physical and social ontology within a naturalistic, though no longer materialist, worldview.

Apart from laying out this metaphysical argument, which is due to others, my own contribution has been to show that it offers traction on some long-standing controversies in social ontology. My "case studies" were issues surrounding the nature of human agents and their relationship to social structures, but these by no means exhaust the applicability of quantum consciousness theory to social science. Thus, in addition to other issues in social ontology beyond the agent–structure problem, it may be useful to highlight what I have *not* done in this book. On the explanatory side, I have not developed a general theory of society like liberalism or Marxism; I have not proposed specific theories of relatively autonomous social systems like the international system; and I have not tested any theories against empirical evidence. And on the normative side, I have not suggested how a quantum approach might affect our thinking about law, human rights, or any of the other issues that make up moral and political philosophy. Insofar as these literatures are based on classical assumptions – or implicit quantum ones that struggle for recognition – my expectation is that bringing the radical conceptual armory of quantum theory to bear on them would be transformative. In short, there is still a long way to go before the potential of a quantum social science is realized.

Night thoughts on epistemology[1]

I do however want to say a few words about another topic that I bracketed in this book, which is social epistemology. As we saw in Part I, unlike in the classical worldview, with its clear separation of subjects and objects, in quantum physics questions of epistemology and ontology are difficult to separate. This is due to the Measurement Problem, in which the observation of sub-atomic phenomena in some way participates in what actually happens, such that we cannot safely assume that the latter is independent of the former. What conclusion to draw from this is a key question in the debate over quantum theory, with some philosophers arguing that it demands an instrumental interpretation of the theory, while others a realist one. By focusing on ontology rather than epistemology I have therefore taken sides right up front on a deeply contested issue. Yet since philosophers of quantum physics are themselves divided, there is no justification I could offer for it (or the reverse) that would satisfy those who think it is a mistake. If pressed I would say that a realist approach is more likely to yield hypotheses, like quantum consciousness theory, that might advance our knowledge down the road. But in the end I see this more as a matter of personal disposition than anything else.

Having said that, however, let me offer a glimpse of how the particular quantum ontology that I have advocated might speak to the Explanation/

---

[1] These are far from worked out in my own mind, and as such are more trial balloons that I hope others will clarify and/or criticize.

Understanding debate in social epistemology.[2] The debate is about "the possibility of naturalism" in social inquiry, or whether a social science on the model of the natural sciences is possible and/or desirable.[3] More specifically, given that the natural sciences aspire to objective, third-person knowledge, the debate is about whether there is any place in a science of society for the second- (and even first-) person analyses characteristic of interpretivist work, which treat the objects of social inquiry as subjects with their own meanings and point of view – as a "You" rather than an "It." That this would be problematic makes sense if our model of natural science is classical. There is no role for second- and first-person perspectives if we are trying to explain rocks or glaciers, which are not subjects and exist whether we observe them or not. In that context, trying to maintain a separation of subject and object makes sense.

In quantum physics objects cannot be said to exist prior to their realization in measurement by an observer. Does this mean we need a participatory epistemology to understand sub-atomic particles? Well, yes and no. On the one hand, I have argued that the collapse of the wave function is a process of temporal symmetry-breaking in which not only a material particle is produced, but also a sub-atomic experience of that process. So yes, in our interaction (sic) with the particle (sic), the second- and first-person (sic) dimensions are always there. On the other hand, the experience of a particle is ephemeral, and "what it is like" to be one is probably beyond the human capacity to understand. Indeed, we face such an epistemic limit even in our dealings with other organisms, like Nagel's bat, whose experiences are not ephemeral but persist in memory. At the highest level of biological organization – apes, chimps, and dogs – we might gain some genuine second-person understanding, but in general when it comes to interpreting meanings we are stuck in a bubble with our fellow human beings. Note, however, that these are limits to *our knowledge*, and as such do not mean that other organisms do not have meaningful experiences, much less that they are mere machines or objects.

As I argued in Chapter 12, however, we are in a very good position to know other human minds, which points to what I see as the most important epistemological implication of this book: subjectivity – and here I mean conscious subjectivity – can and should be "brought back in" to social science. That challenges positivists and interpretivists alike, both of whom have tended to run away from subjectivity, as something either that is not relevant to science or is a Cartesian anxiety that in a post-modern world we can now thankfully leave

---

[2] Quantum theorists who are not wedded to this ontology reach rather different epistemological conclusions; see for example Plotnitsky (1994; 2010) and Barad (2007).

[3] See Bhaskar (1979). The adjective 'physical' would be better than 'natural' here, since, as I argued in Chapter 1, 'physical' is open to both classical and quantum interpretations, which have very different implications for what a "natural" science entails.

behind. In my view, a social science that has no place for subjectivity has no place either for its own subject matter or for its audience, and in that case what's the point? Moreover, if they can get past their hang-ups about it, recovering subjectivity is a task to which interpretivists and positivists can both contribute. Interpretivists, especially in the phenomenological tradition, of course have much to say about how to study subjectivity,[4] but perhaps more surprisingly, so do some positivists who have dealt with individual-level data.[5] This common ground does not mean that first-, second-, and third-person inquiries are the same; quite the contrary, in my own view they are complementary in a quantum sense – individually incomplete, mutually exclusive, but jointly necessary for a complete description.[6] Thus, in contrast to my previous effort to reconcile interpretivism and positivism through a "via media," which assumed a classical either/or choice,[7] from a quantum perspective their relationship should be seen as both/and, such that there will always be "two [or three?] stories to tell."[8] The only question then is which epistemically incomplete stance to take for a given problem.

Speaking of which, the kind of participatory ontology that quantum theory implies gives the problem of what question to ask in our research – von Neumann's "Process 1" – a new and more loaded significance than it has typically had in the past. As graduate students in the social sciences we are taught that it is legitimate to let our values and interests guide us in choosing our questions. But in the conventional, Weberian understanding, that value-laden process ends when we start our actual research; then, the goal of objectivity demands that we keep our values and interests out of the picture. Such a dualism cannot be sustained in quantum physics, and lately is coming into question even in other physical sciences, where non-epistemic values are increasingly seen by philosophers as having a role to play not only in choosing research questions, but also in evaluating their answers.[9] This subject–object "endogeneity" is all the more important in the social sciences, where researchers may be members of the very system they are observing. That does not mean individual social scientists can expect to have a measurable impact on society, although *collectively* we might, analogous to the way in which self-conscious observation of one's own consciousness may change the latter over time.

---

[4] For a good starting point see Zahavi (2005).
[5] See for example Petranker (2003), Kahneman and Krueger (2006), Overgaard et al. (2008), and Lahlou (2011); and for a skeptical view see Irvine (2012).
[6] I'm not sure what to do with first-person experience unmediated by the second-person framework of a researcher, but for some ideas see Rudolph and Rudolph (2003).
[7] See Wendt (1999: Chapter 2).
[8] See Hollis and Smith (1990). Note that not everyone sympathetic to the idea of quantum consciousness agrees that complementarity is its epistemological implication; see for example Jonas (1984: 225–227).
[9] See Elliott and McKaughan (2014) and McAllister (2014).

However, the breakdown of the subject–object dualism in quantum contexts does highlight that in asking a given question: (1) social scientists are collapsing a socially shared wave function, locally for us as researchers, and thereby calling forth a reality rather than just passively observing one given to us as spectators, and (2) our research is therefore responsible, at the micro-level, for helping to create, sustain and/or transform that reality. This is not to say that if social scientists refuse to study something which they disapprove of – as some in IR would have us do with the state – it will disappear, since most social facts are sustained by thousands or millions of lay people "observing" them, not by social scientists. But it does emphasize our complicity in that process, both empirically and politically.

These quantum effects suggest that the hope of social scientific realists – that through the continuing improvement of our theories and measurement techniques we will get closer and closer to truths of social life – is misplaced.[10] If it does sometimes seem like we are getting closer, as in the "democratic peace" (widely seen as the closest thing in IR to a scientific law), then that should be seen as the effect that repeated measurements by lay persons and social scientists alike have in *stabilizing* a certain reality, not of better approximating an independently existing one. However, this does not mean social science does not produce knowledge, or that truth is merely an effect of power. Even in quantum physics, where our power to influence what is observed would seem to be at its maximum, we cannot in fact determine whatever outcome we like. Yet we do not for that reason say physicists' knowledge of sub-atomic particles is purely subjective, only that it is probabilistic. By the same token, even though social scientists cannot completely eliminate their own error terms – not just because of complexity but because that is where agency lies – we can still get some grip on the world around us. So I do not see a quantum social science implying anti-realism so much as what Bernard d'Espagnat calls "open" realism. There is something *there*, independent of us academic observers, even if we cannot subject it to the steel jaws of necessity.[11]

Two last points about epistemology. First, one issue that a quantum perspective suggests positivists and interpretivists alike will need to rethink is causation. As we have seen quantum processes like the collapse of the wave function and action at a distance are not causal, at least in the usual, efficient causal sense. For positivists this challenges the current fashion of trying to explain social phenomena by reference to causal mechanisms, the very language of which exudes a classical worldview. For interpretivists the challenge might seem less, since they were never interested in causation in the first place – but there is still the question of what to do with a quantum naturalism. My own view is that it supports an interpretation of reasons as final or teleological

[10] Cf. Wendt (1999: Chapter 2). [11] See d'Espagnat (2006: 28, 117–118, and passim).

causes, and although the non-local causation found in social structures is not efficient causation, as its name suggests it remains "causation" of a sort, perhaps akin to that in Cambridge Changes. While these are important challenges, however, they are also an opportunity, since quantum theorists themselves do not know how to think about causation.[12] Insofar as sub-atomic phenomena are just like social phenomena, perhaps we can help them figure it out.

Second, an important virtue of quantum consciousness theory is that it can explain the possibility of social science itself, which is something we do consciously. Mastering literature, finding questions, conceptualizing, operationalizing, collecting and analyzing data, and eventually basking in our conclusions – these are all practices that we *experience*, and experience as *freely chosen* as well. The classical materialist naturalism that, at least officially, is the basis for positivism is at a complete loss to explain these experiences – yet it is hard to imagine machines or zombies doing what we do. For their part, interpretivists at least take the consciousness of social scientists as given, but they too cannot explain it. For our activities to make sense, therefore, we need an ontology that can accommodate our subjectivity within a naturalistic worldview. Rather than focusing on our side of this problem, my strategy has been to change the worldview side, in the hopes that this can overcome the implicit dualism within our work.[13]

Too elegant not to be true?

As I indicated in Chapter 1, it is not necessary to believe that the ontology proposed in this book is capital-T true to think that it might be a useful heuristic for doing social science. Indeed, the advocates of some of my key dialectical resources – quantum decision theory, game theory, and semantics – themselves generally take an agnostic position on questions of ontology. That makes sense intellectually, since we do not have to agree on the nature of reality to see if a quantum approach can predict experimental results. And it makes sense strategically, since their work is more likely to be accepted by others if it is not freighted with controversial ontological baggage. My effort to show that a quantum approach sheds new light on debates in social theory can be read in the same, pragmatist spirit. Of course, insofar as these various efforts succeed they will not leave social science where it was before. In particular, they raise the question "if quantum theorizing about X works so well, then why are you still engaged in classical theorizing?" Even as a heuristic, in other words, a successful quantum social science implicitly calls upon those in the

---

[12] For a good introduction see Price and Corry, eds. (2007).

[13] Although I am no student of his work, this was apparently also Michael Polanyi's approach to reconciling the "two cultures"; see Zhenhua (2001–2002) for a useful overview.

mainstream to *justify* something they have always taken for granted. One of the first places where this call may become salient is in graduate methods training, which, if my own experience is any guide, is a touchy subject. But beyond foreshadowing a potential conflict in this area, a non-realist or "as if" reading of quantum social science has the virtue of not forcing those who are interested in exploring these ideas empirically to change their entire worldview first.

Yet in the longer run there are limits to such an agnostic approach, especially in the theoretical domains with which I have been concerned in this book. This is not simply because at the end of the day human beings either are, or are not, walking wave functions. It is because of the politics of science, for many in the classical orthodoxy are *not* agnostic about ontology. In the natural sciences we see this in the dismissal of vitalism as unscientific, in the hostile reception that quantum consciousness theory and until recently panpsychism have received, in the increasingly strident assertions that consciousness and free will are illusions, and – in a domain that I have not addressed in this book – in defenses of evolutionary theory against any and all criticisms. The intensity of these attitudes suggests that these are not just garden-variety scientific disagreements, but spring from a concern to defend a metaphysics that is felt to be under siege. In the social sciences we see the same thing. In many departments the attitude toward interpretivists, post-structuralists, and others who take meaning and subjectivity seriously is not "let a thousand flowers bloom," but "this is not Science and it must not be supported." Don't hire people who practice such arts, if you make that mistake don't tenure them, don't fund their research, and don't publish it in top journals. Crucially, the conception of "science" that is being legislated here is a classical one, in which meaning, subjectivity, and apparently now also free will, all do not belong. So while I encourage those who are intrigued by my argument to remain agnostic about its truth while seeing what it can do on the ground, eventually someone in a position of power may say, "but it *isn't* true" – or perhaps more precisely, "it *can't* be true and therefore it isn't" – and dismiss the results as no more scientific than vitalism.

If you will allow me another military metaphor, therefore, the "as if" approach to quantum social science is akin to guerrilla warfare, attacking unexplained anomalies in the orthodox lines, using local successes like quantum decision theory to build popular support, and training cadres in the new methodological techniques. But the politics of ontology being what they are, if Mao was right then at some point guerrilla warfare must give way to conventional war – to a more realist view of quantum social science that would be a frontal assault on the classical mainstream. Undoubtedly it is too early for such an assault to succeed, not least because if Thomas Kuhn was right, paradigms don't change through decisive battles anyway, but through the gradual generational replacement of those wedded to the old paradigm by those embracing the new.[14] In arguing that human beings *really are* walking wave functions, therefore, my goal has been only to try to foresee what such a confrontation might look like. Given what I have said in this section I cannot urge others to make the same leap, but having gone out on this limb myself I want to conclude this book with my personal justification for a realist view of quantum social science.

The argument would be that it is an inference to the best explanation, or IBE. IBE is a principle of inference for choosing between theories when, for whatever reason, deduction and induction cannot be applied.[15] It has a strong pedigree in the history of science, and is widely used in the law and everyday life as well. The basic idea is that even if we cannot prove a theory to be true, it is still possible to rationally conclude that it is the *best* theory relative to its competitors and as such should be adopted as the most likely to be true. The "relative" is crucial; unlike the principle of inference known as "abduction" or "retroduction," with which often it is conflated, IBE is essentially contrastive in nature, pitting two or more explanations for a given phenomenon against each other.[16] So the question here would be whether, given the CCP, a quantum or a classical ontology provides the best explanation for consciousness and social life.

But how shall we define 'best'? Naïvely one might think it means "most likely to be true given the evidence," but this would be circular and trivialize IBE. The challenge rather is to define 'best' independently of the evidence while showing that, as a consequence of the inference, a theory can be judged as most likely to be true.[17] To do this IBE theorists use "explanatory virtues" to assess competing theories. Inevitably, there is no agreement on what precisely these virtues are, but in an attempt to clarify the literature Adolfas Mackonis has come up with five that are widely used: coherence, depth, breadth, simplicity, and empirical adequacy.[18] However, he ultimately folds the last into the first, and depth is of interest mostly to those with a mechanistic worldview and as such not very useful here. So let me briefly compare our two candidates on the remaining three criteria.

Coherence refers to a theory's fit with relevant background knowledge, by which is primarily meant theories that are already well established, but in Mackonis' view may also include empirical data or tests of the theories in

---

[14] See Kuhn (1962/1996).
[15] The literature on IBE is extensive; see Lipton (2004) for a good overview, and Clayton (1997) offers a succinct treatment specially adapted to theory choice in metaphysics.
[16] Mackonis (2013: 976–978) sees abduction as only the first step in an IBE.
[17] See Glass (2012: 413).
[18] See Mackonis (2013); cf. McAllister (1989), for example, who identifies five criteria: internal consistency, consistency with extant well-corroborated theories, predictive accuracy, predictive scope, and fruitfulness.

question. While often seen as the most important virtue, coherence is tricky to apply to theories that would force a change in the background knowledge itself, as in a Kuhnian paradigm change, which depending on what counts as the "relevant" background may or may not be the case here. If the relevant background is taken to be quantum physics itself – our foundational science of reality – then a quantum social ontology is much more "virtuous" than its classical rival. However, two other ways to define the background point the other way. First, the foundation of my approach, quantum consciousness theory, is not just quantum-physical but panpsychist, which is clearly not coherent with well-established theory. Second, defenders of classical social ontology can point out that quantum phenomena mostly wash out above the molecular level, and so the relevant background is classical physics and neuroscience, not quantum. On the other hand, to the extent that empirical tests are allowed to count as part of the background, then the striking successes of quantum decision theory favor a quantum ontology. So on balance, the coherence jury is either still out, or hung. However, when it comes to the other two virtues I think the ontology developed in this book comes out a clear winner.

Explanatory breadth refers to the extent to which a theory unifies different kinds of facts, the more the better. Normally these facts would all be presumed to be material facts, in which case the quantum formalism – which "far exceeds [the breadth] of any competitor"[19] on its home turf of physics – clearly outperforms classical thinking in the domains of social science to which it has been applied so far. In particular, whereas to my knowledge no one writing from a classical standpoint has suggested that human choice behavior and semantic behavior are in any way connected (and their literatures are totally independent), as we have seen quantum theory can be used to explain both. This is not in itself an argument for quantum *consciousness* theory, although if these behaviors are intentional and therefore presuppose consciousness, then a more direct connection can be made. But even speaking strictly behaviorally, given the gulf that currently separates these domains, it is striking that a single formalism taken straight from physics can unify them – and can subsume classical choice and semantic behavior in the bargain.

Yet that is only the tip of the iceberg, for where there is truly no comparison between the two ontologies is in the ability of quantum consciousness theory to unify utterly different kinds of facts, namely material and phenomenal ones. The latter have stymied the classical worldview for centuries, and so its advocates have now been reduced to arguing that the appearance we have of being conscious is just that, *an appearance*. Quantum consciousness theory, in contrast, can "save" the appearance of consciousness and in so doing make sense of many other apparent phenomenal facts: the apparently meaningful character of

---

[19] See Mackonis (2013: 983), quoting Paul Thagard.

our behavior, our apparently free will, the apparently teleological force of our reasons, our apparent ability to fill in and even change the historical past, and the invisible yet apparently emergent nature of social structures. I cannot *prove* that all these appearances are genuine, but arguing that they are all illusions seems not just like a second-best strategy, but a strategy of last resort. So if we can unify all these appearances under a theory that treats them as real (in a quantum sense), why would we infer otherwise?

Simplicity (or parsimony) is akin to breadth as a virtue, except that whereas breadth is about explaining more facts with the same theoretical resources, simplicity is about explaining the same facts with fewer resources.[20] Ironically, simplicity has several potential dimensions and thus has been difficult to define.[21] However, social scientists are all trained to make their theories as parsimonious as possible, so for my purposes an intuitive, "you know it when you see it" criterion should suffice here.

The simplicity question comes up especially with the theory of quantum consciousness, which, recall, has two parts: quantum brain theory and panpsychism. For the brain to be a quantum computer many things must turn out to be true, which suggests a low simplicity score. On the other hand, it is not obvious a true classical theory of the brain would be any simpler, since the human brain is the most complex system known in the universe. Moreover, and this I think is decisive, the hypothesized effect of the zillions of interactions in a quantum brain is a single organizing principle that is entirely lacking in classical brain theory: quantum coherence, to which everything else in the brain is subservient. So while the details are extraordinarily complex, the result is extremely simple.

This in turn relates to the other half of the theory: consciousness. Here there is almost no contrast to make, since there is no classical theory of consciousness on the horizon, whereas on the quantum side we at least have a candidate. Even a complex quantum explanation would be simpler than no explanation at all. However, the quantum argument is in fact very simple: consciousness is an aspect of matter at the elementary level (panpsychism), which is amplified upward by quantum coherence in the brain. Compare this to the image offered by materialists in their own quest to explain consciousness: an incredibly complex machine, chugging along, neurons firing all over the place, which somehow spits out consciousness. Quantum consciousness theory is speculative, but compared to the alternative its simplicity is hard to beat.

The literature on IBE emphasizes the *explanatory* virtues of theories, because, assuming that science is rational, it should be explanatory power that matters most in judging which of two theories is more likely to be true. I have argued that on this score a quantum social ontology will eventually

---

[20] See Mackonis (2013: 987). [21] See McAllister (1991).

be a clear winner. Nevertheless, given that quantum consciousness theory and quantum social science are both still in their infancy it is difficult to draw this inference convincingly today, which is a good reason for readers to take an "as if" rather than realist attitude toward these ideas. I suppose if I were completely rational then I should do the same, but in my view a quantum social ontology has another kind of virtue from which an IBE can be drawn now: an *aesthetic* one.

Philosophers have long recognized that alongside explanatory power aesthetic considerations play a role in theory choice, and indeed simplicity and breadth are routinely cited as instances of both virtues. Yet concerned to defend the rationality of science, most philosophers think the aesthetic virtues of theories are reducible to the explanatory. This may be less true of scientists, and especially physicists. Philosopher James McAllister offers some choice quotes from the latter, and also makes a strong analytical case that "truth and beauty" are complementary and not always correlated.[22] While I am inclined toward his view, I won't defend it here. Instead, I want to conclude with a bold claim: whatever their current force as explanatory virtues, the coherence, breadth, and simplicity of the quantum hypothesis make it *too elegant not to be true*.

For the price of two simple propositions, quantum consciousness theory offers not just a solution to the mind–body problem, or additionally, to the nature of life and of time, which are mostly beyond the concerns of this book. And it does not just solve the Agent–Structure and Explanation-Understanding problems, or explain quantum decision theory's success in predicting otherwise anomalous behavior. What the theory offers is all of these things and more, and with them a unification of physical and social ontology that gives the human experience a home in the universe. With its elegance, in other words, comes not just extraordinary explanatory power, but extraordinary *meaning*, which at least this situated observer finds utterly lacking in the classical worldview.

You might not share my aesthetic sense and thus be reluctant to believe we really are walking wave functions. That's of course fine. But by arguing it *could* be true I hope I have given you reason to suspend your belief that we really are just classical machines, and thus to suspend your *dis*belief in quantum consciousness long enough to try assuming it in your work. If you do, perhaps you will find your own home in the universe too.

---

[22] See McAllister (1996); note however that although critical of reductionism, he too ultimately wants to defend the rationality of aesthetic judgments. See Montano (2013) for an alternative, more naturalistic view.

Bibliography

Abbott, Derek, Paul Davies, and Arun Pati, eds. (2008) *Quantum Aspects of Life*, London: Imperial College Press.
Abram, David (1996) *The Spell of the Sensuous*, New York: Vintage Books.
Adams, Fred and Kenneth Aizawa (2008) *The Bounds of Cognition*, Oxford: Blackwell.
Adler, Emanuel and Vincent Pouliot (2011) "International Practices," *International Theory*, 3(1), 1–36.
Aerts, Diederik (1998) "The Entity and Modern Physics: The Creation-Discovery View of Reality," in E. Castellani, ed., *Interpreting Bodies*, Princeton University Press, pp. 223–257.
   (2009) "Quantum Particles as Conceptual Entities: A Possible Explanatory Framework for Quantum Theory," *Foundations of Science*, 14(4), 361–411.
   (2010) "Interpreting Quantum Particles as Conceptual Entities," *International Journal of Theoretical Physics*, 49(12), 2950–2970.
Aerts, Diederik and Sven Aerts (1995/6) "Applications of Quantum Statistics in Psychological Studies of Decision Processes," *Foundations of Science*, 1(1), 85–97.
Aerts, Diederik, Jan Broekaert, and Liane Gabora (2011) "A Case for Applying an Abstracted Quantum Formalism to Cognition," *New Ideas in Psychology*, 29(2), 136–146.
Aerts, Diederik, Marek Czachor, and Bart D'Hooghe (2006) "Towards a Quantum Evolutionary Scheme: Violating Bell's Inequalities in Language," in N. Gontier, et al., eds., *Evolutionary Epistemology, Language and Culture*, Dordrecht: Springer, pp. 453–478.
Aerts, Diederik, Bart D'Hooghe, and Emmanuel Haven (2010) "Quantum Experimental Data in Psychology and Economics," *International Journal of Theoretical Physics*, 49(12), 2971–2990.
Affifi, Ramsey (2013) "Learning Plants: Semiosis between the Parts and the Whole," *Biosemiotics*, 6(3), 547–559.
Aharonov, Yakir, Peter Bergmann, and Joel Lebowitz (1964) "Time Symmetry in the Quantum Process of Measurement," *Physical Review*, 134, 1410–1416.
Aharonov, Yakir and Lev Vaidman (1990) "Properties of a Quantum System during the Time Interval between Two Measurements," *Physical Review A*, 41(1), 11–20.
Aharonov, Yakir and M. Suhail Zubairy (2005) "Time and the Quantum: Erasing the Past and Impacting the Future," *Science*, 307, 875–879.
Albert, David (1992) *Quantum Mechanics and Experience*, Cambridge, MA: Harvard University Press.
   (2000) *Time and Chance*, Cambridge, MA: Harvard University Press.

Albert, David and Barry Loewer (1988) "Interpreting the Many Worlds Interpretation," *Synthese*, 77(2), 195–213.
Alfano, Mark (2012) "Wilde Heuristics and Rum Tum Tuggers: Preference Indeterminacy and Instability," *Synthese*, 189(1), 5–15.
Alfinito, Eleonora and Giuseppe Vitiello (2000) "The Dissipative Quantum Model of Brain," *Information Sciences*, 128(3–4), 217–229.
Al-Khalili, Jim and Johnjoe McFadden (2015) *Life on the Edge: The Coming of Age of Quantum Biology*, London: Bantam Press.
Allen, Amy (1998) "Power Trouble: Performativity as Critical Theory," *Constellations*, 5(4), 456–471.
Allen, Colin and Michael Trestman (2014) "Animal Consciousness," *Stanford Encyclopedia of Philosophy* (Summer 2014 edition), Edward N. Zalta (ed.), http://plato.stanford.edu/archives/sum2014/entries/consciousness-animal/.
Allen, Garland (2005) "Mechanism, Vitalism and Organicism in Late Nineteenth and Twentieth-Century Biology," *Studies in History and Philosophy of Biological and Biomedical Sciences*, 36(2), 261–283.
Alter, Torin and Yujin Nagasawa (2012) "What Is Russellian Monism?" *Journal of Consciousness Studies*, 19(9–10), 67–95.
Andersen, P. B., C. Emmeche, N. Finneman, and P. Christiansen, eds. (2000) *Downward Causation: Minds, Bodies, and Matter*, Aarhus University Press.
Ankersmit, Frank (2005) *Sublime Historical Experience*, Stanford University Press.
Apel, Karl-Otto (1984) *Understanding and Explanation*, Cambridge, MA: MIT Press.
Archer, Margaret (1995) *Realist Social Theory: The Morphogenetic Approach*, Cambridge University Press.
  (2007) "The Ontological Status of Subjectivity," in C. Lawson, J. Latsis, and N. Martins, eds., *Contributions to Social Ontology*, London: Routledge, pp. 17–31.
Arenhart, Jonas (2013) "Wither Away Individuals," *Synthese*, 190(16), 3475–3494.
Arfi, Badredine (2005) "Resolving the Trust Predicament in IR: A Quantum Game-Theoretic Approach," *Theory and Decision*, 59(2), 127–174.
  (2007) "Quantum Social Game Theory," *Physica A*, 374(2), 794–820.
Arshavsky, Yuri (2006) "'The Seven Sins' of the Hebbian Synapse: Can the Hypothesis of Synaptic Plasticity Explain Long-Term Memory Consolidation?" *Progress in Neurobiology*, 80(3), 99–113.
Asano, Masanari, Irina Basieva, Andrei Khrennikov, Masanori Ohya, and Yoshiharu Tanaka (2012) "Quantum-Like Dynamics of Decision-Making," *Physica A*, 391(5), 2083–2099.
Asano, Masanari, Masanori Ohya, and Andrei Khrennikov (2011) "Quantum-Like Model for Decision Making Process in Two Players Game," *Foundations of Physics*, 41(3), 538–548.
Astington, Janet Wilde and Eva Filippova (2005) "Language as the Route into Other Minds," in B. Malle and S. Hodges, eds., *Other Minds: How Humans Bridge the Divide between Self and Others*, New York, NY: Guilford Press, pp. 209–222.
Atmanspacher, Harald (2003) "Mind and Matter as Asymptotically Disjoint, Inequivalent Representations with Broken Time-Reversal Symmetry," *Biosystems*, 68(1), 19–30.

(2011) "Quantum Approaches to Consciousness," *Stanford Encyclopedia of Philosophy* (Summer 2011 edition), Edward N. Zalta (ed.), http://plato.stanford.edu/archives/sum2011/entries/qt-consciousness/.

Atmanspacher, Harald and Thomas Filk (2014) "Non-Commutative Operations in Consciousness Studies," *Journal of Consciousness Studies*, 21(3–4), 24–39.

Atmanspacher, Harald, Thomas Filk and Hartmann Römer (2004) "Quantum Zeno Features of Bistable Perception:" *Biological Cybernetics*, 90(1), 33–40.

Atmanspacher, Harald and Hans Primas (2006) "Pauli's Ideas on Mind and Matter in the Context of Contemporary Science," *Journal of Consciousness Studies*, 13(3), 5–50.

Atmanspacher, Harald and Hartmann Römer (2012) "Order Effects in Sequential Measurements of Non-Commuting Psychological Observables," *Journal of Mathematical Psychology*, 56, 274–280.

Atmanspacher, Harald, Hartmann Römer, and Harald Walach (2002) "Weak Quantum Theory: Complementarity and Entanglement in Physics and Beyond," *Foundations of Physics*, 32(3), 379–406.

Baars, Bernaard (2004) "Subjective Experience Is Probably not Limited to Humans," *Consciousness and Cognition*, 14(1), 7–21.

Baars, Bernard and David Edelman (2012) "Consciousness, Biology and Quantum Hypotheses," *Physics of Life Reviews*, 9(3), 285–294.

Baer, John, James Kaufman, and Roy Baumeister, eds. (2008) *Are We Free? Psychology and Free Will*, Oxford University Press.

Baer, Wolfgang (2010) "The Physics of Consciousness," *Journal of Consciousness Studies*, 17(3–4), 165–191.

Bahrami, M. and A. Shafiee (2010) "Postponing the Past: An Operational Analysis of Delayed-Choice Experiments," *Foundations of Physics*, 40(1), 55–92.

Balaguer, Mark (2004) "A Coherent, Naturalistic, and Plausible Formulation of Libertarian Free Will," *Nous*, 38(3), 379–406.

 (2009) "Why There Are No Good Arguments for Any Interesting Version of Determinism," *Synthese*, 168(1), 1–21.

 (2010) *Free Will as an Open Scientific Problem*, Cambridge, MA: MIT Press.

Balazs, Andras (2004) "Internal Measurement: Some Aspects of Quantum Theory in Biology," *Physics Essays*, 17(1), 80–94.

Ball, Philip (2011) "The Dawn of Quantum Biology," *Nature*, 474, 272–274.

Baluska, Frantisek, Dieter Volkmann, and Peter Barlow (2004) "Eukaryotic Cells and Their Cell Bodies: Cell Theory Revised," *Annals of Botany*, 94(1), 9–32.

Banks, Erik (2010) "Neutral Monism Reconsidered," *Philosophical Psychology*, 23(2), 173–187.

Barad, Karen (2003) "Posthumanist Performativity: How Matter Comes to Matter," *Signs*, 28(3), 801–831.

 (2007) *Meeting the Universe Halfway: Quantum Physics and the Entanglement of Matter and Meaning*, Durham, NC: Duke University Press.

Barham, James (2008) "The Reality of Purpose and the Reform of Naturalism," *Philosophia Naturalis*, 44(1), 31–52.

 (2012) "Normativity, Agency, and Life," *Studies in History and Philosophy of Biological and Biomedical Sciences*, 43(1), 92–103.

Barlow, Peter (2008) "Reflections on 'Plant Neurobiology,'" *Biosystems*, 92(2), 132–147.

Barnett, Michael and Raymond Duvall (2005) "Power in International Politics," *International Organization*, 59(1), 39–75.
Barrett, Jeffrey (1999) *The Quantum Mechanics of Minds and Worlds*, Oxford University Press.
  (2006) "A Quantum-Mechanical Argument for Mind–Body Dualism," *Erkenntnis*, 65(1), 97–115.
Bartlett, Gary (2012) "Computational Theories of Conscious Experience: Between a Rock and a Hard Place," *Erkenntnis*, 76(2), 195–209.
Basile, Pierfrancesco (2006) "Rethinking Leibniz: Whitehead, Ward and the Idealistic Legacy," *Process Studies*, 35(2), 207–229.
  (2010) "It Must Be True – But How Can it Be? Some Remarks on Panpsychism and Mental Composition," *Royal Institute of Philosophy Supplement*, 67, 93–112.
Baslow, Morris (2011) "Biosemiosis and the Cellular Basis of Mind," *Biosemiotics*, 4(1), 39–53.
Bass, L. (1975) "A Quantum Mechanical Mind–Body Interaction," *Foundations of Physics*, 5(1), 159–172.
Baumeister, Roy, E. J. Masicampo, and Kathleen Vohs (2011) "Do Conscious Thoughts Cause Behavior?" *Annual Review of Psychology*, 62, 331–361.
Beck, Friedrich and John Eccles (1992) "Quantum Aspects of Brain Activity and the Role of Consciousness," *Proceedings of the National Academy of Sciences*, 89(23), 11357–11361.
  (1998) "Quantum Processes in the Brain: A Scientific Basis of Consciousness," *Cognitive Studies*, 5(2), 95–109.
Becker, Christian and Reiner Manstetten (2004) "Nature as a You: Novalis' Philosophical Thought and the Modern Ecological Crisis," *Environmental Values*, 13(1), 101–118.
Becker, Theodore, ed. (1991) *Quantum Politics: Applying Quantum Theory to Political Phenomena*, New York: Praeger.
Bedau, Mark (1997) "Weak Emergence," *Nous*, 31, 375–399.
  (1998) "Four Puzzles About Life," *Artificial Life*, 4(2), 125–140.
Beim Graben, Peter and Harald Atmanspacher (2006) "Complementarity in Classical Dynamical Systems," *Foundations of Physics*, 36(2), 291–306.
Bekenstein, Jacob (2003) "Information in the Holographic Universe," *Scientific American*, August, 59–65.
Belousek, Darrin (2003) "Non-Separability, Non-Supervenience, and Quantum Ontology," *Philosophy of Science*, 70(4), 791–811.
Benioff, Paul (2002) "Language Is Physical," *Quantum Information Processing*, 1(6), 495–509.
Ben-Jacob, E., D. Coffey, and Alfred Tauber (2005) "Seeking the Foundations of Cognition in Bacteria," *Physica A*, 359(1), 495–524.
Bennett, Jane (2010) *Vibrant Matter: A Political Ecology of Things*, Durham, NC: Duke University Press.
Bennett, Max and Peter Hacker (2003) *Philosophical Foundations of Neuroscience*, Oxford: Blackwell.
Bentley, Arthur (1941) "The Human Skin: Philosophy's Last Line of Defence," *Philosophy of Science*, 8(1), 1–19.

Benton, E. (1974) "Vitalism in Nineteenth-Century Science Thought," *Studies in History and Philosophy of Science*, 5(1), 17–48.
Berkovitz, Joseph (1998) "Aspects of Quantum Non-Locality I," *Studies in History and Philosophy of Modern Physics*, 29(2), 183–222.
  (2008) "On Predictions in Retro-Causal Interpretations of Quantum Mechanics," *Studies in History and Philosophy of Modern Physics*, 39(4), 709–735.
  (2014) "Action at a Distance in Quantum Mechanics," *Stanford Encyclopedia of Philosophy* (Spring 2014 edition), Edward N. Zalta (ed.), http://plato.stanford.edu/archives/spr2014/entries/qm-action-distance/.
Berman, Morris (1981) *The Reenchantment of the World*, Ithaca, NY: Cornell University Press.
Bernecker, Sven (2004) "Memory and Externalism," *Philosophy and Phenomenological Research*, 69(3), 605–632.
Bertau, Marie-Cecile (2014a) "Exploring Language as the 'In-Between,'" *Theory and Psychology*, 24(4), 524–541.
  (2014b) "On Displacement," *Theory and Psychology*, 24(4), 442–458.
Bertelsen, Preben (2011) "Intentional Activity and Free Will as Core Concepts in Criminal Law and Psychology," *Theory and Psychology*, 22(1), 46–66.
Beyler, R. (1996) "Targeting the Organism: The Scientific and Cultural Context of Pascual Jordan's Quantum Biology, 1932–1947," *Isis*, 87(2), 248–273.
Bhargava, Rajeev (1992) *Individualism in Social Science*, Oxford: Clarendon Press.
Bhaskar, Roy (1979) *The Possibility of Naturalism*, New York: Routledge.
  (1982) "Emergence, Explanation, and Emancipation," in P. Secord, ed., *Explaining Human Behavior*, Beverly Hills, CA: Sage Publications, pp. 275–310.
  (1986) *Scientific Realism and Human Emancipation*, London: Verso.
Bierman, Dick (2006) "Empirical Research on the Radical Subjective Solution of the Measurement Problem: Does Time Get Its Direction through Conscious Observation?" in D. Sheehan, ed., *Frontiers of Time, Retrocausation – Experiment and Theory*, Melville, NY: American Institute of Physics, pp. 238–259.
Bigaj, Tomasz (2012) "Ungrounded Dispositions in Quantum Mechanics," *Foundations of Science*, 17(3), 205–221.
Birch, Jonathan (2012) "Robust Processes and Teleological Language," *European Journal for Philosophy of Science*, 2(3), 299–312.
Birksted-Breen, Dana (2003) "Time and the Après-Coup," *The International Journal of Psychoanalysis*, 84(6), 1501–1515.
Bishop, Robert (2013) "Essay Review: Teleology at Work in the World?" *Mind and Matter*, 11(2), 243–255.
Bitbol, Michel (2002) "Science as if Situation Mattered," *Phenomenology and the Cognitive Sciences*, 1(2), 181–224.
  (2007) "Ontology, Matter and Emergence," *Phenomenology and the Cognitive Sciences*, 6(3), 293–307.
  (2008) "Is Consciousness Primary?" *NeuroQuantology*, 6(1), 53–71.
  (2011) "The Quantum Structure of Knowledge," *Axiomathes*, 21(2), 357–371.
  (2012) "Downward Causation without Foundations," *Synthese*, 185(2), 233–255.
Bitbol, Michel, Pierre Kerszberg, and Jean Petitot, eds. (2009) *Constituting Objectivity: Transcendental Perspectives on Modern Physics*, Berlin: Springer.

Bibliography

Bitbol, Michel and Pier Luigi Luisi (2004) "Autopoiesis with or without Cognition: Defining Life at Its Edge," *Journal of the Royal Society Interface*, 1, 99–107.

Bobro, Marc and Kenneth Clatterbaugh (1996) "Unpacking the Monad: Leibniz' Theory of Causality," *The Monist*, 79(3), 408–425.

Bohl, Vivian and Nivedita Gangopadhyay (2014) "Theory of Mind and the Unobservability of Other Minds," *Philosophical Explorations*, 17(2), 203–222.

Bohm, David (1951) *Quantum Theory*, Englewood Cliffs, NJ: Prentice-Hall.

(1980) *Wholeness and the Implicate Order*, London: Routledge.

(1990) "A New Theory of the Relation of Mind and Matter," *Philosophical Psychology*, 3(2), 271–286.

Bohm, David and B. J. Hiley (1993) *The Undivided Universe*, London: Routledge.

Bohr, Niels (1933) "Light and Life," *Nature*, 131, 421–423 and 457–459.

(1937) "Causality and Complementarity," *Philosophy of Science*, 4(3), 289–298.

(1948) "On the Notions of Causality and Complementarity," *Dialectica*, 2, 312–319.

Boi, L. (2004) "Theories of Space-Time in Modern Physics," *Synthese*, 139(3), 429–489.

Bokulich, Alisa (2012) "Distinguishing Explanatory from Nonexplanatory Fictions," *Philosophy of Science*, 79(5), 725–737.

Bolender, John (2001) "An Argument for Idealism," *Journal of Consciousness Studies*, 8(4), 37–61.

Bordley, Robert and Joseph Kadane (1999) "Experiment-Dependent Priors in Psychology and Physics," *Theory and Decision*, 47(3), 213–227.

Bordonaro, Michael and Vasily Ogryzko (2013) "Quantum Biology at the Cellular Level – Elements of the Research Program," *Biosystems*, 112(1), 11–30.

Bortoft, Henri (1985) "Counterfeit and Authentic Wholes," in D. Seamon and R. Mugerauer, eds., *Dwelling, Place and Environment*, Dordrecht: Martinus Nijhoff Publishers, pp. 281–302.

Bouchard, Frédéric and Philippe Huneman, eds. (2013) *From Groups to Individuals: Evolution and Emerging Individuality*, Cambridge, MA: MIT Press.

Bourdieu, Pierre (1990) *The Logic of Practice*, Stanford University Press.

Bourget, David and David Chalmers (2014) "What Do Philosophers Believe?" *Philosophical Studies*, 170(3), 465–500.

Bousso, Raphael (2002) "The Holographic Principle," *Reviews in Modern Physics*, 74(3), 825–874.

Bradley, Raymond (2000) "Agency and the Theory of Quantum Vacuum Interaction," *World Futures*, 55(3), 227–275.

Brainerd, Charles, Zheng Wang, and Valerie Reyna (2013) "Superposition of Episodic Memories: Overdistribution and Quantum Models," *Topics in Cognitive Sciences*, 5(4), 773–799.

Brandenburger, Adam (2010) "The Relationship between Quantum and Classical Correlation in Games," *Games and Economic Behavior*, 69(1), 175–183.

Brandt, Lewis (1973) "The Physics of the Physicist and the Physics of the Psychologist," *International Journal of Psychology*, 8(1), 61–72.

Bratman, Michael (1993) "Shared Intentions," *Ethics*, 104(1), 97–113.

Braun, Kathrin (2012) "From the Body of Christ to Racial Homogeneity: Carl Schmitt's Mobilization of 'Life' against 'the Spirit of Technicity,'" *The European Legacy*, 17(1), 1–17.

Brewer, Bill (2007) "Perception and Its Objects," *Philosophical Studies*, 132(1), 87–97.
Brinkmann, Svend (2006) "Mental Life in the Space of Reasons," *Journal for the Theory of Social Behaviour*, 36(1), 1–16.
Brown, Robin and James Ladyman (2009) "Physicalism, Supervenience and the Fundamental Level," *The Philosophical Quarterly*, 59, 20–38.
Bruza, Peter and Richard Cole (2006) "Quantum Logic of Semantic Space," arXiv:quant-ph/0612178v1.
Bruza, Peter, Kirsty Kitto, Douglas Nelson, and Cathy McEvoy (2009) "Is There Something Quantum-Like about the Human Mental Lexicon?" *Journal of Mathematical Psychology*, 53(5), 362–377.
Bub, Jeffrey (2000) "Indeterminacy and Entanglement: The Challenge of Quantum Mechanics," *British Journal for the Philosophy of Science*, 51(4), 597–615.
Burge, Tyler (1979) "Individualism and the Mental," in P. French, et al., eds., *Midwest Studies in Philosophy*, vol. 4, Minneapolis, MN: University of Minnesota Press, pp. 73–121.
  (1986) "Individualism and Psychology," *The Philosophical Review*, 95(1), 3–45.
  (2005) "Disjunctivism and Perceptual Psychology," *Philosophical Topics*, 33(1), 1–78.
Burns, Jean (1999) "Volition and Physical Laws," *Journal of Consciousness Studies*, 6(10), 27–47.
Burwick, Frederick and Paul Douglass, eds. (1992) *The Crisis in Modernism: Bergson and the Vitalist Controversy*, Cambridge University Press.
Busemeyer, Jerome and Peter Bruza (2012) *Quantum Models of Cognition and Decision*, Cambridge: Cambridge University Press.
Busemeyer, Jerome, Riccardo Franco, Emmanuel Pothos, and Jennifer Trueblood (2011) "A Quantum Theoretical Explanation for Probability Judgment Errors," *Psychological Review*, 118(2), 193–218.
Busemeyer, Jerome, Zheng Wang, and James Townsend (2006) "Quantum Dynamics of Human Decision-Making," *Journal of Mathematical Psychology*, 50(3), 220–241.
Buss, L. (1987) *The Evolution of Individuality*, Princeton, NJ: Princeton University Press.
Butler, Judith (1990) *Gender Trouble: Feminism and the Subversion of Identity*, New York: Routledge.
  (1993) *Bodies that Matter: On the Discursive Limits of 'Sex,'* New York: Routledge.
Butterfield, Jeremy (1995) "Quantum Theory and the Mind," *The Aristotelian Society*, Supplementary Volume LXIX, 112–158.
Buzan, Barry (1991) *People, States, and Fear*, Boulder, CO: Lynne Rienner, 2nd edition.
Callender, Craig and Robert Weingard (1997) "Trouble in Paradise? Problems for Bohm's Theory," *The Monist*, 80(1), 24–43.
Camerer, Colin (2003) *Behavioral Game Theory*, Princeton University Press.
Campbell, Donald T. (1974) "'Downward Causation' in Hierarchically Organised Biological Systems," in F. Ayala and T. Dobzhansky, eds., *Studies in the Philosophy of Biology*, Berkeley, CA: University of California Press, pp. 179–186.
Campbell, Richard (2010) "The Emergence of Action," *New Ideas in Psychology*, 28(3), 283–295.
Campbell, Richard and Mark Bickhard (2011) "Physicalism, Emergence and Downward Causation," *Axiomathes*, 21(1), 33–56.

Čapek, Milič (1992) "Microphysical Indeterminacy and Freedom: Bergson and Peirce," in F. Burwick and P. Douglass, eds., *The Crisis in Modernism: Bergson and the Vitalist Controversy*, Cambridge University Press, pp. 171–189.

Carlin, Laurence (2006) "Leibniz on Final Causes," *Journal of the History of Philosophy*, 44(2), 217–233.

Carminati, G. Galli and F. Martin (2008) "Quantum Mechanics and the Psyche," *Physics of Particles and Nuclei*, 39(4), 560–577.

Carruthers, Peter (2007) "Invertebrate Minds," *The Journal of Ethics*, 11(3), 275–297.

Cartwright, Nancy (1999) *The Dappled World: A Study of the Boundaries of Science*, Cambridge University Press.

Castagnoli, Giuseppe (2009) "The Quantum Speed Up as Advanced Cognition of the Solution," *International Journal of Theoretical Physics*, 48(3), 857–873.

 (2010) "Quantum One Go Computation and the Physical Computation Level of Biological Information Processing," *International Journal of Theoretical Physics*, 49(2), 304–315.

Castellani, Elena, ed. (1998) *Interpreting Bodies: Classical and Quantum Objects in Modern Physics*, Princeton University Press.

 (2002) "Reductionism, Emergence, and Effective Field Theories," *Studies in History and Philosophy of Modern Physics*, 33(2), 251–267.

Chalmers, David (1995) "Facing Up to the Problem of Consciousness," *Journal of Consciousness Studies*, 2(3), 200–219.

 (1996) *The Conscious Mind*, Oxford University Press.

 (1997) "Moving Forward on the Problem of Consciousness," *Journal of Consciousness Studies*, 4(1), 3–46.

 (2010) *The Character of Consciousness*, Oxford University Press.

Cheah, Pheng (2003) *Spectral Nationality*, New York, NY: Columbia University Press.

Chen, Kay-Yut and Tad Hogg (2006) "How Well Do People Play a Quantum Prisoner's Dilemma?" *Quantum Information Processing*, 5(1), 43–67.

Chiereghin, Franco (2011) "Paradoxes of the Notion of Antedating," *Journal of Consciousness Studies*, 18(3–4), 24–43.

Chomsky, Noam (1995) "Language and Nature," *Mind*, 104, 1–61.

Churchland, Patricia Smith (1981) "On the Alleged Backwards Referral of Experiences and its Relevance to the Mind–Body Problem," *Philosophy of Science*, 48(2), 165–181.

Churchland, Paul (1988) *Matter and Consciousness*, Cambridge, MA: MIT Press.

Clark, Andy (2009) "Spreading the Joy? Why the Machinery of Consciousness is (Probably) Still in the Head," *Mind*, 118, 963–993.

Clark, Andy and David Chalmers (1998) "The Extended Mind," *Analysis*, 58(1), 7–19.

Clark, Kevin (2010) "Bose-Einstein Condensates Form in Heuristics Learned by Ciliates Deciding to Signal 'Social' Commitments," *Biosystems*, 99(3), 167–178.

Clarke, Chris (2007) "The Role of Quantum Physics in the Theory of Subjective Consciousness," *Mind and Matter*, 5(1), 45–81.

Clarke, Peter (2014) "Neuroscience, Quantum Indeterminism and the Cartesian Soul," *Brain and Cognition*, 84(1), 109–117.

Clarke, Randolph and Capes, Justin (2014) "Incompatibilist (Nondeterministic) Theories of Free Will," *Stanford Encyclopedia of Philosophy* (Spring 2014 edition), Edward N. Zalta (ed.), http://plato.stanford.edu/archives/spr2014/entries/incompatibilism-theories/.

Clayton, Philip (1997) "Inference to the Best Explanation," *Zygon*, 32(3), 377–391.
  (2006) "Conceptual Foundations of Emergence Theory," in P. Clayton and P. Davies, eds., *The Re-Emergence of Emergence*, Oxford University Press, pp. 1–31.
Cleland, Carol (2012) "Life without Definitions," *Synthese*, 185(1), 125–144.
  (2013) "Is a General Theory of Life Possible?" *Biological Theory*, 7(4), 368–379.
Cleland, Carol and Christopher Chyba (2002) "Defining 'Life'," *Origins of Life and Evolution of the Biosphere*, 32(4), 387–393.
Cochran, Andrew (1971) "Relationships between Quantum Physics and Biology," *Foundations of Physics*, 1(3), 235–250.
Cohen, I. Bernard (1994) "The Scientific Revolution and the Social Sciences," in I. B. Cohen, ed., *The Natural Sciences and the Social Sciences*, Boston, MA: Kluwer, pp. 153–203.
Cole, Andrew (2013) "The Call of Things: A Critique of Object-Oriented Ontologies," *the minnesota review*, 80, 106–118.
Coleman, Sam (2012) "Mental Chemistry: Combination for Panpsychists," *dialectica*, 66(1), 137–166.
  (2014) "The Real Combination Problem: Panpsychism, Micro-Subjects, and Emergence," *Erkenntnis*, 79(1), 19–44.
Compton, Arthur (1935) *The Freedom of Man*, New Haven, CT: Yale University Press.
Conrad, Michael (1996) "Cross-Scale Information Processing in Evolution, Development and Intelligence," *Biosystems*, 38(2–3), 97–109.
Contessa, Gabriele (2010) "Scientific Models and Fictional Objects," *Synthese*, 172(2), 215–229.
Conway, John and Simon Kochen (2006) "The Free Will Theorem," *Foundations of Physics*, 36(10), 1441–1473.
Coole, Diane and Samantha Frost (2010a) "Introducing the New Materialisms," in Coole and Frost, eds., *New Materialisms*, Durham, NC: Duke University Press, pp. 1–43.
  eds. (2010b) *New Materialisms: Ontology, Agency, and Politics*, Durham, NC: Duke University Press.
Cooper, W. Grant (2009) "Evidence for Transcriptase Quantum Processing Implies Entanglement and Decoherence of Superposition Proton States," *Biosystems*, 97(2), 73–89.
Cornejo, Carlos (2004) "Who Says What the Words Say? The Problem of Linguistic Meaning in Psychology," *Theory and Psychology*, 14(1), 5–28.
  (2008) "Intersubjectivity as Co-Phenomenology: From the Holism of Meaning to the Being-in-the-world-with-others," *Integrative Psychological and Behavioral Science*, 42(2), 171–178.
Corradini, Antonella and Timothy O'Connor, eds. (2010) *Emergence in Science and Philosophy*, London: Routledge.
Costa de Beauregard, Olivier (2000) "Efficient and Final Cause as CPT Reciprocals," in E. Agazzi and M. Pauri, eds., *The Reality of the Unobservable*, Dordrecht: Kluwer Academic Publishers, pp. 283–291.
Coulter, Jeff (2001) "Human Practices and the Observability of the 'Macro-Social,'" in T. Schatzki, et al., eds., *The Practice Turn in Contemporary Social Theory*, London: Routledge, pp. 29–41.

Craddock, Travis and Jack Tuszynski (2010) "A Critical Assessment of the Information Processing Capabilities of Neuronal Microtubules Using Coherent Excitations," *Journal of Biological Physics*, 36(1), 53–70.
Cramer, John (1986) "The Transactional Interpretation of Quantum Mechanics," *Reviews of Modern Physics*, 58(3), 647–687.
  (1988) "An Overview of the Transactional Interpretation of Quantum Mechanics," *International Journal of Theoretical Physics*, 27(2), 227–250.
Crane, Tim (2014) "The Problem of Perception," *Stanford Encyclopedia of Philosophy* (Winter 2014 edition), Edward N. Zalta (ed.), http://plato.stanford.edu/archives/win2014/entries/perception-problem/.
Crane, Tim and D. H. Mellor (1990) "There is No Question of Physicalism," *Mind*, 99, 185–206.
Craver, Carl and William Bechtel (2007) "Top-Down Causation without Top-Down Causes," *Biology and Philosophy*, 22(4), 547–563.
Crook, Seth and Carl Gillett (2001) "Why Physics Alone Cannot Define the 'Physical,'" *Canadian Journal of Philosophy*, 31(3), 333–360.
Cudd, Ann (1993) "Game Theory and the History of Ideas about Rationality," *Economics and Philosophy*, 9(1), 101–133.
Cummins, Robert, Martin Roth, and Ian Harmon (2014) "Why It Doesn't Matter to Metaphysics What Mary Learns," *Philosophical Studies*, 167(3), 541–555.
Cunningham, Andrew (2007) "Hume's Vitalism and Its Implications," *British Journal for the History of Philosophy*, 15(1), 59–73.
Currie, Gregory (1984) "Individualism and Global Supervenience," *British Journal for the Philosophy of Science*, 35(4), 345–358.
Cushing, James (1994) *Quantum Mechanics: Historical Contingency and the Copenhagen Hegemony*, University of Chicago Press.
Cvrčková, Fatima, Helena Lipavská, and Viktor Žárský (2009) "Plant Intelligence: Why, Why Not or Where?" *Plant Signaling and Behavior*, 4(5), 394–399.
Dalla Chiara, Maria Luisa, Roberto Giuntini, and Roberto Leporini (2006) "Holistic Quantum Computational Semantics and Gestalt-Thinking," in A. Bassi, et al., eds., *Quantum Mechanics*, Melville, NY:American Institute of Physics, pp. 86–100.
  (2011) "Holism, Ambiguity and Approximation in the Logics of Quantum Computation: A Survey," *International Journal of General Systems*, 40(1), 85–98.
Danto, Arthur (1965) *Analytical Philosophy of History*, Cambridge University Press.
Darby, George (2012) "Relational Holism and Humean Supervenience," *British Journal for the Philosophy of Science*, 63(4), 773–788.
Davidson, Donald (1963) "Actions, Reasons, and Causes," *The Journal of Philosophy*, 60(7), 685–700.
Davies, Paul C. W. (2004) "Does Quantum Mechanics Play a Non-Trivial Role in Life?" *Biosystems*, 78(1–3), 69–79.
  (2006) "The Physics of Downward Causation," in P. Clayton and P. Davies, eds., *The Re-Emergence of Emergence*, Oxford University Press, pp. 35–52.
Davies, Kim (2014) "Emergence from What? A Transcendental Understanding of the Place of Consciousness," *Journal of Consciousness Studies*, 21(5–6), 10–32.
De Barros, J. Acacio (2012) "Quantum-Like Model of Behavioral Response Computation Using Neural Oscillators," *Biosystems*, 110(3), 171–182.

De Barros, J. Acacio, and Patrick Suppes (2009) "Quantum Mechanics, Interference, and the Brain," *Journal of Mathematical Psychology*, 53(5), 306–313.
De Chardin, Teilhard (1959) *The Phenomenon of Man*, New York, NY: Harper & Row.
De Jaegher, Hanne (2009) "Social Understanding Through Direct Perception? Yes, by Interacting," *Consciousness and Cognition*, 18, 535–542.
DeLanda, Manuel (2002) *Intensive Science and Virtual Philosophy*, London: Continuum.
Dellis, A. T. and I. K. Kominis (2012) "The Quantum Zeno Effect Immunizes the Avian Compass against the Deleterious Effects of Exchange and Dipolar Interactions," *Biosystems*, 107(3), 153–157.
Dennett, Daniel (1971) "Intentional Systems," *The Journal of Philosophy*, 68(4), 87–106.
  (1987) *The Intentional Stance*, Cambridge: MIT Press.
  (1991) *Consciousness Explained*, Boston, MA: Little, Brown.
  (1996) "Facing Backwards on the Problem of Consciousness," *Journal of Consciousness Studies*, 3(1), 4–6.
Denton, Michael, Govindasamy Kumaramanickavel, and Michael Legge (2013) "Cells as Irreducible Wholes: The Failure of Mechanism and the Possibility of an Organicist Revival," *Biology and Philosophy*, 28(1), 31–52.
De Regt, Henk and Dennis Dieks (2005) "A Contextual Approach to Scientific Understanding," *Synthese*, 144(1), 137–170.
DeRose, Keith (2009) *The Case for Contextualism*, Oxford University Press.
D'Espagnat, Bernard (1995) *Veiled Reality: An Analysis of Present-Day Quantum Mechanical Concepts*, Reading, MA: Addison-Wesley.
  (2011) "Quantum Physics and Reality," *Foundations of Physics*, 41(11), 1703–1716.
De Quincey, Christian (2002) *Radical Nature: Rediscovering the Soul of Matter*, Montpelier, VT: Invisible Cities Press.
De Souza Vieira, Fabiano and Charbel Niño El-Hani (2008) "Emergence and Downward Determination in the Natural Sciences," *Cybernetics and Human Knowing*, 15(3–4), 101–134.
De Uriarte, Brian (1990) "On the Free Will of Rational Agents in Neoclassical Economics," *Journal of Post Keynesian Economics*, 12(4), 605–617.
Deutsch, David (1999) "Quantum Theory of Probability and Decisions," *Proceedings of the Royal Society*, A455, 3129–3197.
Dewey, John and Arthur Bentley (1949) *Knowing and the Known*, Boston, MA: Beacon Press.
De Witt, Bryce and N. Graham, eds. (1973) *The Many-Worlds Interpretation of Quantum Mechanics*, Princeton University Press.
Dharamsi, Karim (2011) "Re-Enacting in the Second Person," *Journal of the Philosophy of History*, 5(2), 163–178.
D'Hombres, Emmanuel and Soraya Mehdaoui (2012) "'On What Condition is the Equation Organism – Society Valid? Cell Theory and Organicist Sociology in the Works of Alfred Espinas (1870s–80s)," *History of the Human Sciences*, 25(1), 32–51.
Dijksterhuis, Ap and Henk Aarts (2010) "Goals, Attention, and (Un)Consciousness," *Annual Review of Psychology*, 61, 467–490.
Di Paolo, Ezequiel (2005) "Autopoiesis, Adaptivity, Teleology, Agency," *Phenomenology and the Cognitive Sciences*, 4(4), 429–452.

Dobbs, H. A. C. (1951) "The Relation between the Time of Psychology and the Time of Physics (Parts I and II)," *British Journal for the Philosophy of Science*, 2(6 and 7), 122–141 and 177–192.

Domondon, Andrew (2006) "Bringing Physics to Bear on the Phenomenon of Life: The Divergent Positions of Bohr, Delbrück, and Schrödinger," *Studies in History and Philosophy of Biological and Biomedical Sciences*, 37(3), 433–458.

Dorato, Mauro and Michael Esfeld (2010) "GRW as an Ontology of Dispositions," *Studies in History and Philosophy of Modern Physics*, 41(1), 41–49.

Dorato, Mauro and Matteo Morganti (2013) "Grades of Individuality: A Pluralistic View of Identity in Quantum Mechanics and in the Sciences," *Philosophical Studies*, 163(3), 591–610.

D'Oro, Giuseppina and Constantine Sandis (2013) "From Anti-Causalism to Causalism and Back: A History of the Reasons/Causes Debate," in D'Oro and Sandis, eds., *Reasons and Causes*, New York, NY: Palgrave MacMillan, pp. 7–48.

Dorsey, Jonathan (2011) "On the Supposed Limits of Physicalist Theories of Mind," *Philosophical Studies*, 155(2), 207–225.

Dowe, Phil (1997) "A Defense of Backwards in Time Causation Models in Quantum Mechanics," *Synthese*, 112(2), 233–246.

Dray, W. H. (1960) "Historical Causation and Human Free Will," *University of Toronto Quarterly*, 29(3), 357–369.

Drechsler, Wolfgang (2009) "Political Semiotics," *Semiotica*, 173(1–4), 73–97.

Dugić, Miroljub, Milan Ćirković, and Dejan Raković (2002) "On a Possible Physical Metatheory of Consciousness," *Open Systems and Information Dynamics*, 9(2), 153–166.

Dupré, John (1993) *The Disorder of Things*, Cambridge, MA: Harvard University Press.

Eddington, Arthur Stanley (1928) *The Nature of the Physical World*, New York: Macmillan.

Edwards, Jonathan (2005) "Is Consciousness Only a Property of Individual Cells?" *Journal of Consciousness Studies*, 12(4–5), 60–76.

Egg, Matthias (2013) "Delayed-Choice Experiments and the Metaphysics of Entanglement," *Foundations of Physics*, 43(9), 1124–1135.

Ehlers, Jürgen (1997) "Concepts of Time in Classical Physics," in H. Atmanspacher and E. Ruhnau, eds., *Time, Temporality, Now*, Berlin: Springer Verlag, pp. 191–200.

Einstein, Albert, Boris Podolsky, and N. Rosen (1935) "Can Quantum-Mechanical Descriptions of Physical Reality be Considered Complete?," *Physical Review*, 47(10), 777–780.

Eisert, Jens, Martin Wilkens, and Maciej Lewenstein (1999) "Quantum Games and Quantum Strategies," *Physical Review Letters*, 83(15), 3077–3080.

Elder-Vass, Dave (2007) "For Emergence: Refining Archer's Account of Social Structure," *Journal for the Theory of Social Behaviour*, 37(1), 25–44.

(2010a) *The Causal Power of Social Structure*, Cambridge University Press.

(2010b) "The Causal Power of Discourse," *Journal for the Theory of Social Behaviour*, 41(2), 143–160.

El-Hani, Charbel Niño and Claus Emmeche (2000) "On Some Theoretical Grounds for an Organism-Centered Biology: Property Emergence, Supervenience, and Downward Causation," *Theory in Biosciences*, 119(3–4), 234–275.

El-Khoury, Ann (2015) "Alternative Ways of Knowing," Chapter 4 in her *Globalization Development, and Social Justice*, Oxford: Routledge, forthcoming.

Elliott, Kevin and Daniel McKaughan (2014) "Nonepistemic Values and the Multiple Goals of Science," *Philosophy of Science*, 81(1), 1–21.
Elsasser, Walter (1951) "Quantum Mechanics, Amplifying Processes, and Living Matter," *Philosophy of Science*, 18(4), 300–326.
  (1987) *Reflections on a Theory of Organisms: Holism in Biology*, Baltimore, MD: The Johns Hopkins University Press.
Elster, Jon (1983) *Explaining Technical Change*, Cambridge University Press.
Emirbayer, Mustafa (1997) "Manifesto for a Relational Sociology," *American Journal of Sociology*, 103(2), 281–317.
Emmeche, Claus, Simo Koppe and Frederik Stjernfelt (2000) "Levels, Emergence, and Three Versions of Downward Causation," in P. Andersen, et al., eds., *Downward Causation: Minds, Bodies and Matter*, Aarhus University Press, pp. 13–34.
Ephraim, Laura (2013) "Beyond the Two-Sciences Settlement: Giambattista Vico's Critique of the Nature-Politics Opposition," *Political Theory*, 41(5), 710–737.
Epstein, Brian (2009) "Ontological Individualism Reconsidered," *Synthese*, 166(1), 187–213.
Esfeld, Michael (1998) "Holism and Analytic Philosophy," *Mind*, 107, 365–380.
  (1999) "Wigner's View of Physical Reality," *Studies in History and Philosophy of Modern Physics*, 30(1), 145–154.
  (2000) "Is Quantum Indeterminism Relevant to Free Will?" *Philosophia Naturalis*, 37, 177–187.
  (2001) *Holism in Philosophy of Mind and Philosophy of Physics*, Dordrecht: Kluwer.
  (2004) "Quantum Entanglement and a Metaphysics of Relations," *Studies in History and Philosophy of Modern Physics*, 35(4), 625–641.
  (2007) "Mental Causation and the Metaphysics of Causation," *Erkenntnis*, 67, 207–220.
Fahrbach, Ludwig (2005) "Understanding Brute Facts," *Synthese*, 145(3), 449–466.
Farr, Robert (1997) "The Significance of the Skin as a Natural Boundary in the Sub-Division of Psychology," *Journal for the Theory of Social Behaviour*, 27(2/3), 305–323.
Feinberg, Gerald, Shaughan Lavine, and David Albert (1992) "Knowledge of the Past and Future," *The Journal of Philosophy*, 89(12), 607–642.
Feinberg, Todd (2012) "Neuroontology, Neurobiological Naturalism, and Consciousness," *Physics of Life Reviews*, 9(1), 13–34.
Fellingham, John and Doug Schroeder (2006) "Quantum Information and Accounting," *Journal of Engineering and Technology Management*, 23, 33–53.
Fels, Daniel (2012) "Analogy between Quantum and Cell Relations," *Axiomathes*, 22(4), 509–520.
Ferrero, M., V. Gomez Pin, D. Salgado, and J. L. Sanchez-Gomez (2013) "A Further Review of the Incompatibility between Classical Principles and Quantum Postulates," *Foundations of Science*, 18(1), 125–138.
Ferretti, Francesco and Erica Cosentino (2013) "Time, Language and Flexibility of the Mind," *Philosophical Psychology*, 26(1), 24–46.
Filk, Thomas (2013) "Temporal Non-locality," *Foundations of Physics*, 43(4), 533–47.
Filk, Thomas and Albrecht von Müller (2009) "Quantum Physics and Consciousness: The Quest for a Common Conceptual Foundation," *Mind and Matter*, 7(1), 59–79.
Fine, Arthur (1993) "Fictionalism," *Midwest Studies in Philosophy*, 18, 1–18.

Fischer, Joachim (2009) "Exploring the Core Identity of Philosophical Anthropology through the Works of Max Scheler, Helmuth Plessner, and Arnold Gehlen," *Iris*, 1(1), 153–170.
Fitch, W. Tecumseh (2008) "Nano-intentionality: A Defense of Intrinsic Intentionality," *Biology and Philosophy*, 23(2), 157–177.
Flanagan, Brian (2001) "Are Perceptual Fields Quantum Fields?" *Informação e Cognição*, 3(1), 14–41.
  (2007) "Multi-Scaling, Quantum Theory, and the Foundations of Perception," *NeuroQuantology*, 4(1), 404–427.
Fodor, J. A. (1974) "Special Sciences (Or: The Disunity of Science as a Working Hypothesis)," *Synthese*, 28(2), 97–115.
Fodor, Jerry (2000) *The Mind Doesn't Work That Way*, Cambridge, MA: MIT Press.
Fohn, Adeline and Susann Heenen-Wolff (2011) "The Destiny of an Unacknowledged Trauma," *The International Journal of Psychoanalysis*, 92(1), 5–20.
Ford, Alan and F. David Peat (1988) "The Role of Language in Science," *Foundations of Physics*, 18(12), 1233–1242.
Forsdyke, Donald (2009) "Samuel Butler and Human Long Term Memory: Is the Cupboard Bare?" *Journal of Theoretical Biology*, 258(1), 156–164.
Fowler, Carol (1986) "An Event Approach to the Study of Speech Perception from a Direct-Realist Perspective," *Journal of Phonetics*, 14(1), 3–28.
  (1996) "Listeners Do Hear Sounds, not Tongues," *Journal of the Acoustical Society of America*, 99(3), 1730–1741.
Fox Keller, Evelyn (2011) "Towards a Science of Informed Matter," *Studies in History and Philosophy of Biological and Biomedical Sciences*, 42(2), 174–179.
Foxwall, Gordon (2007) "Intentional Behaviorism," *Behavior and Philosophy*, 35, 1–55.
  (2008) "Intentional Behaviorism Revisited," *Behavior and Philosophy*, 36, 113–155.
Francescotti, Robert (2007) "Emergence," *Erkenntnis*, 67, 47–63.
Franck, Georg (2004) "Mental Presence and the Temporal Present," in G. Globus, K. Pribram, and G. Vitiello, eds., *Brain and Being*, Amsterdam: John Benjamins, pp. 47–68.
  (2008) "Presence and Reality: An Option to Specify Panpsychism?" *Mind and Matter*, 6(1), 123–140.
Franco, Riccardo (2009) "The Conjunction Fallacy and Interference Effects," *Journal of Mathematical Psychology*, 53(5), 415–422.
François, Arnaud (2007) "Life and Will in Nietzsche and Bergson," *SubStance*, 36(3), 100–114.
Frank, Manfred (2002) "Self-Consciousness and Self-Knowledge: On Some Difficulties with the Reduction of Subjectivity," *Constellations*, 9(3), 390–408.
Frankfurt, Harry (1971) "Freedom of the Will and the Concept of a Person," *The Journal of Philosophy*, 68(1), 5–20.
Freeman, Walter and Giuseppe Vitiello (2006) "Nonlinear Brain Dynamics as Macroscopic Manifestation of Underlying Many-Body Field Dynamics," *Physics of Life Reviews*, 3, 93–118.
French, Steven (1998) "On the Withering Away of Physical Objects," in E. Castellani, ed., *Interpreting Bodies*, Princeton University Press, pp. 93–113.
  (2002) "A Phenomenological Solution to the Measurement Problem? Husserl and the Foundations of Quantum Mechanics," *Studies in History and Philosophy of Modern Physics*, 33(3), 467–491.

French, Steven and James Ladyman (2003) "Remodelling Structural Realism: Quantum Physics and the Metaphysics of Structure," *Synthese*, 136(1), 31–56.
Freudenburg, William, Scott Frickel and Robert Gramling (1995) "Beyond the Nature/Society Divide: Learning to Think About a Mountain," *Sociological Forum*, 10(3), 361–392.
Freundlieb, Dieter (2000) "Why Subjectivity Matters: Critical Theory and the Philosophy of the Subject," *Critical Horizons*, 1(2), 229–245.
  (2002) "The Return to Subjectivity as a Challenge to Critical Theory," *Idealistic Studies*, 32(2), 171–189.
Friederich, Simon (2011) "How to Spell Out the Epistemic Conception of Quantum States," *Studies in History and Philosophy of Modern Physics*, 42(3), 149–157.
  (2013) "In Defence of Non-Ontic Accounts of Quantum States," *Studies in History and Philosophy of Modern Physics*, 44(2), 77–92.
Friedman, Milton (1953) "The Methodology of Positive Economics," in Friedman, *Essays in Positive Economics*, University of Chicago Press, pp. 3–34.
Friedman, Norman (1997) *The Hidden Domain: Home of the Quantum Wave Function, Nature's Creative Source*, Eugene, OR: Woodbridge Group.
Frisch, Mathias (2010) "Causes, Counterfactuals, and Non-Locality," *Australasian Journal of Philosophy*, 88(4), 655–672.
Fröhlich, Herbert (1968) "Long-Range Coherence and Energy Storage in Biological Systems," *International Journal of Quantum Chemistry*, 2(5), 641–649.
Fry, Iris (2012) "Is Science Metaphysically Neutral?" *Studies in History and Philosophy of Biological and Biomedical Sciences*, 43(3), 665–673.
Fuchs, Christopher and Asher Peres (2000) "Quantum Theory Needs No 'Interpretation,'" *Physics Today*, March, 70–71.
Fuchs, Christopher and Rüdiger Schack (2014) "Quantum Measurement and the Paulian Idea," in H. Atmanspacher and C. Fuchs, eds., *The Pauli-Jung Conjecture*, Exeter: Imprint Academic, pp. 93–107.
Fuchs, Thomas and Hanne De Jaegher (2009) "Enactive Intersubjectivity: Participatory Sense-Making and Mutual Incorporation," *Phenomenology and the Cognitive Sciences*, 8(4), 465–486.
Fusaroli, Riccardo, Nivedita Gangopadhyay, and Kristian Tylén (2014) "The Dialogically Extended Mind: Language as Skilful Intersubjective Engagement," *Cognitive Systems Research*, 29–30, 31–39.
Gabora, Liane (2002) "Amplifying Phenomenal Information," *Journal of Consciousness Studies*, 9(8), 3–29.
Gabora, Liane and Diederik Aerts (2002) "Contextualizing Concepts Using a Mathematical Generalization of the Quantum Formalism," *Journal of Experimental and Theoretical Artificial Intelligence*, 14, 327–358.
Gabora, Liane, Eleanor Rosch, and Diederik Aerts (2008) "Toward an Ecological Theory of Concepts," *Ecological Psychology*, 20(1), 84–116.
Gabora, Liane, Eric Scott, and Stuart Kauffman (2013) "A Quantum Model of Exaptation: Incorporating Potentiality into Evolutionary Theory," *Progress in Biophysics and Molecular Biology*, 113(1), 108–116.
Gallagher, Shaun (2008a) "Inference or Interaction: Social Cognition without Precursors," *Philosophical Explorations*, 11(3), 163–174.
  (2008b) "Direct Perception in the Intersubjective Context," *Consciousness and Cognition*, 17(2), 535–543.

Gallagher, Shaun and Somogy Varga (2014) "Social Constraints on the Direct Perception of Emotions and Intentions," *Topoi*, 33(1), 185–199.
Gallese, Vittorio (2008) "Mirror Neurons and the Social Nature of Language," *Social Neuroscience*, 3(3–4), 317–333.
Gallese, Vittorio and Alvin Goldman (1998) "Mirror Neurons and the Simulation Theory of Mindreading," *Trends in Cognitive Sciences*, 2, 493–501.
Gamez, David (2008) "Progress in Machine Consciousness," *Consciousness and Cognition*, 17(3), 887–910.
Gantt, Edwin and Richard Williams (2014) "Psychology and the Legacy of Newtonianism," *Journal of Theoretical and Philosophical Psychology*, 34(2), 83–100.
Gao, Shan (2008) "A Quantum Theory of Consciousness," *Minds and Machines*, 18(1), 39–52.
  (2011) "Meaning of the Wave Function," *International Journal of Quantum Chemistry*, 111(15), 4124–4138.
  (2013) "A Quantum Physical Argument for Panpsychism," *Journal of Consciousness Studies*, 20(1–2), 59–70.
Garrett, Brian (2006) "What the History of Vitalism Teaches Us about Consciousness and the 'Hard Problem,'" *Philosophy and Phenomenological Research*, 72(3), 576–588.
  (2013) "Vitalism versus Emergent Materialism," in S. Normandin and C. T. Wolfe, eds., *Vitalism and the Scientific Image in Post-Enlightenment Life Science, 1800–2010*, Berlin: Springer Verlag, pp. 127–154.
Gell, Alfred (1992) *The Anthropology of Time*, Oxford: Berg.
Georgiev, Danko (2013) "Quantum No-Go Theorems and Consciousness," *Axiomathes*, 23(4), 683–695.
Germine, Mark (2008) "The Holographic Principle of Mind and the Evolution of Consciousness," *World Futures*, 64, 151–178.
Gerrans, Philip and David Sander (2014) "Feeling the Future: Prospects for a Theory of Implicit Prospection," *Biology and Philosophy*, 29(5), 699–710.
Ghirardi, G. C., Alberto Rimini, and Tullio Weber (1986) "Unified Dynamics for Microscopic and Macroscopic Systems," *Physical Review D*, 34, 470–491.
Ghirardi, Giancarlo (2002) "Making Quantum Theory Compatible with Realism," *Foundations of Science*, 7(1–2), 11–47.
Gibson, James (1979) *The Ecological Approach to Visual Perception*, Boston, MA: Houghton Mifflin.
Giddens, Anthony (1979) *Central Problems in Social Theory*, Berkeley, CA: University of California Press.
  (1984) *The Constitution of Society*, Berkeley, CA: University of California Press.
Giere, Ronald (2009) "Why Scientific Models Should Not Be Regarded as Works of Fiction," in M. Suarez, ed., *Fictions in Science*, London: Routledge, pp. 248–258.
Gigerenzer, Gerd and Wolfgang Gaissmaier (2011) "Heuristic Decision Making," *Annual Review of Psychology*, 62, 451–482.
Gilbert, Margaret (1989) *On Social Facts*, Princeton University Press.
Gilbert, Scott, Jan Sapp, and Alfred Tauber (2012) "A Symbiotic View of Life: We Have Never Been Individuals," *The Quarterly Review of Biology*, 87(4), 325–340.
Gilbert, Scott and Sahotra Sarkar (2000) "Embracing Complexity: Organicism for the 21st Century," *Developmental Dynamics*, 219(1), 1–9.

Gillett, Grant (1989) "Perception and Neuroscience," *British Journal for the Philosophy of Science*, 40(1), 83–103.
Glass, David (2012) "Inference to the Best Explanation: Does It Track Truth?" *Synthese*, 185(3), 411–427.
Glimcher, Paul (2005) "Indeterminacy in Brain and Behavior," *Annual Review of Psychology*, 56, 25–56.
Globus, Gordon (1976) "Mind, Structure, and Contradiction," in G. Globus, G. Maxwell, and I. Savodnik, eds., *Consciousness and the Brain*, New York: Plenum Press, pp. 271–293.
—— (1998) "Self, Cognition, Qualia and World in Quantum Brain Dynamics," *Journal of Consciousness Studies*, 5(1), 34–52.
Glymour, Bruce, Marcelo Sabates, and Andrew Wayne (2001) "Quantum Java: The Upwards Percolation of Quantum Indeterminacy," *Philosophical Studies*, 103(3), 271–283.
Godfrey-Smith, Peter (2009) "Models and Fictions in Science," *Philosophical Studies*, 143(1), 101–116.
Göcke, Benedikt (2009) "What is Physicalism?" *Ratio*, 22(3), 291–307.
—— ed. (2012) *After Physicalism*, South Bend, IN: University of Notre Dame Press.
Goff, Allan (2006) "Quantum Tic-Tac-Toe: A Teaching Metaphor for Superposition in Quantum Mechanics," *American Journal of Physics*, 74(11), 962–973.
Goff, Philip (2006) "Experiences Don't Sum," *Journal of Consciousness Studies*, 13(10–11), 53–61.
—— (2009) "Why Panpsychism Doesn't Help Us Explain Consciousness," *dialectica*, 63(3), 289–311.
Gök, Selvi and Erdinç Sayan (2012) "A Philosophical Assessment of Computational Models of Consciousness," *Cognitive Systems Research*, 17–18, 49–62.
Goldstein, Sheldon (1996) "Review Essay: Bohmian Mechanics and the Quantum Revolution," *Synthese*, 107(1), 145–165.
Gomes, Anil (2011) "Is There a Problem of Other Minds?" *Proceedings of the Aristotelian Society*, 111(3), 353–373.
Gotthelf, Allan (1987) "Aristotle's Conception of Final Causality," in A. Gotthelf and J. Lennox, eds., *Philosophical Issues in Aristotle's Biology*, Cambridge University Press, pp. 204–242.
Grandy, David (2001) "The Otherness of Light: Einstein and Levinas," *Postmodern Culture*, 12(1), 1–20.
—— (2002) "Light as a Solution to Puzzles about Light," *Journal for General Philosophy of Science*, 33(2), 369–379.
—— (2009) *The Speed of Light*, Bloomington, IN: Indiana University Press.
—— (2010) *Everyday Quantum Reality*, Bloomington, IN: Indiana University Press.
—— (2012) "Gibson's Ambient Light and Light Speed Constancy," *Philosophical Psychology*, 25(4), 539–554.
Greco, Monica (2005) "On the Vitality of Vitalism," *Theory, Culture and Society*, 22(1), 15–27.
Gregory, Brad (2008) "No Room for God? History, Science, Metaphysics, and the Study of Religion," *History and Theory*, 47(4), 495–519.
Grethlein, Jonas (2010) "Experientiality and 'Narrative Reference,'" *History and Theory*, 49(3), 315–335.

Greve, Jens (2012) "Emergence in Sociology: A Critique of Nonreductive Individualism," *Philosophy of the Social Sciences*, 42(2), 188–223.
Griffin, David Ray (1998) *Unsnarling the World-Knot: Consciousness, Freedom, and the Mind–Body Problem*, Berkeley, CA: University of California Press.
Griffiths, Paul (2009) "In What Sense Does 'Nothing Make Sense Except in the Light of Evolution'?" *Acta Biotheoretica*, 57, 11–32.
Grimm, Stephen (2006) "Is Understanding a Species of Knowledge?" *British Journal for the Philosophy of Science*, 57(3), 515–535.
Grove, Peter (2002) "Can the Past Be Changed?" *Foundations of Physics*, 32(4), 567–587.
Grush, Rick and Patricia Smith Churchland (1995) "Gaps in Penrose's Toilings," *Journal of Consciousness Studies*, 2(1), 10–29.
Guala, Francesco (2000) "Artefacts in Experimental Economics: Preference Reversals and the Becker-DeGroot-Marschak Mechanism," *Economics and Philosophy*, 16(1), 47–75.
Guo, Hong, Juheng Zhang, and Gary Koehler (2008) "A Survey of Quantum Games," *Decision Support Systems*, 46(1), 318–332.
Gustafson, Don (2007) "Neurosciences of Action and Noncausal Theories," *Philosophical Psychology*, 20(3), 367–374.
Gustafsson, Martin (2010) "Seeing the Facts and Saying What You Like: Retroactive Redescription and Indeterminacy in the Past," *Journal of the Philosophy of History*, 4(3–4), 296–327.
Güzeldere, Güven (1997) "The Many Faces of Consciousness: A Field Guide," in N. Block, et al., eds., *The Nature of Consciousness*, Cambridge, MA: MIT Press, pp. 1–67.
Haber, Matt (2013) "Colonies Are Individuals: Revisiting the Superorganism Revival," in F. Bouchard and P. Huneman, eds., *From Groups to Individuals*, Cambridge, MA: MIT Press, pp. 195–217.
Habermas, Jürgen (2002) "A Conversation about God and World," in E. Mendietta, ed., *Religion and Rationality*, Cambridge: Polity Press, pp. 147–167.
  (2007) "The Language Game of Responsible Agency and the Problem of Free Will," *Philosophical Explorations*, 10(1), 13–50.
Hacking, Ian (1995) *Rewriting the Soul: Multiple Personality and the Sciences of Memory*, Princeton University Press.
Hagan, S., S. Hameroff, and J. Tuszynski (2002) "Quantum Computation in Brain Microtubules: Decoherence and Biological Feasibility," *Physical Review E*, 65, 061901-1 to 061901-11.
Hall, Roland (1995) "The Nature of the Will and Its Place in Schopenhauer's Philosophy," *Schopenhauer-Jahrbuch*, 76, 73–90.
Hameroff, Stuart (1994) "Quantum Coherence in Microtubules: A Neural Basis for Emergent Consciousness?" *Journal of Consciousness Studies*, 1(1), 91–118.
  (1997) "Quantum Vitalism," *Advances: The Journal of Mind–Body Health*, 13(4), 143–22.
  (1998) "Quantum Computation in Brain Microtubules? The Penrose-Hameroff 'Orch OR' Model of Consciousness," *Philosophical Transactions of the Royal Society of London A*, 356, 1869–1896.

(2001a) "Biological Feasibility of Quantum Approaches to Consciousness," in P. Van Loocke, ed., *The Physical Nature of Consciousness*, Amsterdam: John Benjamins, pp. 1–61.

(2001b) "Consciousness, the Brain, and Spacetime Geometry," *Annals of the New York Academy of Sciences*, 929, 74–104.

(2007) "The Brain is Both Neurocomputer and Quantum Computer," *Cognitive Science*, 31(6), 1035–1045.

(2012a) "How Quantum Brain Biology Can Rescue Conscious Free Will," *Frontiers in Integrative Neuroscience*, 6, article 93.

(2012b) "Quantum Brain Biology Complements Neuronal Assembly Approaches to Consciousness: Comment," *Physics of Life Reviews*, 9(3), 303–305.

(2013) "Quantum Mathematical Cognition Requires Quantum Brain Biology," *Behavioral and Brain Sciences*, 36(3), 287–290.

Hameroff, Stuart, Alex Nip, Mitchell Porter, and Jack Tuszynski (2002) "Conduction Pathways in Microtubules, Biological Quantum Computation, and Consciousness," *Biosystems*, 64(1–3), 149–168.

Hameroff, Stuart and Roger Penrose (1996) "Conscious Events as Orchestrated Space-Time Selections," *Journal of Consciousness Studies*, 3(1), 36–53.

(2014a) "Consciousness in the Universe: A Review of the 'Orch OR' Theory," *Physics of Life Reviews*, 11(1), 39–78.

(2014b) "Reply to Criticism of the 'Orch OR Qubit'," *Physics of Life Reviews*, 11, 104–112.

Hamlyn, D. W. (1983) "Schopenhauer on the Will in Nature," *Midwest Studies in Philosophy*, 8, 457–467.

Hanauske, Matthias, Jennifer Kunz, Steffen Bernius, and Wolfgang König (2010) "Doves and Hawks in Economics Revisited: An Evolutionary Quantum Game Theory Based Analysis of Financial Crises," *Physica A*, 389, 5084–5102.

Hannan, Barbara (2009) *The Riddle of the World: A Reconsideration of Schopenhauer's Philosophy*, Oxford University Press.

Hansen, Nathaniel (2011) "Color Adjectives and Radical Contextualism," *Linguistics and Philosophy*, 34(3), 201–221.

Harder, Peter (2003) "The Status of Linguistic Facts," *Mind and Language*, 18(1), 52–76.

Hardy, Lucien (1998) "Spooky Action at a Distance in Quantum Mechanics," *Contemporary Physics*, 39(6), 419–429.

Harré, Rom (2002) "Social Reality and the Myth of Social Structure," *European Journal of Social Theory*, 5(1), 111–123.

Harrington, Anne (1996) *Reenchanted Science: Holism in German Culture from Wilhelm II to Hitler*, Princeton University Press.

Hartshorne, Charles (1972) "The Compound Individual," in Hartshorne, *Whitehead's Philosophy*, Lincoln, NE: University of Nebraska Press, pp. 41–61.

Harvey, Graham (2006) *Animism: Respecting the Living World*, New York: Columbia University Press.

Haven, Emmanuel and Andrei Khrennikov (2013) *Quantum Social Science*, Cambridge University Press.

Healey, Richard (1991) "Holism and Nonseparability," *The Journal of Philosophy*, 88(8), 393–421.

(1994) "Nonseparable Processes and Causal Explanation," *Studies in History and Philosophy of Science*, 25(3), 337–374.

(2011) "Reduction and Emergence in Bose-Einstein Condensates," *Foundations of Physics*, 41(6), 1007–1030.

(2012) "Quantum Theory: A Pragmatist Approach," *British Journal for the Philosophy of Science*, 63(4), 729–771.

(2013) "Physical Composition," *Studies in History and Philosophy of Modern Physics*, 44(1), 48–62.

Heelan, Patrick (1995) "Quantum Mechanics and the Social Sciences: After Hermeneutics," *Science and Education*, 4(2), 127–136.

(2004) "The Phenomenological Role of Consciousness in Measurement," *Mind and Matter*, 2(1), 61–84.

(2009) "The Role of Consciousness as Meaning Maker in Science, Culture, and Religion," *Zygon*, 44(2), 467–486.

Heil, John (1998) "Supervenience Deconstructed," *European Journal of Philosophy*, 6(2), 146–155.

Held, Carsten (1994) "The Meaning of Complementarity," *Studies in History and Philosophy of Science*, 25(6), 871–893.

Hellingwerf, Klaas (2005) "Bacterial Observations: A Rudimentary Form of Intelligence?" *Trends in Microbiology*, 13(4), 152–158.

Helrich, Carl (2007) "Is There a Basis for Teleology in Physics?" *Zygon*, 42(1), 97–110.

Henderson, James (2010) "Classes of Copenhagen Interpretations," *Studies in History and Philosophy of Modern Physics*, 41(1), 1–8.

Henderson, Leah (2014) "Can the Second Law be Compatible with Time Reversal Invariant Dynamics?" *Studies in History and Philosophy of Modern Physics*, 47, 90–98.

Hennig, Boris (2009) "The Four Causes," *The Journal of Philosophy*, 106(3), 137–160.

Henrich, Dieter (2003) "Subjectivity as a Philosophical Principle," *Critical Horizons*, 4(1), 7–27.

Herbert, Nick (1985) *Quantum Reality*, New York, NY: Anchor Books.

Herschbach, Mitchell (2008) "Folk Psychological and Phenomenological Accounts of Social Perception," *Philosophical Explorations*, 11(3), 223–235.

Heylighen, Francis (2007) "The Global Superorganism," *Social Evolution and History*, 6(1), 57–117.

Hildner, Richard, Daan Brinks, Jana Nieder, Richard Cogdell, and Niek van Hulst (2013) "Quantum Coherent Energy Transfer over Varying Pathways in Single Light-Harvesting Complexes," *Science*, 340, 1448–1451.

Hiley, B. J. (1997) "Quantum Mechanics and the Relationship between Mind and Matter," in P. Pylkkänen, et al., eds., *Brain, Mind and Physics*, Amsterdam: IOS Press, pp. 37–53.

Hiley, Basil and Paavo Pylkkänen (1997) "Active Information and Cognitive Science – A Reply to Kieseppa," in P. Pylkkänen, et al., eds., *Brain, Mind and Physics*, Amsterdam: IOS Press, pp. 64–85.

Hiley, B. J. and Paavo Pylkkänen (2001) "Naturalizing the Mind in a Quantum Framework," in P. Pylkkänen and T. Vaden, eds., *Dimensions of Conscious Experience*, Amsterdam: John Benjamins, pp. 119–144.

Hindriks, Frank (2008) "False Models as Explanatory Engines," *Philosophy of the Social Sciences*, 38(3), 334–360.
— (2013) "The Location Problem in Social Ontology," *Synthese*, 190(3), 413–437.
Hinterberger, Thilo and Nikolaus von Stillfried (2013) "The Concept of Complementarity and Its Role in Quantum Entanglement and Generalized Entanglement," *Axiomathes*, 23(3), 443–459.
Ho, Mae-Wan (1996) "The Biology of Free Will," *Journal of Consciousness Studies*, 3(3), 231–244.
— (1997) "Quantum Coherence and Conscious Experience," *Kybernetes*, 26(3), 265–276.
— (1998) *The Rainbow and the Worm: The Physics of Organisms*, Singapore: World Scientific.
— (2012) *Living Rainbow H_2O*, Singapore: World Scientific.
Hodgson, David (2011) "Quantum Physics, Consciousness, and Free Will," in R. Kane, ed., *The Oxford Handbook of Free Will*, Oxford University Press, pp. 57–83.
— (2012) *Rationality + Consciousness = Free Will*, Oxford University Press.
Hodgson, Geoffrey (2002) "Reconstitutive Downward Causation," in E. Fullbrook, ed., *Intersubjectivity in Economics: Agents and Structures*, London: Routledge, pp. 159–180.
Hoefer, Carl (2003) "For Fundamentalism," *Philosophy of Science*, 70(5), 1401–1412.
Hölldobler, B. and Edward O. Wilson (2009) *The Superorganism*, New York, NY: Norton.
Hoffecker, John (2013) "The Information Animal and the Super-Brain," *Journal of Archaeological Method and Theory*, 20(1), 18–41.
Hoffman, J. and G. Rosenkrantz (1984) "Hard and Soft Facts," *The Philosophical Review*, 93(3), 419–434.
Hoffmeyer, Jesper (1996) *Signs of Meaning in the Universe*, Bloomington, IN: Indiana University Press.
— (2010) "A Biosemiotic Approach to the Question of Meaning," *Zygon*, 45(2), 367–390.
Hogarth, R. and H. Einhorn (1992) "Order Effects in Belief Updating," *Cognitive Psychology*, 24(1), 1–55.
Hollis, Martin and Steve Smith (1990) *Explaining and Understanding International Relations*, Oxford: Clarendon Press.
Hollis, Martin and Robert Sugden (1993) "Rationality in Action," *Mind*, 102, 1–35.
Holman, Emmett (2008) "Panpsychism, Physicalism, Neutral Monism and the Russellian Theory of Mind," *Journal of Consciousness Studies*, 15(5), 48–67.
Holton, Richard (2006) "The Act of Choice," *Philosophers' Imprint*, 6(3), 1–15.
Home, Dipankar (1997) *Conceptual Foundations of Quantum Physics*, New York: Plenum Press.
Honderich, Ted (1988) *A Theory of Determinism*, Oxford: Clarendon Press.
Honner, John (1987) *The Description of Nature: Niels Bohr and the Philosophy of Quantum Physics*, Oxford: Clarendon Press.

Honneth, Axel and Hans Joas (1988) *Social Action and Human Nature*, Cambridge University Press.

Hopfield, John (1994) "Physics, Computation, and Why Biology Looks so Different," *Journal of Theoretical Biology*, 171, 53–60.

Horgan, Terence (1993) "From Supervenience to Superdupervenience," *Mind*, 102, 555–586.

Horgan, Terence and Uriah Kriegel (2008) "Phenomenal Intentionality Meets the Extended Mind," *The Monist*, 91(2), 347–373.

Howell, Robert (2009) "Emergentism and Supervenience Physicalism," *Australasian Journal of Philosophy*, 87(1), 83–98.

Hudson, Robert (2000) "Perceiving Empirical Objects Directly," *Erkenntnis*, 52(3), 357–371.

Huebner, Bryce (2011) "Genuinely Collective Emotions," *European Journal for Philosophy of Science*, 1(1), 89–118.

Huebner, Bryce, Michael Bruno, and Hagop Sarkissian (2010) "What Does the Nation of China Think about Phenomenal States?" *Review of Philosophy and Psychology*, 1(2), 225–243.

Hull, George (2013) "Reification and Social Criticism," *Philosophical Papers*, 42(1), 49–77.

Hulswit, Menno (2006) "How Causal Is Downward Causation?" *Journal for General Philosophy of Science*, 36(2), 261–287.

Humphreys, Paul (1985) "Why Propensities Cannot Be Probabilities," *The Philosophical Review*, 94(4), 557–570.

(1997a) "How Properties Emerge," *Philosophy of Science*, 64(March), 1–17.

(1997b) "Emergence, Not Supervenience," *Philosophy of Science*, 64 (Proceedings), S337–S345.

(2008) "Synchronic and Diachronic Emergence," *Minds and Machines*, 18(4), 431–442.

Huneman, Philippe (2006) "From the Critique of Judgment to the Hermeneutics of Nature," *Continental Philosophy Review*, 39(1), 1–34.

Hunt, Tam (2011) "Kicking the Psychophysical Laws into Gear: A New Approach to the Combination Problem," *Journal of Consciousness Studies*, 18(11–12), 96–134.

Hurley, Susan (1998) *Consciousness in Action*, Cambridge, MA: Harvard University Press.

Hut, Piet and Roger Shepard (1996) "Turning 'The Hard Problem' Upside Down and Sideways," *Journal of Consciousness Studies*, 3(4), 313–329.

Hutchins, Edwin (1995) *Cognition in the Wild*, Cambridge, MA: MIT Press.

Hüttemann, Andreas (2005) "Explanation, Emergence, and Quantum Entanglement," *Philosophy of Science*, 72(1), 114–127.

Hutto, Daniel (2004) "The Limits of Spectatorial Folk Psychology," *Mind and Language*, 19(5), 548–573.

(2008) *Folk Psychological Narratives: The Sociocultural Basis of Understanding Reasons*, Cambridge, MA: MIT Press.

Iacoboni, Marco (2008) *Mirroring People: The New Science of How We Connect with Others*, New York, NY: Farrar, Straus and Giroux.

(2009) "Imitation, Empathy, and Mirror Neurons," *Annual Review of Psychology*, 60, 653–670.

Igamberdiev, A. U. (2012) *Physics and Logic of Life*, New York: Nova Science Publishers.

Irvine, Elizabeth (2012) *Consciousness as a Scientific Concept*, Dordrecht: Springer.

Itkonen, Esa (2008) "Concerning the Role of Consciousness in Linguistics," *Journal of Consciousness Studies*, 15(6), 15–33.

Ittelson, William (2007) "The Perception of Nonmaterial Objects and Events," *Leonardo*, 40(3), 279–283.

Jackendoff, Ray (2002) *Foundations of Language*, Oxford University Press.

Jackman, Henry (1999) "We Live Forwards but Understand Backwards," *Pacific Philosophical Quarterly*, 80, 157–177.

— (2005) "Temporal Externalism, Deference, and Our Ordinary Linguistic Practice," *Pacific Philosophical Quarterly*, 86(3), 365–380.

Jackson, Frank (1982) "Epiphenomenal Qualia," *The Philosophical Quarterly*, 32, 127–136.

— (1986) "What Mary Didn't Know," *The Journal of Philosophy*, 83(5), 291–295.

Jackson, Patrick (2008) "Foregrounding Ontology: Dualism, Monism, and IR Theory," *Review of International Studies*, 34(1), 129–153.

Jacob, Pierre (2008) "What Do Mirror Neurons Contribute to Human Social Cognition?" *Mind and Language*, 23(2), 190–223.

— (2011) "The Direct-Perception Model of Empathy: A Critique," *Review of Philosophy and Psychology*, 2(3), 519–540.

— (2014) "Intentionality," *Stanford Encyclopedia of Philosophy* (Winter 2014 edition), Edward N. Zalta (ed.), http://plato.stanford.edu/archives/win2014/entries/intentionality/.

Jacquette, Dale (2005) *The Philosophy of Schopenhauer*, Montreal: McGill-Queens University Press.

Jahn, Robert and Brenda Dunne (2005) "The PEAR Proposition" *Journal of Scientific Exploration*, 19(2), 195–245.

James, William (1890) *The Principles of Psychology*, New York, NY: Henry Holt and Co.

Janaway, Christopher (2004) "Nietzsche and Schopenhauer: Is the Will Merely a Word?" in T. Pink and M. Stone, eds., *The Will and Human Action*, London: Routledge, pp. 173–196.

Jansen, Franz Klaus (2008) "Partial Isomorphism of Superposition in Potentiality Systems of Consciousness and Quantum Mechanics," *NeuroQuantology*, 6(3), 278–288.

Janzen, Greg (2012) "Physicalists Have Nothing to Fear from Ghosts," *International Journal of Philosophical Studies*, 20(1), 91–104.

Jaskolla, Ludwig and Alexander Buck (2012) "Does Panexperiential Holism Solve the Combination Problem?" *Journal of Consciousness Studies*, 19(9–10), 190–199.

Jaynes, Julian (1976) *The Origin of Consciousness in the Breakdown of the Bicameral Mind*, Boston: Houghton Mifflin.

Jenkins, C. S. and Daniel Nolan (2008) "Backwards Explanation," *Philosophical Studies*, 140(1), 103–115.

Ji, Sungchul (1997) "Isomorphism between Cell and Human Languages," *Biosystems*, 44(1), 17–39.

Jibu, Mari and Kunio Yasue (1995) *Quantum Brain Dynamics and Consciousness*, Amsterdam: John Benjamins.
   (2004) "Quantum Brain Dynamics and Quantum Field Theory," in G. Globus, et al., eds., *Brain and Being*, Amsterdam: John Benjamins, pp. 267–290.
John, E. R. (2001) "A Field Theory of Consciousness," *Consciousness and Cognition*, 10(2), 184–213.
Jonas, Hans (1966) *The Phenomenon of Life*, Evanston, IL: Northwestern University Press.
   (1984) "Appendix: Impotence or Power of Subjectivity," in Jonas, *The Imperative of Responsibility*, University of Chicago Press, pp. 205–231.
Jones, Brandon (2014) "Alfred North Whitehead's Flat Ontology," *Journal of Consciousness Studies*, 21(5–6), 174–195.
Jones, Donna (2010) *The Racial Discourses of Life Philosophy*, New York, NY: Columbia University Press.
Jones, Robert (2013) "Science, Sentience, and Animal Welfare," *Biology and Philosophy*, 28(1), 1–30.
Jones, Tessa (2013) "The Constitution of Events," *The Monist*, 96(1), 73–86.
Jonker, Catholun, et al. (2002) "Putting Intentions into Cell Biochemistry," *Journal of Theoretical Biology*, 214(1), 105–134.
Jordan, J. Scott (1998) "Recasting Dewey's Critique of the Reflex Arc Concept via a Theory of Anticipatory Consciousness," *New Ideas in Psychology*, 16(3), 165–187.
Jorgensen, Andrew (2009) "Holism, Communication, and the Emergence of Public Meaning," *Philosophia*, 37(1), 133–147.
Josephson, Brian and Fotini Pallikari-Viras (1991) "Biological Utilization of Quantum Nonlocality," *Foundations of Physics*, 21(2), 197–207.
Judisch, Neal (2008) "Why 'Non-Mental' Won't Work: On Hempel's Dilemma and the Characterization of the 'Physical'," *Philosophical Studies*, 140(3), 299–318.
Kadar, Endre and Judith Effken (1994) "Heideggerian Meditations on an Alternative Ontology for Ecological Psychology," *Ecological Psychology*, 6(4), 297–341.
Kadar, Endre and Robert Shaw (2000) "Toward an Ecological Field Theory of Perceptual Control of Locomotion," *Ecological Psychology*, 12(2), 141–180.
Kahneman, Daniel and Alan Krueger (2006) "Developments in the Measurement of Subjective Well-Being," *Journal of Economic Perspectives*, 20(1), 3–24.
Kaidesoja, Tuukka (2009) "Bhaskar and Bunge on Social Emergence," *Journal for the Theory of Social Behaviour*, 39(3), 300–322.
Kane, Robert (1996) *The Significance of Free Will*, Oxford University Press.
   ed. (2011) *The Oxford Handbook of Free Will*, Oxford University Press.
Karakostas, Vassilios (2009) "From Atomism to Holism: The Primacy of Non-Supervenient Relations," *NeuroQuantology*, 7(4), 635–656.
Karsten, Siegfried (1990) "Quantum Theory and Social Economics," *The American Journal of Economics and Sociology*, 49(4), 385–399.
Kastner, Ruth (1999) "Time-Symmetrized Quantum Theory, Counterfactuals and 'Advanced Action,'" *Studies in History and Philosophy of Modern Physics*, 30(2), 237–259.
   (2008) "The Transactional Interpretation, Counterfactuals, and Weak Values in Quantum Theory," *Studies in History and Philosophy of Modern Physics*, 39(4), 806–818.

Kawade, Yoshimi (2009) "On the Nature of the Subjectivity of Living Things," *Biosemiotics*, 2(2), 205–220.
  (2013) "The Origin of Mind: The Mind-Matter Continuity Thesis," *Biosemiotics*, 6(3), 367–378.
Kesebir, Selin (2012) "The Superorganism Account of Human Sociality," *Personality and Social Psychology Review*, 16(3), 233–261.
Kessler, Oliver (2007) "From Agents and Structures to Minds and Bodies: Of Supervenience, Quantum, and the Linguistic Turn," *Journal of International Relations and Development*, 10(3), 243–271.
Ketterle, Wolfgang (1999) "Experimental Studies of Bose-Einstein Condensation," *Physics Today*, December, 30–35.
Khalifa, Kareem (2013) "The Role of Explanation in Understanding," *British Journal for the Philosophy of Science*, 64(1), 161–187.
Khalil, Elias (2003) "The Context Problematic, Behavioral Economics and the Transactional View," *Journal of Economic Methodology*, 10(2), 107–130.
Khandker, Wahida (2013) "The Idea of Will and Organic Evolution in Bergson's Philosophy of Life," *Continental Philosophy Review*, 46(1), 57–74.
Khoshbin-e-Khoshnazar, Mohammadreza and Rita Pizzi (2014) "Quantum Superposition in the Retina: Evidence and Proposals," *NeuroQuantology*, 12(1), 97–101.
Khrennikov, Andrei (2010) *Ubiquitous Quantum Structure: From Psychology to Finance*, Berlin: Springer.
  (2011) "Quantum-Like Model of Processing of Information in the Brain Based on Classical Electromagnetic Field," *Biosystems*, 105(3), 251–259.
Khrennikov, Andrei and Emmanuel Haven (2009) "Quantum Mechanics and Violations of the Sure-Thing Principle," *Journal of Mathematical Psychology*, 53(5), 378–388.
Khrennikova, Polina, Emmanuel Haven, and Andrei Khrennikov (2014) "An Application of the Theory of Open Quantum Systems to Model the Dynamics of Party Governance in the US Political System," *International Journal of Theoretical Physics*, 53(4), 1346–1360.
Kieseppa, I. A. (1997) "Is David Bohm's Notion of Active Information Useful in Cognitive Science?" in P. Pylkkänen, et al., eds., *Brain, Mind and Physics*, Amsterdam: IOS Press, pp. 54–63.
Kim, Jaegwon (1974) "Noncausal Connections," *Nous*, 8(1), 41–52.
  (1990) "Supervenience as a Philosophical Concept," *Metaphilosophy*, 21(1–2), 1–27.
  (1998) *Mind in a Physical World*, Cambridge, MA: MIT Press.
  (1999) "Making Sense of Emergence," *Philosophical Studies*, 95(1–2), 3–36.
  (2000) "Making Sense of Downward Causation," in P. Andersen, et al., eds., *Downward Causation*, Aarhus University Press, pp. 305–321.
  (2006) "Emergence: Core Ideas and Issues," *Synthese*, 151(3), 547–559.
King, Chris (1997) "Quantum Mechanics, Chaos and the Conscious Brain," *The Journal of Mind and Behavior*, 18(2/3), 155–170.
King, Gary, Robert Keohane, and Sidney Verba (1994) *Designing Social Inquiry*, Princeton University Press.
Kirk, Robert (1997) "Consciousness, Information, and External Relations," *Communication and Cognition*, 30(3/4), 249–272.

Kirschner, Marc, John Gerhart, and Tim Mitchison (2000) "Molecular 'Vitalism'," *Cell*, 100(1), 79–88.

Kitto, Kirsty and R. Daniel Kortschak (2013) "Contextual Models and the Non-Newtonian Paradigm," *Progress in Biophysics and Molecular Biology*, 113(1), 97–107.

Kitto, Kirsty, Brentyn Ramm, Laurianne Sitbon, and Peter Bruza (2011) "Quantum Theory Beyond the Physical: Information in Context," *Axiomathes*, 21(2), 331–345.

Klemm, David and William Klink (2008) "Consciousness and Quantum Mechanics: Opting from Alternatives," *Zygon*, 43(2), 307–327.

Kojevnikov, Alexei (1999) "Freedom, Collectivism, and Quasiparticles: Social Metaphors in Quantum Physics," *Historical Studies in the Physical and Biological Sciences*, 29(2), 295–331.

Kolak, Daniel (2004) *I Am You: The Metaphysical Foundations for Global Ethics*, Berlin: Springer.

Koons, Robert and George Bealer, eds. (2010) *The Waning of Materialism*, Oxford University Press.

Kriegel, Uriah (2003) "Is Intentionality Dependent upon Consciousness?" *Philosophical Studies*, 116(3), 271–307.

(2004) "Consciousness and Self-Consciousness," *The Monist*, 87(2), 185–209.

Kronz, Frederick and Justin Tiehen (2002) "Emergence and Quantum Mechanics," *Philosophy of Science*, 69(2), 324–347.

Krueger, Joel (2012) "Seeing Mind in Action," *Phenomenology and the Cognitive Sciences*, 11, 149–173.

Kuhn, Thomas (1962/1996) *The Structure of Scientific Revolutions*, University of Chicago Press, 3rd edition.

Kull, Kalevi (2000) "An Introduction to Phytosemiotics: Semiotic Botany and Vegetative Sign Systems," *Sign Systems Studies*, 28, 326–350.

Kuttner, Ran (2011) "The Wave/Particle Tension in Negotiation," *Harvard Negotiation Law Review*, 16(1), 331–366.

Ladyman, James (2008) "Structural Realism and the Relationship between the Special Sciences and Physics," *Philosophy of Science*, 75(5), 744–755.

Lahlou, Saadi (2011) "How Can We Capture the Subject's Perspective?" *Social Science Information*, 50(3–4), 607–655.

Lakatos, Imre (1970) "Falsification and the Methodology of Scientific Research Programmes," in I. Lakatos and A. Musgrave, eds., *Criticism and the Growth of Knowledge*, Cambridge University Press, pp. 91–196.

Laloe, F. (2001) "Do We Really Understand Quantum Mechanics?" *American Journal of Physics*, 69(6), 655–701.

Lambert-Mogiliansky, Ariane and Jerome Busemeyer (2012) "Quantum Type Indeterminacy in Dynamic Decision-Making," *Games*, 3(2), 97–118.

Lambert-Mogiliansky, Ariane, Shmuel Zamir, and Herve Zwirn (2009) "Type Indeterminacy: A Model of the KT (Kahneman-Tversky)-Man," *Journal of Mathematical Psychology*, 53(5), 349–361.

La Mura, Pierfrancesco (2009) "Projective Expected Utility," *Journal of Mathematical Psychology*, 53(5), 408–414.

Larrain, Antonia and Andres Haye (2014) "A Dialogical Conception of Concepts," *Theory and Psychology*, 24(4), 459–478.

Lasersohn, Peter (2012) "Contextualism and Compositionality," *Linguistics and Philosophy*, 35(2), 171–189.
Lash, Scott (2005) "Lebenssoziologie: Georg Simmel in the Information Age," *Theory, Culture and Society*, 22(3), 1–23.
Laszlo, Ervin (1995) *The Interconnected Universe*, Singapore: World Scientific.
Latour, Bruno (2005) *Reassembling the Social: An Introduction to Actor-Network Theory*, Oxford University Press.
Lau, Joe and Max Deutsch (2014) "Externalism about Mental Content," *Stanford Encyclopedia of Philosophy* (Summer 2014 edition), Edward N. Zalta (ed.), http://plato.stanford.edu/archives/sum2014/entries/content-externalism/.
Laughlin, Victor (2013) "Sketch This: Extended Mind and Consciousness Extension," *Phenomenology and the Cognitive Sciences*, 12(1), 41–50.
Lavelle, Jane Suilin (2012) "Theory-Theory and the Direct Perception of Mental States," *Review of Philosophy and Psychology*, 3(2), 213–230.
Lawson, Tony (2012) "Ontology and the Study of Social Reality," *Cambridge Journal of Economics*, 36(2), 345–385.
Le Boutillier, Shaun (2013) "Emergence and Reduction," *Journal for the Theory of Social Behaviour*, 43(2), 205–225.
Lehner, Christoph (1997) "What It Feels Like to be in a Superposition, and Why," *Synthese*, 110(2), 191–216.
Lemons, John, Kristin Shrader-Frechette, and Carl Cranor (1997) "The Precautionary Principle: Scientific Uncertainty and Type I and Type II Errors," *Foundations of Science*, 2(2), 207–236.
Lenoir, Timothy (1982) *The Strategy of Life: Teleology and Mechanics in Nineteenth Century German Biology*, Boston, MA: Kluwer.
Leudar, Ivan and Alan Costall (2004) "On the Persistence of the 'Problem of Other Minds' in Psychology," *Theory and Psychology*, 14(5), 601–621.
Levine, Donald (1995) "The Organism Metaphor in Sociology," *Social Research*, 62(2), 239–265.
Levine, Joseph (1983) "Materialism and Qualia: The Explanatory Gap," *Pacific Philosophical Quarterly*, 64, 354–361.
— (2001) *Purple Haze: The Puzzle of Consciousness*, Oxford University Press.
Levy, Neil (2005) "Libet's Impossible Demand," *Journal of Consciousness Studies*, 12(12), 67–76.
Lewis, Peter (2005) "Interpreting Spontaneous Collapse Theories," *Studies in History and Philosophy of Modern Physics*, 36(1), 165–180.
Lewtas, Patrick (2013a) "What It Is Like to Be a Quark," *Journal of Consciousness Studies*, 20(9–10), 39–64.
— (2013b) "Emergence and Consciousness," *Philosophy*, 88(4), 527–553.
— (2014) "The Irrationality of Physicalism," *Axiomathes*, 24(3), 313–341.
Libet, Benjamin (1985) "Unconscious Cerebral Initiative and the Role of Conscious Will in Voluntary Action," *The Behavioral and Brain Sciences*, 8(4), 529–566.
— (2004) *Mind Time*, Cambridge, MA: Harvard University Press.
Linell, Per (2013) "Distributed Language Theory, With or Without Dialogue," *Language Sciences*, 40, 168–173.
Lipari, Lisbeth (2014) "On Interlistening and the Idea of Dialogue," *Theory and Psychology*, 24(4), 504–523.

Lipton, Peter (2004) *Inference to the Best Explanation*, London: Routledge, 2nd edition.
    (2009) "Understanding without Explanation," in H. de Regt, S. Leonelli and K. Eigner, eds., *Scientific Understanding*, University of Pittsburgh Press, pp. 43–63.
List, Christian and Kai Spiekermann (2013) "Methodological Individualism and Holism in Political Science," *American Political Science Review*, 107(4), 629–643.
Litt, Abninder, et al. (2006) "Is the Brain a Quantum Computer?" *Cognitive Science*, 30(3), 593–603.
Lloyd, Seth (2011) "Quantum Coherence in Biological Systems," *Journal of Physics: Conference Series* 302, article 012037.
Lockwood, Michael (1989) *Mind, Brain, and the Quantum*, Oxford: Blackwell.
    (1996) "'Many Minds' Interpretations of Quantum Mechanics," *British Journal for the Philosophy of Science*, 47(2), 159–188.
Lodge, Paul and Marc Bobro (1998) "Stepping Back inside Leibniz's Mill," *The Monist*, 81(4), 553–572.
Loewer, Barry (1996) "Freedom from Physics: Quantum Mechanics and Free Will," *Philosophical Topics*, 24(2), 91–112.
London, Fritz and Edmond Bauer (1939/1983) "The Theory of Observation in Quantum Mechanics," in J. Wheeler and W. Zurek, eds., *Quantum Theory and Measurement*, Princeton University Press, pp. 217–259.
Look, Brandon (2002) "On Monadic Domination in Leibniz' Metaphysics," *British Journal for the History of Philosophy*, 10(3), 379–399.
Luisi, Pier Luigi (1998) "About Various Definitions of Life," *Origins of Life and Evolution of the Biosphere*, 28(4–6), 613–622.
Maas, Harro (1999) "Mechanical Rationality: Jevons and the Making of Economic Man," *Studies in History and Philosophy of Science*, 30(4), 587–619.
Machery, Edouard (2012) "Why I Stopped Worrying about the Definition of Life... and Why You Should as Well," *Synthese*, 185(1), 145–164.
MacKenzie, Donald (2006) "Is Economics Performative? Option Theory and the Construction of Derivatives Markets," *Journal of the History of Economic Thought*, 28(1), 29–55.
Mackonis, Adolfas (2013) "Inference to the Best Explanation, Coherence and Other Explanatory Virtues," *Synthese*, 190(6), 975–995.
Mainville, Sebastién (2015) "The International System and Its Environment: Evolutionary and Developmental Perspectives on Change in World Politics," Ph.D. dissertation, Ohio State University.
Majorek, Marek (2012) "Does the Brain Cause Conscious Experience?" *Journal of Consciousness Studies*, 19(3–4), 121–144.
Malin, Shimon (2001) *Nature Loves to Hide: Quantum Physics and the Nature of Reality, a Western Perspective*, Oxford University Press.
Malpas, Jeff (2002) "The Weave of Meaning: Holism and Contextuality," *Language and Communication*, 22(4), 403–419.
Manousakis, Efstratios (2006) "Founding Quantum Theory on the Basis of Consciousness," *Foundations of Physics*, 36(6), 795–838.
Mantzavinos, C. (2012) "Explanations of Meaningful Actions," *Philosophy of the Social Sciences*, 42(2), 224–238.

Manzotti, Riccardo (2006) "A Process Oriented View of Conscious Perception," *Journal of Consciousness Studies*, 13(6), 7–41.
Marcer, Peter (1995) "A Proposal for a Mathematical Specification for Evolution and the Psi Field," *World Futures*, 44(2), 149–159.
Marcer, Peter and Walter Schempp (1997) "Model of the Neuron Working by Quantum Holography," *Informatica*, 21, 519–534.
  (1998) "The Brain as a Conscious System," *International Journal of General Systems*, 27(1–3), 231–248.
Marchettini, Nadia, et al. (2010) "Water: A Medium Where Dissipative Structures Are Produced by a Coherent Dynamics," *Journal of Theoretical Biology*, 265(4), 511–516.
Marcin, Raymond (2006) *In Search of Schopenhauer's Cat*, Washington, DC: Catholic University Press of America.
Margenau, Henry (1967) "Quantum Mechanics, Free Will, and Determinism," *The Journal of Philosophy*, 64(21), 714–725.
Margulis, Lynn (2001) "The Conscious Cell," *Annals of the New York Academy of Sciences*, 929, 55–70.
Margulis, Lynn and Dorion Sagan (1995) *What Is Life?*, New York, NY: Simon & Schuster.
Marijuan, Pedro, Raquel del Moral, and Jorge Navarro (2013) "On Eukaryotic Intelligence: Signaling System's Guidance in the Evolution of Multicellular Organization," *Biosystems*, 114(1), 8–24.
Marin, Juan Miguel (2009) "'Mysticism in Quantum Mechanics: The Forgotten Controversy," *European Journal of Physics*, 30(4), 807–822.
Markoš, Anton and Fatima Cvrčková (2013) "The Meanings(s) of Information, Code... and Meaning," *Biosemiotics*, 6(1), 61–75.
Markosian, Ned (1995) "The Open Past," *Philosophical Studies*, 79(1), 95–105.
  (2004) "A Defense of Presentism," in D. Zimmerman, ed., *Oxford Studies in Metaphysics: Volume I*, Oxford University Press, pp. 47–82.
Marr, David (1982) *Vision: A Computational Investigation into the Human Representation and Processing of Visual Information*, New York, NY: W. H. Freeman and Company.
Marshall, I. N. (1989) "Consciousness and Bose-Einstein Condensates," *New Ideas in Psychology*, 7(1), 73–83.
Marston, Sallie, John Paul Jones III, and Keith Woodward (2005) "Human Geography without Scale," *Transactions of the Institute of British Geographers*, 30(4), 416–423.
Martin, F., F. Carminati, and G. Galli Carminati (2010) "Quantum Information, Oscillations and the Psyche," *Physics of Particles and Nuclei*, 41(3), 425–451.
Martinez-Martinez, Ismael (2014) "A Connection between Quantum Decision Theory and Quantum Games," *Journal of Mathematical Psychology*, 58, 33–44.
Mather, Jennifer (2008) "Cephalopod Consciousness: Behavioural Evidence," *Consciousness and Cognition*, 17(1), 37–48.
Mathews, Freya (2003) *For Love of Matter: A Contemporary Panpsychism*, Albany, NY: SUNY Press.
Mathiesen, Kay (2005) "Collective Consciousness," in D. Smith and A. Thomasson, eds., *Phenomenology and Philosophy of Mind*, Oxford: Clarendon Press, pp. 235–250.

Matson, Floyd (1964) *The Broken Image: Man, Science and Society*, New York: G. Braziller.
Matsuno, Koichiro (1993) "Being Free from Ceteris Paribus: A Vehicle for Founding Physics on Biology Rather than the Other Way Around," *Applied Mathematics and Computation*, 56(2–3), 261–279.
Matthiessen, Hannes Ole (2010) "Seeing and Hearing Directly," *Review of Philosophy and Psychology*, 1(1), 91–103.
Matzkin, A. (2002) "Realism and the Wavefunction," *European Journal of Physics*, 23(3), 285–294.
Maudlin, Tim (1998) "Part and Whole in Quantum Mechanics," in E. Castellani, ed., *Interpreting Bodies*, Princeton University Press, pp. 46–60.
  (2003) "Distilling Metaphysics from Quantum Physics," in M. Loux and D. Zimmerman, eds., *Oxford Handbook of Metaphysics*, Oxford University Press, pp. 461–487.
  (2007) *The Metaphysics within Physics*, Oxford University Press.
Maul, Andrew (2013) "On the Ontology of Psychological Attributes," *Theory and Psychology*, 23(6), 752–769.
Mavromatos, Nick (2011) "Quantum Coherence in (Brain) Microtubules and Efficient Energy and Information Transport," *Journal of Physics: Conference Series*, 329(1), 12026–12056.
Maynard, Douglas and Thomas Wilson (1980) "On the Reification of Social Structure," in S. McNall and G. Howe, eds., *Current Perspectives in Social Theory*, vol. 1, Greenwich, CT: JAI Press, pp. 287–322.
Mayr, Ernst (1982) "Teleological and Teleonomic: A New Analysis," in H. Plotkin, ed., *Learning, Development, and Culture*, New York: John Wiley & Sons, pp. 17–38.
  (1992) "The Idea of Teleology," *Journal of the History of Ideas*, 53(1), 117–135.
McAllister, James (1989) "Truth and Beauty in Science Reason," *Synthese*, 78(1), 25–51.
  (1991) "The Simplicity of Theories: Its Degree and Form," *Journal for General Philosophy of Science*, 22(1), 1–14.
  (1996) *Beauty and Revolution in Science*, Ithaca, NY: Cornell University Press.
  (2014) "Methodological Dilemmas and Emotion in Science," *Synthese*, 191(13), 3143–3158.
McClure, John (2011) "Attributions, Causes, and Actions: Is the Consciousness of Will a Perceptual Illusion?" *Theory and Psychology*, 22(4), 402–419.
McDaniel, Jay (1983) "Physical Matter as Creative and Sentient," *Environmental Ethics*, 5(4), 291–317.
McDermid, Douglas (2001) "What Is Direct Perceptual Knowledge? A Fivefold Confusion," *Grazer Philosophische Studien*, 62(1), 1–16.
McFadden, Johnjoe (2001) *Quantum Evolution*, New York: Norton.
  (2007) "Conscious Electromagnetic Field Theory," *NeuroQuantology*, 5(3), 262–270.
McGinn, Colin (1989) "Can We Solve the Mind–Body Problem?" *Mind*, 98, 349–366.
  (1995) "Consciousness and Space," *Journal of Consciousness Studies*, 2(3), 220–230.
  (1999) *The Mysterious Flame*, New York: Basic Books.
McIntyre, Lee (2007) "Emergence and Reduction in Chemistry: Ontological or Epistemological Concepts?" *Synthese*, 155(3), 337–343.

McKemmish, Laura, Jeffrey Reimers, Ross McKenzie, Alan Mark, and Noel Hush (2009) "Penrose-Hameroff Orchestrated Objective-Reduction Proposal for Human Consciousness is not Biologically Feasible," *Physical Review E*, 80(2), 021912–1 to 021912–6.

McLaughlin, Brian (1992) "The Rise and Fall of British Emergentism," in A. Beckermann, H. Flohr, and J. Kim, eds., *Emergence or Reduction?*, Berlin: Walter de Gruyter, pp. 49–93.

McSweeney, Brendan (2000) "Looking Forward to the Past," *Accounting, Organizations and Society*, 25(8), 767–786.

McTaggart, J. M. E. (1908) "The Unreality of Time," *Mind*, 17, 456–473.

McTaggart, Lynne (2002) *The Field*, New York, NY: Quill.

Megill, Jason (2013) "A Defense of Emergence," *Axiomathes*, 23(4), 597–615.

Melkikh, Alexey (2014) "Congenital Programs of the Behavior and Nontrivial Quantum Effects in the Neurons Work," *Biosystems*, 119, 10–19.

Menard, Claude (1988) "The Machine and the Heart: An Essay on Analogies in Economic Reasoning," *Social Concept*, 5, 81–95.

Mensky, M. (2005) "Concept of Consciousness in the Context of Quantum Mechanics," *Physics Uspekhi*, 48(4), 389–409.

Mercer, Jonathan (2010) "Emotional Beliefs," *International Organization*, 64(1), 1–31.
    (2014) "Feeling like a State: Social Emotion and Identity," *International Theory*, 6(3), 515–535.

Mesquita, Marcus et al. (2005) "Large-Scale Quantum Effects in Biological Systems," *International Journal of Quantum Chemistry*, 102(6), 1116–1130.

Meyer, David (1999) "Quantum Strategies," *Physical Review Letters*, 82(5), 1052–1055.

Meyer, John and Ronald Jepperson (2000) "The 'Actors' of Modern Society: The Cultural Construction of Social Agency," *Sociological Theory*, 18(1), 100–120.

Meyering, Theo (2000) "Physicalism and Downward Causation in Psychology and the Special Sciences," *Inquiry*, 43(2), 181–202.

Michell, Joel (2005) "The Logic of Measurement: A Realist Overview," *Measurement*, 38(4), 285–294.

Midgley, David (2006) "Intersubjectivity and Collective Consciousness," *Journal of Consciousness Studies*, 13(5), 99–109.

Millar, Boyd (2014) "The Phenomenological Directness of Perceptual Experience," *Philosophical Studies*, 170(2), 235–253.

Miller, Dale (1990) "Biological Systems and the Rumored Animate-Sentient Like Aspect of Physical Phenomena," *Journal of Biological Physics*, 17(3), 145–150.
    (1992) "Agency as a Quantum-theoretic Parameter: Synthetic and Descriptive Utility for Theoretical Biology," *Nanobiology*, 1, 361–371.

Milovanovic, Dragan (2014) *Quantum Holographic Criminology*, Durham, NC: Carolina Academic Press.

Mingers, John (1995) "Information and Meaning," *Information Systems Journal*, 5(4), 285–306.

Miranker, Willard (2000) "Consciousness is an Information State," *Neural, Parallel and Scientific Computations*, 8, 83–104.
    (2002) "A Quantum State Model of Consciousness," *Journal of Consciousness Studies*, 9(3), 3–14.

Mirowski, Philip (1988) *Against Mechanism: Protecting Economics from Science*, Totowa, NJ: Rowman and Littlefield.
   (1989) "The Probabilistic Counter-Revolution, or How Stochastic Concepts Came to Neoclassical Economic Theory," *Oxford Economic Papers*, 41(1), 217–235.
Mitchell, Edgar and Robert Staretz (2011) "The Quantum Hologram and the Nature of Consciousness," in R. Penrose, S. Hameroff, and S. Kak, eds., *Consciousness and the Universe*, Cambridge, MA: Cosmology Science Publishers, pp. 933–965.
Mitchell, Jeff and Mirella Lapata (2010) "Composition in Distributional Models of Semantics," *Cognitive Science*, 34(8), 1388–1429.
Monk, Nicholas (1997) "Conceptions of Space-Time: Problems and Possible Solutions," *Studies in History and Philosophy of Modern Physics*, 28(1), 1–34.
Montano, Ulianov (2013) "Beauty in Science: A New Model of the Role of Aesthetic Evaluations in Science," *European Journal for Philosophy of Science*, 3(2), 133–156.
Montero, Barbara (1999) "The Body Problem," *Nous*, 33(2), 183–200.
   (2001) "Post-Physicalism," *Journal of Consciousness Studies*, 8(2), 61–80.
   (2003) "Varieties of Causal Closure," in S. Walter and H.-D. Heckmann, eds., *Physicalism and Mental Causation*, London: Imprint Academic, pp. 173–187.
   (2009) "What Is the Physical?" in B. McLaughlin, et al., eds., *The Oxford Handbook of Philosophy of Mind*, Oxford University Press, pp. 173–188.
Moore, David (2002) "Measuring New Types of Question-Order Effects," *Public Opinion Quarterly*, 66(1), 80–91.
Moore, J. (2013) "Mentalism as a Radical Behaviorist Views It – Part I," *The Journal of Mind and Behavior*, 34(2), 133–164.
Moreno, Alvaro and Jon Umerez (2000) "Downward Causation at the Core of Living Organization," in P. Andersen, et al., eds., *Downward Causation*, Aarhus University Press, pp. 99–117.
Morganti, Matteo (2009) "A New Look at Relational Holism in Quantum Mechanics," *Philosophy of Science*, 76(5), 1027–1038.
Morris, Suzanne, John Taplin, and Susan Gelman (2000) "Vitalism in Naïve Biological Thinking," *Developmental Psychology*, 36(5), 582–595.
Mortimer, Geoffrey (2001) "Did Contemporaries Recognize a 'Thirty Years War'?" *English Historical Review*, 116, 124–136.
Mould, Richard (1995) "The Inside Observer in Quantum Mechanics," *Foundations of Physics*, 25(11), 1621–1629.
   (2003) "Quantum Brain States," *Foundations of Physics*, 33(4), 591–612.
Mozersky, M. Joshua (2011) "Presentism," in C. Callender, ed., *The Oxford Handbook of Philosophy of Time*, Oxford University Press, pp. 122–144.
Munro, William Bennett (1928) "Physics and Politics – An Old Analogy Revised," *American Political Science Review*, 22(1), 1–11.
Mureika, J. R. (2007) "Implications for Cognitive Quantum Computation and Decoherence Limits in the Presence of Large Extra Dimensions," *International Journal of Theoretical Physics*, 46(1), 133–145.
Nachtomy, Ohad (2007) "Leibniz on Nested Individuals," *British Journal for the History of Philosophy*, 15(4), 709–728.
Nadeau, Robert and Menas Kafatos (1999) *The Non-Local Universe: The New Physics and Matters of the Mind*, Oxford University Press.

Nagel, Alexandra (1997) "Are Plants Conscious?" *Journal of Consciousness Studies*, 4(3), 215–230.
Nagel, Thomas (1974) "What Is It Like to Be a Bat?" *The Philosophical Review*, 83(4), 435–450.
  (1979) "Panpsychism," in Nagel, *Mortal Questions*, Cambridge University Press, pp. 181–195.
  (2012) *Mind and Cosmos: Why the Materialist Neo-Darwinian Conception of Nature is Almost Certainly False*, Oxford University Press.
Nakagomi, Teruaki (2003a) "Mathematical Formulation of Leibnizian World: A Theory of Individual-Whole or Interior-Exterior Reflective Systems," *Biosystems*, 69(1), 15–26.
  (2003b) "Quantum Monadology: A Consistent World Model for Consciousness and Physics," *Biosystems*, 69(1), 27–38.
Narby, Jeremy (2005) *Intelligence in Nature*, New York, NY: Penguin.
Nelson, Douglas, Cathy McEvoy, and Lisa Pointer (2003) "Spreading Activation or Spooky Action at a Distance?" *Journal of Experimental Psychology*, 29(1), 42–**52.
Neuman, Yair (2008) "The Polysemy of the Sign: From Quantum Computing to the Garden of Forking Paths," *Semiotica*, 169(6), 155–168.
Neurath, Otto (1932/1959) "Sociology and Physicalism," in A. J. Ayer, ed., *Logical Positivism*, Glencoe, IL: Free Press, pp. 282–317.
Ney, Alyssa and David Albert, eds. (2013) *The Wave Function: Essays on the Metaphysics of Quantum Mechanics*, Oxford University Press.
Ni, Peimin (1992) "Changing the Past," *Nous*, 26(3), 349–359.
Nicholson, Daniel (2010) "Biological Atomism and Cell Theory," *Studies in History and Philosophy of Biological and Biomedical Sciences*, 41(3), 202–211.
  (2013) "Organisms ≠ Machines," *Studies in History and Philosophy of Biological and Biomedical Sciences*, 44(4), 669–678.
Noë, Alva, ed. (2002) "Is the Visual World a Grand Illusion?" *Journal of Consciousness Studies*, 9(5–6), special issue.
  (2004) *Action in Perception*, Cambridge, MA: MIT Press.
Noë, Alva and Evan Thompson (2004) "Are There Neural Correlates of Consciousness?" *Journal of Consciousness Studies*, 11(1), 3–28.
Normandin, Sebastian and Charles Wolfe, eds. (2013) *Vitalism and the Scientific Image in Post-Enlightenment Life Science, 1800–2010*, Berlin: Springer Verlag.
Norris, Christopher (1998) "On the Limits of 'Undecidability': Quantum Physics, Deconstruction, and Anti-Realism," *Yale Journal of Criticism*, 11(2), 407–432.
Nunn, Chris (2013) "On Taking Monism Seriously," *Journal of Consciousness Studies*, 20(9–10), 77–89.
Ochs, Elinor (2012) "Experiencing Language," *Anthropological Theory*, 12(2), 142–160.
O'Connor, Timothy (2000) *Persons and Causes: The Metaphysics of Free Will*, Oxford University Press.
  (2014) "Free Will," *Stanford Encyclopedia of Philosophy* (Summer 2014 edition), Edward N. Zalta (ed.), http://plato.stanford.edu/archives/fall2014/entries/freewill/.
O'Connor, Timothy and Hong Yu Wong (2005) "The Metaphysics of Emergence," *Nous*, 39(4), 658–678.

(2012) "Emergent Properties," *Stanford Encyclopedia of Philosophy* (Spring 2014 edition), Edward N. Zalta (ed.), http://plato.stanford.edu/archives/spr2012/entries/properties-emergent/.

Okasha, Samir (2011) "Biological Ontology and Hierarchical Organization: A Defense of Rank Freedom," in B. Calcott and K. Sterelny, eds., *The Major Transitions in Evolution Revisited*, Cambridge, MA: MIT Press, pp. 53–64.

Omnès, Roland (1995) "A New Interpretation of Quantum Mechanics and Its Consequences in Epistemology," *Foundations of Physics*, 25(4), 605–629.

O'Neill, John, ed. (1973) *Modes of Individualism and Collectivism*, New York, NY: St. Martin's Press.

Orlandi, Nicoletta (2013) "Embedded Seeing: Vision in the Natural World," *Nous*, 47(4), 727–747.

Orlov, Yuri (1982) "The Wave Logic of Consciousness: A Hypothesis," *International Journal of Theoretical Physics*, 21(1), 37–53.

Ortner, Sherry (2001) "Specifying Agency: The Comaroffs and Their Critics," *Interventions*, 3(1), 76–84.

(2005) "Subjectivity and Cultural Critique," *Anthropological Theory*, 5(1), 31–52.

Overgaard, Morten, Shaun Gallagher, and Thomas Ramsoy (2008) "An Integration of First-Person Methodologies in Cognitive Science," *Journal of Consciousness Studies*, 15(5), 100–120.

Overgaard, Soren (2004) "Exposing the Conjuring Trick: Wittgenstein on Subjectivity," *Phenomenology and the Cognitive Sciences*, 3(3), 263–286.

Oyama, Susan (2010) "Biologists Behaving Badly: Vitalism and the Language of Language," *History and Philosophy of the Life Sciences*, 32(2–3), 401–423.

Pacherie, Elisabeth (2014) "Can Conscious Agency Be Saved?" *Topoi*, 33(1), 33–45.

Packham, Catherine (2002) "The Physiology of Political Economy: Vitalism and Adam Smith's Wealth of Nations," *Journal of the History of Ideas*, 63(3), 465–481.

Padgett, John, Doowan Lee, and Nick Collier (2003) "Economic Production as Chemistry," *Industrial and Corporate Change*, 12(4), 843–878.

Pagin, Peter (1997) "Is Compositionality Compatible with Holism?" *Mind and Language*, 12(1), 11–33.

(2006) "Meaning Holism," in E. Lepore and B. Smith, eds., *The Oxford Handbook of Philosophy of Language*, Oxford: Clarendon Press, pp. 213–232.

Palmer, Stephen (1999) *Vision Science: Photons to Phenomenology*, Cambridge, MA: MIT Press.

Papineau, David (2001) "The Rise of Physicalism," in C. Gillett and B. Loewer, eds., *Physicalism and Its Discontents*, Cambridge University Press, pp. 3–36.

(2009) "Physicalism and the Human Sciences," in C. Mantzavinos, ed., *Philosophy of the Social Sciences*, Cambridge University Press, pp. 103–123.

(2011) "What Exactly is the Explanatory Gap?" *Philosophia*, 39, 5–19.

Parsons, Stephen (1991) "Time, Expectations and Subjectivism," *Cambridge Journal of Economics*, 15(4), 405–423.

Paternoster, Alfredo (2007) "Vision Science and the Problem of Perception," in M. Marraffa, M. De Caro, and F. Ferretti, eds., *Cartographies of the Mind*, Berlin: Springer Verlag, pp. 53–64.

Paty, Michel (1999) "Are Quantum Systems Physical Objects with Physical Properties?" *European Journal of Physics*, 20(6), 373–388.

Pauen, Michael (2012) "The Second-Person Perspective," *Inquiry*, 55(1), 33–49.
Peacock, Kent (1998) "On the Edge of a Paradigm Shift: Quantum Nonlocality and the Breakdown of Peaceful Coexistence," *International Studies in the Philosophy of Science*, 12(2), 129–149.
Peijnenburg, Jeanne (2006) "Shaping Your Own Life," *Metaphilosophy*, 37(2), 240–253.
Penrose, Roger (1994) *Shadows of the Mind: A Search for the Missing Science of Consciousness*, Oxford University Press.
Pepper, John and Matthew Herron (2008) "Does Biology Need an Organism Concept?" *Biological Review*, 83(4), 621–627.
Pereira, Alfredo (2003) "The Quantum Mind/Classical Brain Problem," *NeuroQuantology*, 1(1), 94–118.
Perlman, Mark (2004) "The Modern Philosophical Resurrection of Teleology," *The Monist*, 87(1), 3–51.
Perus, Mitja (2001) "Image Processing and Becoming Conscious of Its Result," *Informatica*, 25, 575–592.
Pestana, Mark (2001) "Complexity Theory, Quantum Mechanics and Radically Free Self Determination," *Journal of Mind and Behavior*, 22(4), 365–388.
Peterman, William (1994) "Quantum Theory and Geography: What Can Dr. Bertlmann Teach Us?" *Professional Geographer*, 46(1), 1–9.
Petranker, Jack (2003) "Inhabiting Conscious Experience: Engaged Objectivity in the First-Person Study of Consciousness," *Journal of Consciousness Studies*, 10(12), 3–23.
Pettit, Philip (1993a) *The Common Mind*, Oxford University Press.
  (1993b) "A Definition of Physicalism," *Analysis*, 53(4), 213–223.
Pettit, Philip and Christian List (2011) *Group Agents*, Oxford University Press.
Phelan, Mark and Adam Waytz (2012) "The Moral Cognition/Consciousness Connection," *Review of Philosophy and Psychology*, 3(3), 293–301.
Pickering, Martin and Simon Garrod (2013) "An Integrated Theory of Language Production and Comprehension," *Behavioral and Brain Sciences*, 36(4), 329–347.
Piotrowski, Edward and Jan Sladkowski (2003) "An Invitation to Quantum Game Theory," *International Journal of Theoretical Physics*, 42(5), 1089–1099.
Piro, Francesco (1997) "Is It Possible to Co-operate without Interaction?" *Synthesis Philosophica*, 12(2), 433–444.
Plankar, Matej, Simon Brezan, and Igor Jerman (2013) "The Principle of Coherence in Multi-level Brain Information Processing," *Progress in Biophysics and Molecular Biology*, 111(1), 8–29.
Platt, John (1956) "Amplification Aspects of Biological Response and Mental Activity," *American Scientist*, 44(2), 180–197.
Plotnitsky, Arkady (1994) *Complementarity: Anti-Epistemology after Bohr and Derrida*, Durham, NC: Duke University Press.
  (2010) *Epistemology and Probability: Bohr, Heisenberg, Schrödinger, and the Nature of Quantum-Theoretical Thinking*, New York, NY: Springer.
Pockett, Susan (2002) "On Subjective Back-Referral and How Long It Takes to Become Conscious of a Stimulus," *Consciousness and Cognition*, 11(2), 144–161.
Poland, Jeffrey (1994) *Physicalism: The Philosophical Foundations*, Oxford University Press.

Polanyi, Michael (1968) "Life's Irreducible Structure," *Science*, 160, 1308–1312.
Polonioli, Andrea (2014) "Blame It on the Norm: The Challenge from 'Adaptive Rationality'," *Philosophy of the Social Sciences*, 44(2), 131–150.
Porpora, Douglas (1989) "Four Concepts of Social Structure," *Journal for the Theory of Social Behaviour*, 19(2), 195–221.
  (2006) "Methodological Atheism, Methodological Agnosticism and Religious Experience," *Journal for the Theory of Social Behaviour*, 36(1), 57–75.
Portmore, Douglas (2011) "The Teleological Conception of Practical Reasons," *Mind*, 120(477), 117–153.
Poser, Hans (1992) "The Notion of Consciousness in Schrödinger's Philosophy of Nature," in J, Götschl, ed., *Erwin Schrödinger's World View*, Dordrecht: Kluwer, pp. 153–168.
Pothos, Emmanuel and Jerome Busemeyer (2009) "A Quantum Probability Explanation for Violations of 'Rational' Decision Theory," *Proceedings of the Royal Society B*, 276, 2171–78.
  (2013) "Can Quantum Probability Provide a New Direction for Cognitive Modeling?" *Behavioral and Brain Sciences*, 36(3), 255–327 (includes open peer commentary).
  (2014) "In Search for a Standard of Rationality," *Frontiers in Psychology*, 5, article 49.
Power, Sean Enda (2010) "Perceiving External Things and the Time-Lag Argument," *European Journal of Philosophy*, 21(1), 94–117.
Pradeu, Thomas (2010) "What Is an Organism? An Immunological Answer," *History and Philosophy of Life Science*, 32(2–3), 247–268.
Pradhan, Rajat (2012) "Psychophysical Interpretation of Quantum Theory," *NeuroQuantology*, 10(4), 629–654.
Pratten, Stephen (2013) "Critical Realism and the Process Account of Emergence," *Journal for the Theory of Social Behaviour*, 43(3), 251–279.
Predelli, Stefano (2005) "Painted Leaves, Context, and Semantic Analysis," *Linguistics and Philosophy*, 28(3), 351–374.
Pribram, Karl (1971) *Languages of the Brain*, Englewood Cliffs, NJ: Prentice-Hall.
  (1986) "The Cognitive Revolution and Mind/Brain Issues," *American Psychologist*, 41(5), 507–520.
Price, Huw (1996) *Time's Arrow and Archimedes' Point*, Oxford University Press.
  (2012) "Does Time-Symmetry Imply Retrocausality? How the Quantum World Says 'Maybe,'" *Studies in History and Philosophy of Modern Physics*, 43(2), 75–83.
Price, Huw and Richard Corry, eds. (2007) *Causation, Physics, and the Constitution of Reality*, Oxford University Press.
Primas, Hans (1992) "Time-Asymmetric Phenomena in Biology," *Open Systems and Information Dynamics*, 1(1), 3–34.
  (2003) "Time-Entanglement between Mind and Matter," *Mind and Matter*, 1(1), 81–119.
  (2007) "Non-Boolean Descriptions for Mind-Matter Problems," *Mind and Matter*, 5(1), 7–44.
  (2009) "Complementarity of Mind and Matter," in H. Atmanspacher and H. Primas, eds., *Recasting Reality*, Berlin: Springer Verlag, pp. 171–209.
Prosser, Simon (2012) "Emergent Causation," *Philosophical Studies*, 159(1), 21–39.

Puryear, Stephen (2010) "Monadic Interaction," *British Journal for the History of Philosophy*, 18(5), 763–795.
Putnam, Hilary (1975) *Mind, Language and Reality*, Cambridge University Press.
Pylkkänen, Paavo (1995) "Mental Causation and Quantum Ontology," *Acta Philosophica Fennica*, 58, 335–348.
  (2004) "Can Quantum Analogies Help Us to Understand the Process of Thought?" in G. Globus, K. Pribram, and G. Vitiello, eds., *Brain and Being*, Amsterdam: John Benjamins, pp. 165–195.
  (2007) *Mind, Matter and the Implicate Order*, Berlin: Springer.
Pyyhtinen, Olli (2009) "Being-With: Georg Simmel's Sociology of Association," *Theory, Culture and Society*, 26(5), 108–128.
Radder, Hans and Gerben Meynen (2012) "Does the Brain 'Initiate' Freely Willed Processes? A Philosophy of Science Critique of Libet-type Experiments and Their Interpretation," *Theory and Psychology*, 23(1), 3–21.
Rahnama, Majid, Vahid Salari, and Jack Tuszynski (2009) "How Can the Visual Quantum Information Be Transferred to the Brain Intact, Collapsing There and Causing Consciousness?" *NeuroQuantology*, 7(4), 491–499.
Ram, Vimal (2009) "Meanings Attributed to the Term 'Consciousness': An Overview," *Journal of Consciousness Studies*, 16(5), 9–27.
Read, Rupert (2008) "The 'Hard' Problem of Consciousness Is Continually Reproduced and Made Harder by All Attempts to Solve It," *Theory, Culture and Society*, 25(2), 51–86.
Recanati, François (2002) "Does Linguistic Communication Rest on Inference?" *Mind and Language*, 17(1–2), 105–126.
  (2005) "Literalism and Contextualism," in G. Preyer and G. Peter, eds., *Contextualism in Philosophy*, Oxford: Clarendon Press, pp. 171–196.
Reddy, Vasudevi and Paul Morris (2004) "Participants Don't Need Theories: Knowing Minds in Engagement," *Theory and Psychology*, 14(5), 647–665.
Redman, Deborah (1997) *The Rise of Political Economy as a Science*, Cambridge: MIT Press.
Reed, Edward (1983) "Two Theories of the Intentionality of Perceiving," *Synthese* 54(1), 85–94.
Rehberg, Karl-Siegbert (2009) "Philosophical Anthropology from the End of World War I to the 1940s in Current Perspective," *Iris*, 1(1), 131–152.
Reill, Peter-Hanns (2005) *Vitalizing Nature in the Enlightenment*, Berkeley, CA: University of California Press.
Reimers, Jeffrey, Laura McKemmish, Ross McKenzie, Alan Mark, and Noel Hush (2009) "Weak, Strong, and Coherent Regimes of Fröhlich Condensation and their Applications to Terahertz Medicine and Quantum Consciousness," *PNAS*, 106(11), 4219–4224.
Reynolds, Andrew (2007) "The Theory of the Cell State and the Question of Cell Autonomy in Nineteenth and Early Twentieth-Century Biology," *Science in Context*, 20(1), 71–95.
  (2008) "Ernst Haeckel and the Theory of the Cell State," *History of Science*, 46(2), 123–152.
  (2010) "The Redoubtable Cell," *Studies in History and Philosophy of Biological and Biomedical Sciences*, 41(3), 194–201.

Reynolds, David (2003) "The Origins of the Two 'World Wars': Historical Discourse and International Politics," *Journal of Contemporary History*, 38(1), 29–44.

Ricciardi, L. and H. Umezawa (1967) "Brain and Physics of Many-Body Problems," *Kybernetik*, 4(2), 44–48.

Rieskamp, Jorg, Jerome Busemeyer, and Barbara Mellers (2006) "Extending the Bounds of Rationality: Evidence and Theories of Preferential Choice," *Journal of Economic Literature*, 44(3), 631–661.

Ringen, Jon (1999) "Radical Behaviorism: B.F. Skinner's Philosophy of Science," in W. O'Donohue and R. Kitchener, eds., *Handbook of Behaviorism*, New York, NY: Academic Press, pp. 159–177.

Risjord, Mark (2004) "Reasons, Causes, and Action Explanation," *Philosophy of the Social Sciences*, 35(3), 294–306.

Robb, David and John Heil (2014) "Mental Causation," *Stanford Encyclopedia of Philosophy* (Spring 2014 edition), Edward N. Zalta (ed.), http://plato.stanford.edu/archives/spr2014/entries/mental-causation/.

Robbins, Stephen (2002) "Semantics, Experience and Time," *Cognitive Systems Research*, 3(3), 301–337.

  (2006) "Bergson and the Holographic Theory of Mind," *Phenomenology and the Cognitive Sciences*, 5(3–4), 365–394.

Robinson, Andrew and Christopher Southgate (2010) "A General Definition of Interpretation and Its Application to Origin of Life Research," *Biology and Philosophy*, 25(2), 163–181.

Robinson, Howard (2012) "Qualia, Qualities, and the Our Conception of the Physical World," in B. Göcke, ed., *After Physicalism*, South Bend, IN: University of Notre Dame Press, pp. 231–263.

Robinson, William (2005) "Zooming In on Downward Causation," *Biology and Philosophy*, 20(1), 117–136.

Rogeberg, Ole and Morten Nordberg (2005) "A Defence of Absurd Theories in Economics," *Journal of Economic Methodology*, 12(4), 543–562.

Romero-Isart, Oriol, Mathieu Juan, Romain Quidant, and Ignacio Cirac (2010) "Toward Quantum Superposition of Living Organisms," *New Journal of Physics*, 12(3), article 033015.

Romijn, Herms (2002) "Are Virtual Photons the Elementary Carriers of Consciousness?" *Journal of Consciousness Studies*, 9(1), 61–81.

Rosa, Luiz and Jean Faber (2004) "Quantum Models of the Mind: Are They Compatible with Environment Decoherence?" *Physical Review E*, 70(3), 031902.

Rosen, Steven (2008) *The Self-Evolving Cosmos: A Phenomenological Approach to Nature's Unity-in-Diversity*, Singapore: World Scientific.

Rosenberg, Gregg (2004) *A Place for Consciousness*, Oxford University Press.

Rosenblueth, Arturo, Norbert Wiener, and Julian Bigelow (1943) "Behavior, Purpose and Teleology," *Philosophy of Science*, 10(1), 18–24.

Rosenblum, Bruce and Fred Kuttner (1999) "Consciousness and Quantum Mechanics," *Journal of Mind and Behavior*, 20(1), 229–256.

  (2002) "The Observer in the Quantum Experiment," *Foundations of Physics*, 32(8), 1273–1293.

  (2006) *Quantum Enigma: Physics Encounters Consciousness*, Oxford University Press.

Roth, Paul (2012) "The Pasts," *History and Theory*, 51(3), 313–339.
Rovane, Carol (2004) "Alienation and the Alleged Separateness of Persons," *The Monist*, 87(4), 554–572.
Rudolph, Lloyd and Susanne Rudolph (2003) "Engaging Subjective Knowledge: How Amar Singh's Diary Narratives of and by the Self Explain Identity Formation," *Perspectives on Politics*, 1(4), 681–694.
Ruetsche, Laura (2002) "Interpreting Quantum Theories," in P. Machamer and M. Silberstein, eds., *The Blackwell Guide to the Philosophy of Science*, Oxford: Blackwell, pp. 199–226.
Ruiz-Mirazo, Kepa, Arantza Etxeberria, Alvaro Moreno, and Jesús Ibáñez (2000) "Organisms and Their Place in Biology," *Theory in Biosciences*, 119(3–4), 209–233.
Rupert, Robert (2009) *Cognitive Systems and the Extended Mind*, Oxford University Press.
Ryle, Gilbert (1949) *The Concept of Mind*, London: Hutchinson.
Rysiew, Patrick (2011) "Epistemic Contextualism," *Stanford Encyclopedia of Philosophy* (Winter 2011 edition), Edward N. Zalta (ed.), http://plato.stanford.edu/archives/win2011/entries/contextualism-epistemology/.
Sahu, Satyajit, Subrata Ghosh, Daisuke Fujita, and Anirban Bandyopadhyay (2011) "Computational Myths and Mysteries That Have Grown Around Microtubule in the Last Half a Century and Their Possible Verification," *Journal of Computational and Theoretical Nanoscience*, 8(3), 509–515.
Sahu, Satyajit, Subrata Ghosh, Batu Ghosh, Krishna Aswani, Kazuto Hirata, Daisuke Fujita, and Anirban Bandyopadhyay (2013a) "Atomic Water Channel Controlling Remarkable Properties of a Single Brain Microtubule," *Biosensors and Bioelectronics*, 47(15), 141–148.
Sahu, Satyajit, Subrata Ghosh, Kazuto Hirata, Daisuke Fujita, and Anirban Bandyopadhyay (2013b) "Multi-Level Memory Switching Properties of a Single Brain Microtubule," *Applied Physics Letters*, 102(12), 123701–123704.
Samuel, Arthur (2011) "Speech Perception," *Annual Review of Psychology*, 62, 49–72.
Sappington, A. A. (1990) "Recent Psychological Approaches to the Free Will versus Determinism Issue," *Psychological Bulletin*, 108(1), 19–29.
Satinover, Jeffrey (2001) *The Quantum Brain*, New York, NY: Wiley.
Savage, L. J. (1954) *The Foundations of Statistics*, New York, NY: John Wiley & Sons.
Savitt, Steven (1996) "The Direction of Time," *British Journal for the Philosophy of Science*, 47(3), 347–370.
Sawyer, R. Keith (2002) "Durkheim's Dilemma: Toward a Sociology of Emergence," *Sociological Theory*, 20(2), 227–247.
  (2005) *Social Emergence: Societies as Complex Systems*, Cambridge University Press.
Sayes, Edwin (2014) "Actor-Network Theory and Methodology: Just What Does It Mean to Say that Nonhumans Have Agency?" *Social Studies of Science*, 44(1), 134–149.
Schäfer, Lothar (2006) "Quantum Reality and the Consciousness of the Universe," *Zygon*, 41(3), 505–532.
Schatzki, Theodore (2002) *The Site of the Social*, University Park, PA: Pennsylvania State University Press.

(2005) "The Sites of Organizations," *Organization Studies*, 26(3), 465–484.
(2006) "On Organizations as They Happen," *Organization Studies*, 27(12), 1863–1873.
Schatzki, Theodore, Karin Knorr-Cetina, and Eike von Savigny, eds. (2001) *The Practice Turn in Contemporary Social Theory*, London: Routledge.
Schilbach, Leonhard, Bert Timmermans, Vasudevi Reddy, Alan Costall, Gary Bente, Tobias Schlicht, and Kai Vogeley (2013) "Toward a Second-Person Neuroscience," *Behavioral and Brain Sciences*, 36(4), 393–414.
Schindler, Samuel (2014) "Explanatory Fictions – For Real?" *Synthese*, 191(8), 1741–1755.
Schlosser, Markus (2014) "The Neuroscientific Study of Free Will: A Diagnosis of the Controversy," *Synthese*, 191(2), 245–262.
Schmid, Hans Bernhard (2014) "Plural Self-Awareness," *Phenomenology and the Cognitive Sciences*, 13(1), 7–24.
Schmidt, R. C. (2007) "Scaffolds for Social Meaning," *Ecological Psychology*, 19(2), 137–151.
Schneider, Jean (2005) "Quantum Measurement Act as a Speech Act," in R. Buccheri, et al., eds., *Endophysics, Time, Quantum and the Subjective*, Singapore: World Scientific, pp. 345–354.
Schrödinger, Erwin (1944) *What Is Life?*, Cambridge University Press.
(1959) *Mind and Matter*, Cambridge University Press.
Schroeder, Severin (2001) "Are Reasons Causes? A Wittgensteinian Response to Davidson," in Schroeder, ed., *Wittgenstein and Contemporary Philosophy of Mind*, New York: Palgrave, pp. 150–170.
Schubert, Glendon (1983) "The Evolution of Political Science: Paradigms of Physics, Biology, and Politics," *Politics and the Life Sciences*, 1(2), 97–124.
Schueler, G. F. (2003) *Reasons and Purposes*, Oxford University Press.
Schwartz, Jeffrey, Henry Stapp, and Mario Beauregard (2005) "Quantum Physics in Neuroscience and Psychology," *Philosophical Transactions of the Royal Society B*, 360, 1309–1327.
Schwartz, Sanford (1992) "Bergson and the Politics of Vitalism," in F. Burwick and P. Douglass, eds., *The Crisis in Modernism*, Cambridge University Press, pp. 277–305.
Scott, Joan (1991) "The Evidence of Experience," *Critical Inquiry*, 17(4), 773–797.
Seager, William (1995) "Consciousness, Information and Panpsychism," *Journal of Consciousness Studies*, 2(3), 272–288.
(2009) "Panpsychism," in B. McLaughlin, et al., eds., *The Oxford Handbook of Philosophy of Mind*, Oxford University Press, pp. 206–219.
(2010) "Panpsychism, Aggregation, and Combinatorial Infusion," *Mind and Matter*, 8(2), 167–184.
(2012) "Emergentist Panpsychism," *Journal of Consciousness Studies*, 19(9–10), 19–39.
(2013) "Classical Levels, Russellian Monism and the Implicate Order," *Foundations of Physics*, 43(4), 548–567.
Searle, John (1991) "Intentionalistic Explanations in the Social Sciences," *Philosophy of the Social Sciences*, 21(3), 332–344.
(1992) *The Rediscovery of the Mind*, Cambridge, MA: MIT Press.

(1995) *The Construction of Social Reality*, New York: Free Press.

(2001) *Rationality in Action*, Cambridge, MA: MIT Press.

Seevinck, M. P. (2004) "Holism, Physical Theories and Quantum Mechanics," *Studies in History and Philosophy of Modern Physics*, 35(4), 693–712.

Sehon, Scott (2005) *Teleological Realism: Mind, Agency, and Explanation*, Cambridge, MA: MIT Press.

Sending, Ole (2002) "Constitution, Choice and Change: Problems with the 'Logic of Appropriateness' and Its Use in Constructivist Theory," *European Journal of International Relations*, 8(4), 443–470.

Seth, Anil, Bernaard Baars, and David Edelman (2005) "Criteria for Consciousness in Humans and Other Mammals," *Consciousness and Cognition*, 14(1), 119–139.

Sevush, Steven (2006) "Single-Neuron Theory of Consciousness," *Journal of Theoretical Biology*, 238(3), 704–725.

Shani, Itay (2010) "Mind Stuffed with Red Herrings: Why William James' Critique of the Mind-Stuff Theory Does not Substantiate a Combination Problem for Panpsychism," *Acta Analytica*, 25(4), 413–434.

Shanks, Niall (1993) "Quantum Mechanics and Determinism," *The Philosophical Quarterly*, 43, 20–37.

Shannon, C. (1949) "The Mathematical Theory of Communication," in C. Shannon and W. Weaver, *The Mathematical Theory of Communication*, Urbana, IL: University of Illinois Press, pp. 3–91.

Shapiro, J. A. (2007) "Bacteria Are Small But Not Stupid," *Studies in History and Philosophy of Biological and Biomedical Sciences*, 38(4), 807–819.

Shaw, Robert, Endre Kadar, and Jeffrey Kinsella-Shaw (1994) "Modelling Systems with Intentional Dynamics: A Lesson from Quantum Mechanics," in K. Pribram, ed., *Origins: Brain and Self Organization*, Hillsdale, NJ: Lawrence Erlbaum, pp. 54–101.

Sheehy, Paul (2006) *The Reality of Social Groups*, Aldershot: Ashgate.

Sheets-Johnstone, Maxine (1998) "Consciousness: A Natural History," *Journal of Consciousness Studies*, 5(3), 260–294.

(2009) "Animation: The Fundamental, Essential, and Properly Descriptive Concept," *Continental Philosophy Review*, 42(3), 375–400.

Sheperd, Joshua (2013) "The Apparent Illusion of Conscious Deciding," *Philosophical Explorations*, 16(1), 18–30.

Shimony, Abner (1963) "Role of the Observer in Quantum Theory," *American Journal of Physics*, 31(10), 755–773.

(1978) "Metaphysical Problems in the Foundations of Quantum Mechanics," *International Philosophical Quarterly*, 18(1), 2–17.

Shoemaker, Paul (1982) "The Expected Utility Model: Its Variants, Purposes, Evidence and Limitations," *Journal of Economic Literature*, 20(2), 529–563.

Short, T. L. (2007) "Final Causation," chapter 5 in Short, *Peirce's Theory of Signs*, Cambridge University Press, pp. 117–150.

Shulman, Robert and Ian Shapiro (2009) "Reductionism in the Human Sciences," in C. Mantzavinos, ed., *Philosophy of the Social Sciences*, Cambridge University Press, pp. 124–129.

Siegfried, Tom (2000) *The Bit and the Pendulum: From Quantum Computing to M Theory*, New York, NY: Wiley.

Sieroka, Norman (2007) "Weyl's 'Agens Theory' of Matter and the Zurich Fichte," *Studies in History and Philosophy of Science*, 38(1), 84–107.
  (2010) "Geometrization versus Transcendent Matter: A Systematic Historiography of Theories of Matter Following Weyl," *British Journal for the Philosophy of Science*, 61(4), 769–802.
Siewert, Charles (1998) *The Significance of Consciousness*, Princeton University Press.
  (2011) "Consciousness and Intentionality," *Stanford Encyclopedia of Philosophy* (Fall 2011 edition), Edward N. Zalta (ed.), http://plato.stanford.edu/archives/fall2011/entries/consciousness-intentionality/.
Silberstein, Michael (2002) "Reduction, Emergence and Explanation," in P. Machamer and M. Silberstein, eds., *The Blackwell Guide to the Philosophy of Science*, Oxford: Blackwell, pp. 80–107.
  (2009) "Essay Review: Why Neutral Monism is Superior to Panpsychism," *Mind and Matter*, 7(2), 239–248.
Silberstein, Michael and John McGeever (1999) "The Search for Ontological Emergence," *The Philosophical Quarterly*, 49, 182–200.
Sinnott-Armstrong, Walter and Lynn Nadel, eds. (2011) *Conscious Will and Responsibility: A Tribute to Benjamin Libet*, Oxford University Press.
Sitte, Peter (1992) "A Modern Concept of the 'Cell Theory,'" *International Journal of Plant Science*, 153(3), S1–S6.
Sklar, Lawrence (2003) "Dappled Theories in a Uniform World," *Philosophy of Science*, 70(2), 424–441.
Skrbina, David (2005) *Panpsychism in the West*, Cambridge, MA: MIT Press.
Sloan, Phillip (2012) "How Was Teleology Eliminated in Early Molecular Biology?" *Studies in History and Philosophy of Biological and Biomedical Sciences*, 43(1), 140–151.
Slovic, Paul (1995) "The Construction of Preference," *American Psychologist*, 50(5), 364–371.
Smith, C. U. M. (2001) "Renatus Renatus: The Cartesian Tradition in British Neuroscience and the Neurophilosophy of John Carew Eccles," *Brain and Cognition*, 46(3), 364–372.
Smith, Joel (2010a) "The Conceptual Problem of Other Bodies," *Proceedings of the Aristotelian Society*, 110(2), 201–217.
  (2010b) "Seeing Other People," *Philosophy and Phenomenological Research*, 81(3), 731–748.
Smith, John (2012) "The Endogenous Nature of the Measurement of Social Preferences," *Mind and Society*, 11(2), 235–256.
Smolin, Lee (2001) *Three Roads to Quantum Gravity*, New York, NY: Basic Books.
Sole, Albert (2013) "Bohmian Mechanics without Wave Function Ontology," *Studies in History and Philosophy of Modern Physics*, 44(4), 365–378.
Sollberger, Michael (2008) "Naïve Realism and the Problem of Causation," *Disputatio*, 3, 1–19.
  (2012) "Causation in Perception: A Challenge to Naïve Realism," *Review of Philosophy and Psychology*, 3(4), 581–595.
Sozzo, Sandro (2014) "A Quantum Probability Explanation in Fock Space for Borderline Contradictions," *Journal of Mathematical Psychology*, 58(1), 1–12.

Spaulding, Shannon (2012) "Mirror Neurons are Not Evidence for the Simulation Theory," *Synthese*, 189(3), 515–534.
Sprigge, T. L. S. (1983) *The Vindication of Absolute Idealism*, Edinburgh University Press.
Squires, Euan (1990) *Conscious Mind in the Physical World*, Philadelphia, PA: Institute of Physics.
——— (1994) *The Mystery of the Quantum World*, Philadelphia, PA: Institute of Physics.
Stapp, Henry (1972/1997) "The Copenhagen Interpretation," *Journal of Mind and Behavior*, 18(2–3), 127–154.
——— (1993) *Mind, Matter, and Quantum Mechanics*, Berlin: Springer Verlag.
——— (1996) "The Hard Problem: A Quantum Approach," *Journal of Consciousness Studies*, 3(3), 194–210.
——— (1999) "Attention, Intention, and Will in Quantum Physics," *Journal of Consciousness Studies*, 6(8–9), 143–164.
——— (2001) "Quantum Theory and the Role of Mind in Nature," *Foundations of Physics*, 31(10), 1465–1499.
——— (2005) "Quantum Interactive Dualism: An Alternative to Materialism," *Journal of Consciousness Studies*, 12(11), 43–58.
——— (2006) "Quantum Interactive Dualism, II: The Libet and Einstein-Podolsky-Rosen Causal Anomalies," *Erkenntnis*, 65(1), 117–142.
Stawarska, Beata (2008) "'You' and 'I,' 'Here' and 'Now': Spatial and Social Situatedness in Deixis," *International Journal of Philosophical Studies*, 16(3), 399–418.
——— (2009) *Between You and I: Dialogical Phenomenology*, Athens, OH: Ohio University Press.
Stazicker, James (2011) "Attention, Visual Consciousness and Indeterminacy," *Mind and Language*, 26(2), 156–184.
Steffensen, Sune Vork (2009) "Language, Languaging, and the Extended Mind Hypothesis," *Pragmatics and Cognition*, 17(3), 677–697.
Steffensen, Sune Vork and Stephen Cowley (2010) "Signifying Bodies and Health: A Non-Local Aftermath," in S. Cowley, et al., eds., *Signifying Bodies*, Braga: Portuguese Catholic University Press, pp. 331–355.
Steinberg, S. H. (1947) "The Thirty Years War: A New Interpretation," *History*, 32, 89–102.
Stella, Marco and Karel Kleisner (2010) "Uexküllian Umwelt as Science and as Ideology," *Theory in Biosciences*, 129(1), 39–51.
Stephan, Achim, Sven Walter, and Wendy Wilutzky (2014) "Emotions beyond Brain and Body," *Philosophical Psychology*, 27(1), 65–81.
Stewart, John (1996) "Cognition = Life: Implications for Higher-Level Cognition," *Behavioural Processes*, 35(1–3), 311–326.
Stout, Rowland (1996) *Things That Happen Because They Should: A Teleological Approach to Action*, Oxford: Clarendon Press..
Strawson, Galen (2004) "Real Intentionality," *Phenomenology and the Cognitive Sciences*, 3(3), 287–313.
——— (2006) "Realistic Monism: Why Physicalism Entails Panpsychism," *Journal of Consciousness Studies*, 13(10–11), 3–31.
Strehle, Stephen (2011) "The Nazis and the German Metaphysical Tradition of Voluntarism," *The Review of Metaphysics*, 65(1), 113–137.

Stuart, C., Y. Takahashi, and H. Umezawa (1978) "On the Stability and Non-Local Properties of Memory," *Journal of Theoretical Biology*, 71(4), 605–618.

Stubenberg, Leopold (2014) "Neutral Monism," *Stanford Encyclopedia of Philosophy* (Fall 2014 edition), Edward N. Zalta (ed.), http://plato.stanford.edu/archives/fall2014/entries/neutral-monism/.

Stueber, Karsten (2002) "The Psychological Basis of Historical Explanation: Reenactment, Simulation, and the Fusion of Horizons," *History and Theory*, 41(1), 25–42.

Suárez, Antoine and Peter Adams, eds. (2013) *Is Science Compatible with Free Will? Exploring Free Will and Consciousness in the Light of Quantum Physics and Neuroscience*, Berlin: Springer Verlag.

Suarez, Mauricio (2007) "Quantum Propensities," *Studies in History and Philosophy of Modern Physics*, 38(2), 418–438.

ed. (2009) *Fictions in Science*, London: Routledge.

Suddendorf, Thomas and Michael Corballis (2007) "The Evolution of Foresight: What Is Mental Time Travel, and Is It Unique to Humans?" *Behavioral and Brain Sciences*, 30(3), 299–351.

Suddendorf, Thomas, Donna Rose Addis, and Michael Corballis (2009) "Mental Time Travel and the Shaping of the Human Mind," *Philosophical Transactions of the Royal Society B*, 364, 1317–1324.

Sullivan, Gregory (2011) "The Instinctual Nation-State: Non-Darwinian Theories, State Science and Ultra-Nationalism in Oka Asajiro's Evolution and Human Life," *Journal of the History of Biology*, 44(3), 547–586.

Svozil, Karl and Ron Wright (2005) "Statistical Structures Underlying Quantum Mechanics and Social Science," *International Journal of Theoretical Physics*, 44(7), 1067–1086.

Swann, William, Angel Gomez, D. Conor Seyle, J. Francisco Morales, and Carmen Huici (2009) "Identity Fusion: The Interplay of Personal and Social Identities in Extreme Group Behavior," *Journal of Personality and Social Psychology*, 96(5), 995–1011.

Swenson, Rod (1999) "Epistemic Ordering and the Development of Space-Time: Intentionality as a Universal Entailment," *Semiotica*, 127(1–4), 567–597.

Swinburne, Richard (2013) *Mind, Brain, and Free Will*, Oxford University Press.

Sylvester, Christine (2012) "War Experiences/War Practices/War Theory," *Millennium*, 40(3), 483–503.

Sytsma, Justin (2009) "Phenomenological Obviousness and the New Science of Consciousness," *Philosophy of Science*, 76(5), 958–969.

Szanto, Thomas (2014) "How to Share a Mind: Reconsidering the Group Mind Hypothesis," *Phenomenology and the Cognitive Sciences*, 13(1), 99–120.

Tabaczek, Mariusz (2013) "The Metaphysics of Downward Causation," *Zygon*, 48(2), 380–404.

Talbot, Michael (1991) *The Holographic Universe*, New York: Harper Collins.

Tanesini, Alessandra (2006) "Bringing About the Normative Past," *American Philosophical Quarterly*, 43(3), 191–206.

(2014) "Temporal Externalism: A Taxonomy, an Articulation, and a Defence," *Journal of the Philosophy of History*, 8(1), 1–19.

Tanney, Julia (1995) "Why Reasons May Not Be Causes," *Mind and Language*, 10(1/2), 105–128.

Tantillo, Astrida (2002) *The Will to Create: Goethe's Philosophy of Nature*, University of Pittsburgh Press.

Tarde, Gabriel (1895/2012) *Monadology and Sociology*, Melbourne: re.press.

Tauber, Alfred (1994) *The Immune Self: Theory or Metaphor?* Cambridge University Press.

— (2013) "Immunology's Theories of Cognition," *History and Philosophy of the Life Sciences*, 35(2), 239–264.

Tegmark, Max (2000a) "Importance of Quantum Decoherence in Brain Processes," *Physical Review E*, 61(4), 4194–4206.

— (2000b) "Why the Brain Is Probably Not a Quantum Computer," *Information Sciences*, 128(3), 155–179.

— (2014) "Consciousness as a State of Matter," arXiv:1401.1219v2.

Teller, Paul (1986) "Relational Holism and Quantum Mechanics," *British Journal for the Philosophy of Science*, 37(1), 71–81.

— (1998) "Quantum Mechanics and Haecceities," in E. Castellani, ed., *Interpreting Bodies*, Princeton University Press, pp. 114–141.

Temby, Owen (2013) "What Are Levels of Analysis and What Do They Contribute to International Relations Theory?" *Cambridge Review of International Affairs*, online.

Teufel, Thomas (2011) "Wholes that Cause their Parts: Organic Self-Reproduction and the Reality of Biological Teleology," *Studies in History and Philosophy of Biological and Biomedical Sciences*, 42(2), 252–260.

Theiner, Georg, Colin Allen, and Robert Goldstone (2010) "Recognizing Group Cognition," *Cognitive Systems Research*, 11(4), 378–395.

Thomas, Michael, Harry Purser, and Denis Mareschal (2012) "Is the Mystery of Thought Demystified by Context-Dependent Categorisation?" *Mind and Language*, 27(5), 595–618.

Thompson, Evan (2007) *Mind in Life: Biology, Phenomenology, and the Sciences of the Mind*, Cambridge, MA: Harvard University Press.

Todd, Patrick (2013) "Soft Facts and Ontological Dependence," *Philosophical Studies*, 164(3), 829–844.

Toepfer, Georg (2012) "Teleology and Its Constitutive Role for Biology as the Science of Organized Systems in Nature," *Studies in History and Philosophy of Biological and Biomedical Sciences*, 43(1), 113–119.

Togeby, Ole (2000) "Anticipated Downward Causation and the Arch Structure of Texts," in P. Andersen, et al., eds., *Downward Causation: Minds, Bodies and Matter*, Aarhus University Press, pp. 261–277.

Tollaksen, Jeff (1996) "New Insights from Quantum Theory on Time, Consciousness, and Reality," in S. Hameroff, A. Kaszniak, and A. Scott, eds., *Toward a Science of Consciousness*, Cambridge, MA: MIT Press, pp. 551–567.

Tononi, Giulio (2008) "Consciousness as Integrated Information: A Provisional Manifesto," *Biological Bulletin*, 215(3), 216–242.

Trewavas, Anthony (2003) "Aspects of Plant Intelligence," *Annals of Botany*, 92(1), 1–20.

— (2008) "Aspects of Plant Intelligence: Convergence and Evolution," in S. Conway Morris, ed., *The Deep Structure of Biology*, West Conshohocken, PA: Templeton Foundation Press, pp. 68–110.

Trout, J. D. (2002) "Scientific Explanation and the Sense of Understanding," *Philosophy of Science*, 69(2), 212–233.
Trueblood, Jennifer and Jerome Busemeyer (2011) "A Quantum Probability Account of Order Effects in Inference," *Cognitive Science*, 35(8), 1518–1552.
Turausky, Keith (2014) "Conference Report: 'The Most Interesting Problem in the Universe,'" *Journal of Consciousness Studies*, 21(7–8), 220–240.
Tuszynksi, Jack, ed. (2006) *The Emerging Physics of Consciousness*, Berlin: Springer Verlag.
Tuszynski, J., J. Brown, and P. Hawrylak (1997) "Dielectric Polarization, Electrical Conduction, Information Processing, and Quantum Computation in Microtubules," *Philosophical Transactions of the Royal Society of London A*, 356, 1897–1926.
Tversky, Amos and Daniel Kahneman (1983) "Extensional versus Intuitive Reasoning: The Conjunctive Fallacy in Probability Judgment," *Psychological Review*, 90(4), 293–315.
Tversky, Amos, Paul Slovic, and Daniel Kahneman (1990) "The Causes of Preference Reversal," *American Economic Review*, 80(1), 204–217.
Tylén, Kristian, Ethan Weed, Mikkel Wallentin, Andreas Roepstorff, and Chris Frith (2010) "Language as a Tool for Interacting Minds," *Mind and Language*, 25(1), 3–29.
Uzan, Pierre (2012) "A Quantum Approach to the Psychosomatic Phenomenon: Co-Emergence and Time Entanglement of Mind and Matter," *KronoScope*, 12(2), 219–244.
Vaihinger, Hans (1924) *The Philosophy of 'As If'*, New York: Harcourt Brace.
Valenza, Robert (2008) "Possibility, Actuality, and Free Will," *World Futures*, 64(2), 94–108.
Van Camp, Wesley (2014) "Explaining Understanding (or Understanding Explanation)," *European Journal for Philosophy of Science*, 4(1), 95–114.
Vandenberghe, Frederic (2002) "Reconstructing Humans: A Humanist Critique of Actant-Network Theory," *Theory, Culture and Society*, 19(5–6), 51–67.
Van Dijk, Ludger and Rob Withagen (2014) "The Horizontal Worldview: A Wittgensteinian Attitude towards Scientific Psychology," *Theory and Psychology*, 24(1), 3–18.
Van Duijn, Marc and Sacha Bem (2005) "On the Alleged Illusion of Conscious Will," *Philosophical Psychology*, 18(6), 699–714.
Van Gulick, Robert (2001) "Reduction, Emergence and Other Recent Options on the Mind–Body Problem: A Philosophical Overview," *Journal of Consciousness Studies*, 8(9–10), 1–34.
Van Putten, Cornelis (2006) "Changing the Past: Retrocausality and Narrative Construction," *Metaphilosophy*, 37(2), 254–258.
Vannini, Antonella (2008) "Quantum Models of Consciousness," *Quantum Biosystems*, 2, 165–184.
Varga, Somogy (2011) "Existential Choices: To What Degree is Who We Are a Matter of Choice?" *Continental Philosophy Review*, 44(1), 65–79.
Vedral, Vlatko (2010) *Decoding Reality: The Universe as Quantum Information*, Oxford University Press.
Velmans, Max (2000) *Understanding Consciousness*, London: Routledge.

(2002) "Making Sense of Causal Interactions between Consciousness and Brain," *Journal of Consciousness Studies*, 9(11), 69–95.
   (2003) "Preconscious Free Will," *Journal of Consciousness Studies*, 10(12), 42–61.
   (2008) "Reflexive Monism," *Journal of Consciousness Studies*, 15(2), 5–50.
Verheggen, Claudine (2006) "How Social Must Language Be?," *Journal for the Theory of Social Behaviour*, 36(2), 203–219.
Vermersch, Pierre (2004) "Attention between Phenomenology and Experimental Psychology," *Continental Philosophy Review*, 37(1), 45–81.
Vicente, Agustin (2006) "On the Causal Completeness of Physics," *International Studies in the Philosophy of Science*, 20(2), 149–171.
   (2011) "Current Physics and 'the Physical,'" *British Journal for the Philosophy of Science*, 62(2), 393–416.
Vimal, Ram (2009) "Subjective Experience Aspect of Consciousness, Parts I and II," *Neuroquantology*, 7(3), 390–410 and 411–434.
Vitiello, Giuseppe (2001) *My Double Unveiled: The Dissipative Quantum Model of Brain*, Amsterdam: John Benjamins.
Von Lucadou, Walter (1994) "Wigner's Friend Revitalized?" in H. Atmanspacher and G. Dalenoort, eds., *Inside Versus Outside*, Berlin: Springer Verlag, pp. 369–388.
Von Neumann, John and Oskar Morgenstern (1944) *Theory of Games and Economic Behavior*, Princeton University Press.
Von Uexküll, Jakob (1982[1940]) "The Theory of Meaning," *Semiotica*, 42(1), 25–82.
Von Wright, Georg (1971) *Explanation and Understanding*, Ithaca, NY: Cornell University Press.
Walach, Harald and Nikolaus von Stillfried (2011) "Generalised Quantum Theory – Basic Idea and General Intuition," *Axiomathes*, 21(2), 185–209.
Walker, Evan Harris (1970) "The Nature of Consciousness," *Mathematical Biosciences*, 7(1–2), 131–178.
Wallace, Alan (2000) *The Taboo of Subjectivity*, Oxford University Press.
Wallin, Annika (2013) "A Peace Treaty for the Rationality Wars?" *Theory and Psychology*, 23(4), 458–478.
Walsh, D. M. (2006) "Organisms as Natural Purposes," *Studies in History and Philosophy of Biological and Biomedical Sciences*, 37(4), 771–791.
Walsh, Denis (2012) "Mechanism and Purpose: A Case for Natural Teleology," *Studies in History and Philosophy of Biological and Biomedical Sciences*, 43(1), 173–181.
Walter, Sven (2014a) "Willusionism, Epiphenomenalism, and the Feeling of Conscious Will," *Synthese*, 191(10), 2215–2238.
   (2014b) "Situated Cognition: A Field Guide to Some Open Conceptual and Ontological Issues," *Review of Philosophy and Psychology*, 5(2), 241–263.
Waltz, Kenneth (1979) *Theory of International Politics*, Boston: Addison-Wesley.
Wang, Zheng and Jerome Busemeyer (2013) "A Quantum Question Order Model Supported by Empirical Tests of an a priori and Precise Prediction," *Topics in Cognitive Science*, 5(4), 689–710.
Wang, Zheng, Jerome Busemeyer, Harald Atmanspacher, and Emmanuel Pothos (2013) "The Potential of Using Quantum Theory to Build Models of Cognition," *Topics in Cognitive Science*, 5(4), 672–688.
Ward, Barry (2014) "Is There a Link between Quantum Mechanics and Consciousness?" in C. U. M. Smith and H. Whitaker, eds., *Brain, Mind and Consciousness in the History of Neuroscience*, Berlin: Springer Verlag, pp. 273–302.

Ward, Dave (2012) "Enjoying the Spread: Conscious Externalism Reconsidered," *Mind*, 121, 731–751.
Warfield, Ted (2003) "Compatibilism and Incompatibilism," in M. Loux and D. Zimmerman, eds., *The Oxford Handbook of Metaphysics*, Oxford University Press, pp. 613–30.
Warren, William (2005) "Direct Perception: The View from Here," *Philosophical Topics*, 33(1), 335–361.
Waters, Christopher and Bonnie Bassler (2005) "Quorum Sensing: Cell-to-Cell Communication in Bacteria," *Annual Review of Cell and Development*, 21, 319–346.
Weber, Andreas and Francisco Varela (2002) "Life after Kant: Natural Purposes and the Autopoietic Foundations of Biological Individuality," *Phenomenology and the Cognitive Sciences*, 1(2), 97–125.
Weber, Marcel (2005) "Genes, Causation and Intentionality," *History and Philosophy of the Life Sciences*, 27(3–4), 407–420.
Weberman, David (1997) "The Nonfixity of the Historical Past," *Review of Metaphysics*, 50(4), 749–768.
Weekes, Anderson (2009) "Whitehead's Unique Approach to the Topic of Consciousness," in M. Weber and A. Weekes, eds., *Process Approaches to Consciousness in Psychology, Neuroscience, and Philosophy of Mind*, Albany, NY: SUNY Press, pp. 137–172.
  (2012) "The Mind–Body Problem and Whitehead's Non-Reductive Monism," *Journal of Consciousness Studies*, 19(9–10), 40–66.
Wegner, Daniel (2002) *The Illusion of Conscious Will*, Cambridge, MA: MIT Press.
Wegner, Daniel, et al. (2004) "Précis of The Illusion of Conscious Will, with Commentaries," *Behavioral and Brain Sciences*, 27(5), 649–692.
Weisskopf, Walter (1979) "The Method is the Ideology: From a Newtonian to a Heisenbergian Paradigm in Economics," *Journal of Economic Issues*, 13(4), 869–884.
Wendt, Alexander (1987) "The Agent–Structure Problem in International Relations Theory," *International Organization*, 41(3), 335–370.
  (1998) "On Constitution and Causation in International Relations," *Review of International Studies*, 24 (special issue), 101–117.
  (1999) *Social Theory of International Politics*, Cambridge University Press.
  (2003) "Why a World State Is Inevitable," *European Journal of International Relations*, 9(4), 491–542.
  (2004) "The State as Person in International Theory," *Review of International Studies*, 30(2), 289–316.
  (2006) "Social Theory as Cartesian Science: An Auto-Critique from a Quantum Perspective," in S. Guzzini and A. Leander, eds., *Constructivism in International Relations: Alexander Wendt and his Critics*, London: Routledge, pp. 181–219.
  (2010) "Flatland: Quantum Mind and the International Hologram," in M. Albert, L.-E. Cederman, and A. Wendt, eds., *New Systems Theories of World Politics*, New York, NY: Palgrave, pp. 279–310.
Wendt, Alexander and Raymond Duvall (2008) "Sovereignty and the UFO," *Political Theory*, 36(4), 607–644.

Wheeler, John Archibald (1978) "The 'Past' and the 'Delayed-Choice' Double-Slit Experiment," in A. Marlow, ed., *Mathematical Foundations of Quantum Theory*, New York: Academic Press, pp. 9–48.
  (1988) "World as System Self-Synthesized by Quantum Networking," in E. Agazzi, ed., *Probability in the Sciences*, Dordrecht: Kluwer Academic Publishers, pp. 103–129.
  (1990) "Information, Physics, Quantum: The Search for Links," in W. Zurek, ed., *Complexity, Entropy, and the Physics of Information*, Reading, MA: Addison Wesley, pp. 3–28.
  (1994) "Delayed-Choice Experiments and the Bohr-Einstein Dialogue," in J. Wheeler, *At Home in the Universe*, Woodbury, NY: American Institute of Physics, pp. 112–131.
Whitford, Josh (2002) "Pragmatism and the Untenable Dualism of Means and Ends," *Theory and Society*, 31(3), 325–363.
Widdows, Dominic (2004) *Geometry and Meaning*, Stanford, CA: CSLI Publications.
Wight, Colin (2006) *Agents, Structures and International Relations*, Cambridge University Press.
Wigner, Eugene (1962) "Remarks on the Mind–Body Question," in I. J. Good, ed., *The Scientist Speculates*, London: Heinemann, pp. 284–302.
  (1964) "Two Kinds of Reality," *The Monist*, 48(2), 248–264.
  (1970) "Physics and the Explanation of Life," *Foundations of Physics*, 1(1), 35–45.
Williams, Meredith (2000) "Wittgenstein and Davidson on the Sociality of Language," *Journal for the Theory of Social Behaviour*, 30(3), 299–318.
Wilson, David Sloan (2002) *Darwin's Cathedral: Evolution, Religion and the Nature of Society*, University of Chicago Press.
Wilson, David Sloan and Elliott Sober (1989) "Reviving the Superorganism," *Journal of Theoretical Biology*, 136(3), 337–356.
Wilson, Jessica (2006) "On Characterizing the Physical," *Philosophical Studies*, 131(1), 61–99.
Wilson, Robert (2001) "Group-Level Cognition," *Philosophy of Science*, 68 (Proceedings), S262–S273.
Wimsatt, William (2006) "Reductionism and Its Heuristics: Making Methodological Reductionism Honest," *Synthese*, 151(3), 445–475.
Winsberg, Eric and Arthur Fine (2003) "Quantum Life: Interaction, Entanglement, and Separation," *Journal of Philosophy*, 100(2), 80–97.
Witte, F. M. C. (2005) "Quantum 2-Player Gambling and Correlated Pay-Off," *Physica Scripta*, 71(2), 229–232.
Wolf, Fred Alan (1989) "On the Quantum Physical Theory of Subjective Antedating," *Journal of Theoretical Biology*, 136(1), 13–19.
  (1998) "The Timing of Conscious Experience," *Journal of Scientific Exploration*, 12(4), 511–542.
Wolters, Gereon (2001) "Hans Jonas' Philosophical Biology," *Graduate Faculty Philosophy Journal*, 23(1), 85–98.
Wong, Hong Yu (2006) "Emergents from Fusion," *Philosophy of Science*, 73(3), 345–367.
Woodward, Keith, John Paul Jones III, and Sallie Marston (2012) "The Politics of Autonomous Space," *Progress in Human Geography*, 36(2), 204–224.

Woolf, Nancy and Stuart Hameroff (2001) "A Quantum Approach to Visual Consciousness," *Trends in Cognitive Sciences*, 5(11), 472–478.
Worden, R. P. (1999) "Hybrid Cognition," *Journal of Consciousness Studies*, 6(1), 70–90.
Worgan, S. F. and R. K. Moore (2010) "Speech as the Perception of Affordances," *Ecological Psychology*, 22(4), 327–343.
Wrong, Dennis (1961) "The Oversocialized Conception of Man in Modern Sociology," *American Sociological Review*, 26(2), 183–193.
Yearsley, James and Emmanuel Pothos (2014) "Challenging the Classical Notion of Time in Cognition: A Quantum Perspective," *Proceedings of the Royal Society B*, 281 (1781), article 20133056.
Ylikoski, Petri (2013) "Causal and Constitutive Explanation Compared," *Erkenntnis*, 78(2), 277–297.
Young, Arthur (1976) *The Reflexive Universe*, New York, NY: Delacorte Press.
Yu, Shan and Danko Nikolic (2011) "Quantum Mechanics Needs No Consciousness," *Annalen der Physik*, 523(11), 931–938.
Yukalov, Vyacheslav and Didier Sornette (2009a) "Physics of Risk and Uncertainty in Quantum Decision Making," *European Physical Journal B*, 71(4), 533–548.
  (2009b) "Processing Information in Quantum Decision Theory," *Entropy*, 11(4), 1073–1120.
  (2011) "Decision Theory with Prospect Interference and Entanglement," *Theory and Decision*, 70(3), 283–328.
  (2014) "Conditions for Quantum Interference in Cognitive Sciences," *Topics in Cognitive Science*, 6(1), 79–90.
Zahavi, Dan (2005) *Subjectivity and Selfhood: Investigating the First-Person Perspective*, Cambridge, MA: MIT Press.
  (2008) "Simulation, Projection and Empathy," *Consciousness and Cognition*, 17(2), 514–522.
Zahavi, Dan and Shaun Gallagher (2008) "The (In)visibility of Others: A Reply to Herschbach," *Philosophical Explorations*, 11(3), 237–244.
Zahle, Julie (2014) "Practices and the Direct Perception of Normative States: Part I," *Philosophy of the Social Sciences*, 43(4), 493–518.
Zaman, L. Frederick (2002) "Nature's Psychogenic Forces: Localized Quantum Consciousness," *Journal of Mind and Behavior*, 23(4), 351–374.
Zammito, John (2006) "Teleology Then and Now: The Question of Kant's Relevance for Contemporary Controversies over Function in Biology," *Studies in History and Philosophy of Biological and Biomedical Sciences*, 37(4), 748–770.
Zeilinger, Anton (1999) "Experiment and the Foundations of Quantum Physics," *Reviews of Modern Physics*, 71(2), S288–S297.
Zeki, S. (2003) "The Disunity of Consciousness," *Trends in Cognitive Sciences*, 7(5), 214–218.
Zhenhua, Yu (2001–2002) "Two Cultures Revisited: Michael Polanyi on the Continuity between the Natural Science and the Study of Man," *Tradition and Discovery*, 28(3), 6–19.
Zhu, Jing (2003) "Reclaiming Volition: An Alternative Interpretation of Libet's Experiment," *Journal of Consciousness Studies*, 10(11), 61–77.
  (2004a) "Understanding Volition," *Philosophical Psychology*, 17(2), 247–273.

(2004b) "Intention and Volition," *Canadian Journal of Philosophy*, 34(2), 175–194.
Ziman, John (2003) "Emerging Out of Nature into History: The Plurality of the Sciences," *Philosophical Transactions of the Royal Society of London A*, 361, 1617–1633.
Zimmerman, Michael (1988) "Quantum Theory, Intrinsic Value, and Panentheism," *Environmental Ethics*, 10(1), 3–30.
Zlatev, Jordan (2008) "The Dependence of Language on Consciousness," *Journal of Consciousness Studies*, 15(6), 34–62.
Zohar, Danah (1990) *The Quantum Self*, New York, NY: Quill.
Zohar, Danah and Ian Marshall (1994) *The Quantum Society: Mind, Physics and a New Social Vision*, New York, NY: Morrow.
Zovko, Jure (2008) "Metaphysics as Interpretation of Conscious Life: Some Remarks on D. Henrich's and D. Kolak's Thinking," *Synthese*, 162(3), 425–438.
Zukav, Gary (1979) *The Dancing Wu Li Masters: An Overview of the New Physics*, New York: Morrow.

Index

A-Series time 126–130, 199–200. *See also* time-symmetry
absolute space and time 59, 65–66, 88–89. *See also* temporal non-locality
action at a distance 51, 53–54, 219–220. *See also* quantum entanglement
active information 88
Addition Effect, temporal non-locality 202–203
advanced action 174
  relation to reasons as causes 179–181
  and temporal non-locality 187
  and will 221
Aerts, Diederik 110, 215, 216, 217–220
agency 174–175, 182–188
  performative model of 163, 172. *See also* will
agent-causation 184
agent–structure problem 6–7, 32, 33–34, 210, 243–246, 255–260. *See also* social structure
agents, matter as 122
Albert, David 80
analogy
  humans as quantum systems 3–4
  mental and quantum domains 91–92
Analytical Philosophy of History (Danto) 192
anomalous phenomena 34
  consciousness 14
  social structure 22–23
anthropological perspectives 36–37, 103
après-coup 203–204
Archer, Margaret 247
Aristotle 64
  causation 63–64, 65, 119–123, 177, 262–263
arrow of time 126–130, 199–200. *See also* time-symmetry
as if explanations 26–28, 35–36
assumptions, classical 11–14, 29, 51–52, 59, 88–89, 111–112. *See also* absolute space and time; atomism; determinism; materialism; mechanism; subject–object distinction
Atmanspacher, Harald 126, 200
atomism 59, 60–62, 88–89, 244–245

B-Series time 126–130, 199–200
backwards causation 65, 179–181
bacterial cognition 117–118. *See also* single-celled organisms
bank teller example, representativeness heuristic 159–161
Barad, Karen (and intra-action) 172, 238
Bauer, Edmond 82
Beck, Friedrich 97
behaviorism 22, 28
Bell, John 40
  Experiments 51–54
Bennett, Jane 133, 146–147
Bergson, Henri 132
Bhaskar, Roy 247
bifurcation of nature 67
biological perspectives
  definition of life 133–134
  materialist–vitalist controversy 133
  quantum biology 107, 135–136
biosemiotics 141–142
bird migration 135
Bitbol, Michel 91–92
black box radiation problem 44
body problem 30. *See also* mind–body problem
Bohm, David (Bohm Interpretation) 85–89
Bohr, Niels 48, 49, 50–54, 73–76, 135, 217
Bose-Einstein condensates (BECs) 98–101
Broglie, Louis 45, 87
Bruza, Peter 160
Burge, Tyler 225, 253
Busemeyer, Jerome 158, 159, 160–161
Butler, Judith 163

Cambridge Changes 195, 287–288
Carminati, G. Galli 202

345

Cartwright, Nancy 8
causal closure of physics (CCP) 7–11
   causal closure of classical physics (CCCP) 11
   definition 7–8
   and *élan vital* 21
   emergentist–reductionist debate 247–248
   language 211–212
   problem of mental causation 175–176
   and quantum coherence 144
   and social science 11–14, 252, 257–258
   and temporal locality 190–191
   unscientific fictions 27–28
causal exclusion 261
causation
   Aristotelian 63–64, 65, 119–123, 177, 262–263
   backwards 65, 179–181
   classical worldviews 64, 177
   counterfactual model 64
   direct perception 224–225
   downward 182, 248, 252, 260–266
   epistemology 287–288
   final or teleological 177–179, 181
   hard problem of time 129
   human nature 152
   problem of mental causation 175–176, 182
cell theory 279–280. *See also* single-celled organisms
Chalmers, David 15, 16, 21, 146, 189, 276
   The Conscious Mind 113
changing the past 191–205. *See also* temporal non-locality
Clark, Andy 276
Clarke, Chris 131
classical physics. *See* physics
classical worldview
   assumptions. *See* assumptions (classical)
   causal exclusion 261
   causation 177
   changing the past 193–194
   direct perception 222, 224–225
   emergence 249
   free will 182, 183–184, 186
   human nature 151–153
   language 212–215
   mind-reading 231, 233, 237, 239
   paradigm shifts 58–59
   and quantum theory 40
   separability 208–209
   social science/social structures 11–14, 22, 245
   supervenience 254–255
   temporal non-locality 202–203

cognition; *See* quantum cognition; quantum consciousness theory
coherence 164–165, 167–168, 290–291.
   See also quantum coherence
collapse of wave functions
   in all matter 119–123
   and Bohm Interpretation 87
   and consciousness 189
   and determinism 62–63
   downward causation 263–265
   experiments 46–48, 54–57
   interpretations of quantum theory 72
   language 216–217
   observation, role 81–85
   quantum coherence 138, 139–141, 145–146
   quantum emergence 256–257
   semantic non-locality 235–236
   subject–object distinction 67–68
   superpositions 278
   temporal non-locality 200–202
   and will 109–111, 120, 121, 174–175, 179–181
collective consciousness 275–281
combination problem
   definition 92
   and panpsychism 92, 123–124, 131–132
commutativity
   and order effects 157–158
   probability 159–160, 161
complementarity, principle of 35, 64, 74, 75, 135
   Two-Slit Experiment 48–49
compositionalism
   composition vs. context 212–215
   language 212
computational theory of mind. *See* quantum brain theory
concept interference, language 217–218
conjunction fallacy, quantum decision theory 159–161
"The Conscious Cell" (Margulis) 118
The Conscious Mind (Chalmers) 113
consciousness 14
   collective 275–281
   definition 15, 276–277
   features of 115, 116
   hard problem of 14–18, 21, 147
   idealist vs. materialist interpretations 89
   illusion of 17–18, 220, 249, 283, 289, 292
   and language 220–221
   in all living things 136, 137, 139, 145–146
   in matter 120–121
   Many Minds Interpretation 80–81
   metaphysics of 109–111
   mind–body problem 14–18

in plants 118
in single-celled organisms 117–118, 279–280
temporal non-locality 189–191
vitalism 21–22, 146–147
and will 18–20, 174
see also experience; quantum consciousness theory; subjectivity
constitutive interactions 85
contextualism
  composition vs. context 212–215
  language 212
  quantum 215–221
Conway, John 122
Copenhagen Interpretation 72, 73–76, 86
correspondence view of rationality 164–165
counterfactual model of causation 64
Cramer, John 179, 180, 187–188

Danto, Arthur 192–193
Davidson, Donald 176
decisions 181. *See also* quantum decision theory
decoherence 3, 96, 104
Delayed-Choice Experiment 54–57, 127
delayed choice, temporal non-locality 201–204
Dennett, Daniel 21
Descartes, René 83
d'Espagnat, Bernard 74, 287
determinism 59, 62–63, 88–89
  free will 183–184, 186
  and quantum theory 40–41, 47–48, 50–54
Devil 27
DeWitt, Bryce 78
dialogue. *See* language
Dirac choice 68–69, 83–84
direct perception 222–228, 230–233, 242
disjunction effect, probability 160–161
downward causation 182, 248, 252, 260–266
Driesch, Hans 132
dualism 10, 17, 28–29, 189
  emergence 248
  mind-reading 233
  panpsychism 31
  and Subjectivist Interpretation 85
  see also mind–body problem
Durkheim, Emile 244

Eccles, John 97
efficient causation, Aristotelian view 119–123, 177
ego 84–85
Einstein, Albert 44, 50–54, 75
Eisert, Jens 169
élan vital. *See* life force

eliminative materialism 22, 28
emergence 33
  classical 249
  of consciousness 16, 124–125
  Fröhlich tradition 98
  quantum 33, 255–260, 263, 265–266
  and social structures 244, 247–250
enactivism 277
entanglement, quantum. *See* quantum entanglement
epiphenomenalism 256
epistemology 93
  Copenhagen Interpretation 75
  changing the past 192–193, 199–200
  emergence 248–249, 251–252
  social 284–288
EPR paper (Einstein, Podolsky and Rosen) 50–54
Esfeld, Michael 175
event-causation 184
Everett, Hugh 78
exclusion argument, mental causation 256–257
expected utility theory (EUT) 155–156, 162, 167
experience 33, 109–111, 141–143. *See also* consciousness; subjectivity
experiments, quantum mechanics 43
  Bell Experiments 50–54
  Delayed-Choice Experiment 54–57
  Two-Slit Experiment 43–49
explanation vs. understanding debate 5–6
explanatory breadth, quantum consciousness theory 291
explanatory power, quantum consciousness theory 292–293
extended mind hypothesis 276. *See also* collective consciousness
externalism, and supervenience 252–255, 259–260

fascism 143–144
feminist perspectives 163, 189
Feynman, Richard 40
fictionalism (*as if* explanations) 26–28, 35–36
final causation 177–179, 181
first-person perspective 109–111. *See also* subjectivity
flat ontology 33, 244, 247–250, 255, 264–265, 268–269, 277
Fodor, Jerry 17, 212, 215
folk psychology 34, 138, 181, 184
formal causation, Aristotelian view 177
Foucault, Michel 19
free will 28, 175
  classical debate about 62–63, 122, 152, 153

free will (*cont.*)
  human nature 152
  illusion of 62, 183–184, 186, 289, 292
  and quantum theory 182–188
  theorem 122, 184–185
  "The Free Will Theorem" (Conway and Kochen) 122
French, Steven 41, 84–85
Fröhlich tradition, quantum brain theory 97, 98–101
Fuchs, Christopher 75

game theory 32–33, 154, 169–173
Garrett, Brian 21
Gell-Mann, Murray 75
generalized (weak) quantum theory 5, 97
Giddens, Anthony 243–244
Gigerenzer, Gerd 165
Gisin, Nicolas 53
global supervenience 254
Glymour, Bruce 63
God 10–11
Godfrey-Smith, Peter 27
Grandy, David 227
group consciousness 275–281
GRW Interpretation (Ghirardi, Rimini, and Weber) 76–77

Haber, Matt 274
Habermas, Jürgen 19
Hacking, Ian 192–193
Hameroff, Stuart 30, 99, 187
Hanauske, Matthias 171
hard problem
  of consciousness 14–18, 21, 147
  of life 134
  of time 126–130, 199
Heisenberg, Werner 49, 73
Heisenberg choice 67–68, 83–84
Hiley, Basil 88
Ho, Mae-Wan, 131, 137
holism 3, 33, 62, 128
  Copenhagen Interpretation 74
  epistemological 74
  light 228
  mind-reading 233
  organicism 144
  in quantum physics 62, 167, 216
  separability 242
  social 244–245, 249–250, 258–259
  temporal 196
  will 140–141
holograms/holographic projection 228–230, 239
holographic state 271–273

horizontal question, social structures 244–245
human nature 150–153
humans, quantum model 3–4, 149–153
Hume, David 132
Humphreys, Paul 256
Husserl, Edmund 84

idealism 31, 81–86, 89. *See also* panpsychism; realism
indeterminacy of the past 192
indirect perception, Time-Lag Argument 239
individualism 249–250
  mind-reading 233
  ontology 251–252
  social structure 259
  supervenience 250–251
inference to the best explanation (IBE) 290, 292–293
information
  Bohm Interpretation 88
  neutral monism 125–126
  subjectivity in living things 117–118, 141–143
  theoretic ontology 125–126
instrumentalism, Copenhagen Interpretation 75
intentionality 13–14, 28, 175
  consciousness 18–20
  as if explanations 26–28
  social structure 252
  see also will
interaction
  quantum game theory 172
  semantic non-locality 237–238
  see also quantum entanglement
interconnnectedness vs. individuality. *See* separability
interference patterns, light 44, 46
interference, quantum 160, 161
internalism, and supervenience 252–255
interpretations of quantum theory 58–59, 70–73
  Bohmian 85–89
  Copenhagen 72, 73–76, 86
  GWR 76–77
  Many Minds 78, 80–81, 181
  Many Worlds 77–81
  Subjectivist 81–85
  time-symmetric/Transactional 70, 179, 187–188, 229–230
  see also panpsychism; realism
interpretivism 28–29
  anomaly of consciousness 14
  causal closure of physics 8, 10–11
  classical social science 12–14
  intentionality 18–19

Index   349

methodological atheism 10–11
quantum consciousness theory 35
intersubjectivity 74
intrinsic properties 252
introspection 84, 110. *See also* subjectivity
Iraq War 269–270
irreducibility 248
it from bit 83–84, 125–126

Jackson, Frank 16
Jonas, Hans 282
Jordan, Pascual 135
Jung, C. G. 128

Kahneman, Daniel 156, 159
Kahneman-Tversky effects 34, 164, 166–167
Kant, Immanuel 114
Kim, Jaegwon 26
Kochen, Simon 122
Kolak, Daniel 242
Kuttner, Fred 69

Lambert-Mogiliansky, Ariane 162
language 33, 210–212
  composition vs. context 212–215
  and experience 190
  mind-reading 222–223, 233–237
  quantum contextualism 215–221
Lapata, Mirella 215
Latour, Bruno 133
leaders 270
Leibniz, Gottfried Wilhelm von 269
Levine, Joseph 15
Lewenstein, Maciej 169
Lewis, David 64
Libet, Benjamin 185–188
life
  definitions of 131–132, 133–134, 137–138, 147
  hard problem of 134
  quantum vitalism 131–132
  and subjectivity 116–118, 136, 137, 139
life force (*élan vital*) 21–22, 27, 32, 92
  quantum coherence 137–138, 267. *See also* vitalism
light
  Bell Experiments 51–54
  dual nature of 43–49, 226–228
  visual perception 222–223, 226–228
linguistics. *See* language
Litt, Abninder 104–105
living things. *See* life
locality, assumption of 51–52, 54, 63–65. *See also* non-locality
location problem, in social ontology 25

Lockwood, Michael 80
Loewer, Barry 80
London, Fritz 82

machines, humans as 153. *See also* mechanism
MacKenzie, Alexander 192
Mackonis, Adolfas 290
macroscopic systems 68–69, 131
Malin, Shimon 46
Many Minds Interpretation 78, 80–81, 181
Many Worlds Interpretation 77–81
Marcer, Peter 229
Margulis, Lynn 117–118
Martin, F. 202
material causation, Aristotelian view 177
materialism 59–60, 88–89, 181
  Bohm Interpretation 86
  conflation with physicalism 9
  consciousness 89
  definition 59, 60
  emergence 247–248
  and experience 189
  human nature 151
  language 211–212
  materialist–idealist debate 6
  materialist–vitalist debate 132–137
  mind–body problem 14, 111–112, 114
  New Materialism 132, 146, 147, 267
  realism 76–81
  social structure 267
  supervenience 251
  See also physicalism
matter (inanimate), distinction from living things 123–124
Mayr, Ernst 140
McEvoy, Cathy 219
McGeever, John 256
McGinn, Colin 17
McKemmish, Laura 105–107
McTaggart, J. M. E. 126–130, 199–200
meaning, language composition vs. context 212–215
measurement
  collapse of wave functions 217
  Delayed-Choice Experiment 56–57
  EPR paper 50–54
  order effects 157–158
  semantic non-locality 235–236
  subject–object distinction 66–69
  Two-Slit Experiment 46
  See also observation
mechanism 59, 63–65, 88–89
memory
  language 217
  temporal non-locality 200–202
  Umezawa tradition 101–102

mental causation
  exclusion argument 256–257
  problem of 175–176, 182
mental time travel 182, 240–241
mentality 85–88, 89, 125. *See also*
  consciousness; No Fundamental
  Mentality; panpsychism
Mercer, Jonathan 277
meta-interpretive frameworks 71–73. *See also*
  interpretations of quantum theory
metaphysics, new 109–111, 289
meta-theory, international relations (IR)
  discipline 1
Meyer, David 169
microtubules 99–101, 104–107
mind
  interpretations of quantum theory 72
  in all living things 136–137
  problem of other 33, 116, 230
  see also consciousness; panpsychism
Mind in Life (Thompson) 136
mind–body dualism 80, 83
mind–body problem 14–18, 29, 30, 72, 93
  dualist and materialist views 111–112
  experience 116
  materialism 114
  neutral monism 125–126
  quantum brain theory 97
  subjective world of experience 111
mind–matter interactions 83, 85
mind-reading 222–223, 230–242. *See also*
  direct perception
minded matter 112. *See also* panpsychism
Mirowski, Philip 4
Mitchell, Edgar 272
Mitchell, Jeff 215
monads/monadology 269–271
Monk, Nicholas 66
Montero, Barbara 9, 30
Munro, William Bennett 4

Nakagomi, Teruaki 269
Nelson, Douglas 219
Neo-Vitalism 145–146. *See also* New
  Materialism
Neuron Doctrine 96, 99, 103, 109
neuroscientific perspectives, free will 183, 185–188
neutral monism 93, 111, 124–130, 179
New Materialism 132, 146, 147, 267
Newton, Sir Isaac 43
No Fundamental Mentality 9, 31, 76, 251. *See also* realism
non-commutativity 157–158

non-conscious matter, distinction from living
  things 123–124
non-locality
  assumptions of locality 51–52, 54, 63–65
  causation 54, 64
  problem of memory 101–102
  semantic 230–242
  temporal 54–57, 180, 182, 189–205
non-separability. *See* separability vs.
  non-separability

object–subject distinction. *See* subject–object
  distinction
observation, role in quantum phenomena
  36–37, 67
  Bohm Interpretation 87
  Subjectivist Interpretation 81–85
  See also measurement
Occam's Razor, applied to Many World
  Interpretation 79
Okasha, Samir 274
ontology 93
  Copenhagen Interpretation 75
  emergence 248–250, 255–260, 264–265
  of events (changing the past) 193–198
  flat 33, 244, 247–250, 255, 264–265, 268–269, 277
  hierarchical 93, 244
  individualism 251–252
  quantum emergence 258–259
  social 245, 280
  supervenience 251
  vitalist sociology 267–268
Open Individualism 242
order effects 157–158
ordered water 100, 102, 106
organicism 144
organisms 273–275
organismic purposiveness 177–178
other minds, problem of 33, 116, 230
*The Oxford Handbook of Philosophy of
  Language* 220–221

pain, experience of 117
panpsychism 31, 35, 70, 92, 111–114
  Bohm Interpretation 85–89
  combination problem and quantum
    coherence 123–124
  definitions 114–116
  and free will 184–185
  and neutral monism 124–130
  subjectivity in all matter 119–123
  subjectivity in living things 116–118
Panpsychism in the West (Skrbina) 112
pan-social ontology 280

Index

Papineau, David 17
paradox of Wigner's friend 82–83
parsimony, quantum consciousness theory 292
particle theory of light 43–49
past, possibility of changing 191–205. *See also* temporal non-locality
Peijnenburg, Jeanne 193, 196–197
Penrose, Roger 30, 99, 186
perception
    direct 222–228, 230–233, 242
    holographic projection 228–230, 239
    problem of 223–226
Peres, Asher 75
performative model of agency 163, 172
Perus, Mitja 229
Pettit, Philip 244, 250–251
phenomenological approach 84–85
philosophical anthropology 36–37, 103
photoelectric effect 44–45
photosynthesis 105, 135
physical basis of language 211–212
physicalism 76, 251
    ambiguity of contemporary definitions 88
    definitions 8–9, 10
    No Fundamental Mentality 9, 31, 76, 251
    See also materialism; realism
physics
    classical 2, 151–153
    dual nature of light 226–228
    of changing the past 198–205
Pia and the Painted Leaves story 213
pilot-wave model 87
Planck, Max 44
Planck's constant 44
plants
    photosynthesis 105, 135
    quantum effects 104–105, 107
    subjectivity in 118
Podolsky, Boris, EPR paper 50–54
Pointer, Lisa 219
polarization of light, Bell Experiments 52–54
politics of vitalist sociology 281–282
Polonioli, Andrea 165
positivism 6, 18–20, 28–29
    anomaly of consciousness 14
    causal closure of physics 8, 10
    classical social science 12–13
    intentionality 18
    methodological atheism 10–11
    quantum consciousness theory 35
post-modernism 74
power, structural 34–35
practice turn 264–265
Pradhan, Rajat 229
pragmatics, linguistic 211, 214

preference reversals, quantum decision theory 161–164
Pribram, Karl 229
Price, Huw 65
Primas, Hans 126, 200
Prisoner's Dilemma, quantum game theory 172
probability judgment 159–161, 164
probability, quantum 40–41, 46–48, 159–161
Process 1 286
Process and Reality (Whitehead) 113
projection postulate 82, 200
proximity principle, causality 225
psyche, definitions 114–116
psycho-physical duality 84
purposiveness 177–178
Putnam, Hilary 253
Pylkkänen, Paavo 88

QFT (quantum field theory) 101–102
quantum biology 107, 135–136
quantum brain theory 15, 30–31, 35, 92, 95, 97
    current debates 102–108
    Fröhlich tradition 98–101
    holographic projection 230
    memory 200–202
    quantum coherence 144–145
    and quantum decision theory 155
    Umezawa tradition 101–102
    see also quantum decision theory
quantum cognition 32–33, 154–156
    probability judgment 159–161, 164
    rational choice theory 33, 155–156, 164–169
    see also quantum consciousness theory; quantum decision theory
quantum coherence 123–124
    cognition 139
    as definition of life 131–132, 137–138, 147
    experience 141–143, 145–146
    Fröhlich tradition 98–101
    as life force 267
    materialist–vitalist controversy 132–137
    quantum vitalism 143–147
    superorganisms 274
    will 139–141
quantum contextualism 215–221
quantum consciousness theory 5, 30–36, 91–94, 154–156
    elegance/aesthetics 293
    explanatory breadth/power 291, 292–293
    and quantum game theory 169
    quantum vitalism 146–147
    social science 288

quantum consciousness theory (*cont.*)
  see also panpsychism; quantum brain theory; quantum cognition
quantum decision theory 4–5, 6, 12, 32–33, 154, 157
  Kahneman-Tversky effects 34
  order effects 157–158
  preference reversals 161–164
  probability 159–161
  and quantum brain theory 155
quantum emergence 33, 255–260, 263, 265–266
quantum entanglement 53–54
  and atomism 61
  emergence 256–257, 263
  language 218–220
  and quantum game theory 169–173
quantum field theory (QFT) 101–102
quantum game theory 32–33, 154, 169–173
quantum interference 160, 161
quantum potential 86, 87–88
quantum probability 40–41, 46–48, 159–161
The Quantum Society (Zohar) 2
quantum theory 29, 39–42, 147, 182–188
quantum tunneling 100
quantum vitalism. *See* vitalism
Quantum Zeno Effect 164

rational choice theory 33, 155–156, 164–169
readiness potential (RP) 185–188
realism 72–73
  assumption 51–52, 54
  idealist interpretations 81–89
  materialist interpretations 76–81
  naïve 223
  social structure 258–259
  see also Copenhagen Interpretation
reasons as cause of actions (problem of mental causation) 175–176, 182
reductionism 33
  and atomism 61
  causal closure of physics 8
  will 140–141
reification of social structure 23, 25, 26, 27, 28, 244
Reimers, Jeffrey 99
relativism 74
relativity theory 65, 227
Replacement Effect, temporal non-locality 202–203
representational models of perception 223, 224
representativeness heuristic 159
retro-causality (changing the past) 191–205.
  See also temporal non-locality
Ricciardi, Luigi 97, 101–102
Rosen, Nathan, EPR paper 50–54

Rosenblum, Bruce 69
RP (readiness potential) 185–188
Ruetsche, Laura 77, 79
Russell, Bertrand 113, 114, 126

Sachs, Mendel 228
Saussure, Ferdinand de 210
Sawyer, Keith 251
Schempp, Walter 229
Schmid, Hans Bernhard 277
Schmitt, Carl 281
Schneider, Jean 235
Schopenhauer, Arthur 110
Schrödinger, Erwin 66, 135
  cat thought experiment 68–69, 76–77
  equation 47, 92
  What Is Life? 135
Seager, William 113
Searle, John 20, 63, 113
Sebeok, Thomas 142
self-consciousness 15
self-organization in organisms 136
semantics
  linguistic 211
  non-locality 230–242
  quantum 215–221
sensory perception, living things 143
separability vs. non-separability
  classical worldviews 208–209
  human mind 33, 149–150, 152
  light 228
  mind-reading 242
  problem of perception 224–225
  rational choice theory 166
Silberstein, Michael 256
simplicity (parsimony), quantum consciousness theory 292
Simulation Theory, theory of mind debate 231, 232
single-celled organisms
  learning in 135
  subjectivity/consciousness 117–118, 136, 279–280
skeletal past 194–196, 199–200
Skinner, B. F. 22
Sklar, Lawrence 8
Skrbina, David 112
Smith, Adam 132
Smolin, Lee 66
social epistemology 284–288
social holism 244–245, 249–250, 258–259
social organisms 273–275
social science
  classical worldviews 11–14
  quantum consciousness theory 288
social structure 22–23, 33–34

definition 243–244
downward causation 260–266
quantum emergence 258–260
state, role of 23–25
threat of reification 23, 25, 26, 27, 28, 244
see also agent–structure problem
Socrates 194, 270
Sollberger, Michael 225
space
  absolute 59, 65–66, 88–89
  experience of 190
speed of light 227
spooky action at a distance 51, 219–220.
  See also quantum entanglement
spreading activation model, language 219–220
Squires, Euan 87
Stapp, Henry 73, 84, 186
Staretz, Robert 272
state
  collective consciousness 275–281
  holographic 271–273
  as organism 273–275
  social structure 23–25
strong argument, quantum brain theory 97.
  See also Fröhlich tradition; Umezawa tradition
structural power 34–35
structure, social. *See* social structure
sub-atomic particles, cloud chamber 119
subject–object distinction 59, 66–69, 88–89
  interpretations of quantum theory 72, 73, 87
  light 228
  temporal non-locality 189
Subjectivist Interpretation 81–85
subjectivity 28, 93, 109–111
  anomaly 7
  definitions 114–116
  in living things 116–118
  in matter 119–124
  see also consciousness
superorganism, state as 273–275
superpositions 47, 52, 85, 162–163, 258–259, 278
  definition 32–33
  see also collapse of wave functions
supervenience 80, 250–255, 256
  definition 61
  externalism 246, 250–255, 259–260
sure thing principle 160–161
swarm intelligence 275

Tegmark, Max 104
teleological causation 177–179, 181
teleology
  living things 140
  reasons as causes 175–176, 182

telepathy 232
Teller, Paul 61
temporal non-locality 54–57, 180, 182, 189–205
temporal symmetry. *See* time-symmetry
theory of mind debate 230–233
  Simulation Theory 231, 232
  Theory-Theory 231
theory of relativity 65, 227
thick past 194–196
Thompson, Evan 136
time
  absolute 59, 65–66, 88–89
  hard problem of 126–130, 199
  speed of light 227
  see also temporal non-locality
Time-Lag argument, indirect perception 225, 239
time-reversal invariance 127
time-symmetry 126–130, 179–181, 189
Tononi, Giulio 146
Transactional Interpretations 70, 179, 187–188, 229–230
trauma, après-coup 203–204
Trewavas, Anthony 118
tunneling, quantum 100
Tversky, Amos 34, 159, 164, 166–167
Twenty Questions game, quantum 67–68
Twin Earth thought experiments 253
Two-Slit Experiment 43–49, 161, 217–218
type indeterminacy, KT man 162

Umezawa, Hiroomi 97, 101–102
unbounded rationality 164–169
Uncertainty Principle 49
  atomism 61–62
  Bohm Interpretation 86
  materialism 60
  supervenience 251
understanding vs. explanation debate 5–6
unobservable social structures anomaly 7
unscientific fictions/fictionalized models 26–28, 35–36
utility maximization 165, 166. *See also* rational choice theory

Vaihinger, Hans 27
van Gulick, Robert 261
van Putten, Cornelis 197, 202–203
Velmans, Max 228, 229
vertical question, social structures 244
visual perception. *See* perception
vitalism 21–22, 32, 33–34, 92, 131–134
  cognition 136, 139
  experience 137, 141–143, 145–146

Index

vitalism (cont.)
  materialist–vitalist controversy 132–137
  politics of 281–282
  quantum 137–138, 267, 268
  terminological usage 143–147
  will 136, 139–141
  see also life force
vitalist sociology 276
volition 175. See also will
voluntarism 282
von Lucadou, Walter 81
von Neumann chain 68–69, 84
  Subjectivist Interpretation 82–83
von Uexküll, Jakob 142, 281

Wang, Zheng 158
war veteran example 202–203
wave front reconstruction
  holographic state 272–273
wave functions 3, 29, 31, 283
  interpretations of quantum theory 71–72.
  See also collapse of wave functions
wave theory of light
  Two-Slit Experiment 43–49
wave-particle duality
  complementarity, principle of 48–49
weak quantum theory 5, 97
"weak" objectivity 74

Weberman, David 193–198
Weyl, Hermann 122
Wheeler, John 55, 67–68, 82, 83–84
Whitehead, Alfred North
  Process and Reality 113
Wigner, Eugene 82
Wilkens, Martin 169
will 33
  in all living things 136, 139–141
  in all matter 121–122
  collapse of wave functions 263–265
  features of subjectivity 116
  free will 182–188
  plants 118
  problem of mental causation 182
  single-celled organisms 117–118
Wilson, David Sloan 273
Witte, F. M. C. 170
Wittgenstein, Ludwig 19, 214, 232
Wolf, Fred Alan 187–188
World War I 204–205
World War II 204–205

Young, Thomas
  Two-Slit Experiment 44

Zahle, Julie 225
Zaman, Frederick 173
zombies 153

For EU product safety concerns, contact us at Calle de José Abascal, 56–1°, 28003 Madrid, Spain or eugpsr@cambridge.org.

www.ingramcontent.com/pod-product-compliance
Ingram Content Group UK Ltd.
Pitfield, Milton Keynes, MK11 3LW, UK
UKHW020305140625
459647UK00005B/43